DATE			

Annual Reviews Monograph

TELESCOPES FOR THE 1980s

Production Editor E. P. BROWER

Annual Reviews Monograph

TELESCOPES FOR THE 1980s

Edited by G. Burbidge and A. Hewitt

ANNUAL REVIEWS INC. 4139 EL CAMINO WAY PALO ALTO, CALIFORNIA 94306 USA

ANNUAL REVIEWS INC.
Palo Alto, California, USA

International Standard Book Number: 0-8243-2902-3
Library of Congress Catalog Card Number: 81-65966

PRINTED AND BOUND IN THE UNITED STATES OF AMERICA

BST
R

CONTENTS

PREFACE

This monograph contains accounts of the history, planning, and completion of the latest generation of telescopes to be built in the US—a radio telescope, an optical/infrared telescope, an X-ray telescope, and the first large space telescope. These new instruments deserve particular scrutiny because astronomy is an observational, and not an experimental science. This means that we cannot design experiments in which the boundary conditions are under our control. We can only passively observe the various types of radiation and particles that celestial bodies emit. Our effectiveness depends on finding efficient methods for collecting and analyzing photons over the whole electromagnetic spectrum.

Until the 1940s practically all of observational astronomy was restricted to what astronomers normally call optical wavelengths—between about 3000 and 7000 angstroms. The advances that were made in optical astronomy in the first half of the century nearly all stemmed from 1. the construction of large reflectors in good astronomical climates, and 2. the steady improvement of the quality of auxiliary equipment, and in particular photographic emulsions.

Over a period of about fifty years optical astronomy was transformed with the creation of the Lick Observatory in the latter part of the 19th century, the building of the Mount Wilson 60-inch and 100-inch telescopes in the first two decades of the 20th century, the building of the Palomar 200-inch telescope in the 1920s and 1930s (completed in 1948), and the construction of the McDonald Observatory with its 82-inch reflector in the 1930s. All of these telescopes were placed in good climates on mountains in the southwestern United States. The new telescopes, together with the development of auxiliary equipment, gave unparalleled opportunities for research and led to the modern framework of our understanding of the universe.

Following the hiatus due to World War II (1939–1945), the next thirty-five years saw the construction of many more large reflecting telescopes (and many smaller ones as well), mostly in good astronomical climates. In the US telescopes with apertures of 2–4 meters were constructed at the Lick Observatory, at the University of Arizona, at McDonald Observatory, on Mauna Kea in Hawaii (two, one optimized for the infrared), the Wyoming Infrared Observatory (Jelm Mountain), and at the Kitt Peak National Observatory (two). Outside the US telescopes with apertures of ~ 2 meters had earlier been built in Australia (Mount Stromlo), in South Africa (Radcliffe Observatory, Pretoria, later moved to Sutherland), and in Canada (David Dunlap Observatory,

Toronto). In Northern Europe it became increasingly obvious that, to compete, it would be necessary to build large reflectors in good climates, i.e. abroad. The British started building first, but they made a mistake by building a 100-inch telescope (The Isaac Newton Telescope) practically at sea level in Southern England. Nearly 30 years later they have tacitly acknowledged their mistake by moving that telescope to La Palma in the Canary Islands. In the 1970s, Great Britain and Australia jointly built the Anglo-Australia telescope, a superb instrument (3.9 meters), at Siding Spring in New South Wales. Germany, France, Holland, Belgium, Denmark, and Sweden banded together in the same period (the 1970s) to build a large reflector (3.6 meters) at the European Southern Observatory on La Silla in Chile. Canada and France combined to build a comparable reflector (3.8 meters) on Mauna Kea, Hawaii. Germany is constructing a National Observatory in Spain (Calar Alto) where they will have two major reflectors (2.2 meters and 3.5 meters). US astronomers, realizing that they had so far left the Southern Hemisphere untouched, constructed two major telescopes, the 4-meter telescope at Cerro Tololo and the 2.5-meter telescope on Las Campanas.

In the same period three major reflectors were built in the USSR, a 6-meter telescope in the Caucasus Mountains and two 2.5-meter telescopes, one in the Crimea and the other in Byurakan. All of these telescopes, and many others (this is not meant to be a definitive list), together with steadily improving instrumentation, vastly increased our ability to study the universe in optical wavelengths.

Infrared astronomy also became a full-fledged branch of astronomy in this period. However, the major steps forward first were made through the development of infrared detectors, rather than by the building of special purpose large telescopes. A number of infrared telescopes with apertures of ~ 1.5 meters were built, but the only very large telescopes that have been built with the infrared predominantly in mind are those at the University of Wyoming (Jelm Mountain), the NASA Infrared Telescope, and the United Kingdom Infrared Telescope, both of the latter on Mauna Kea, Hawaii.

In the same period, many telescopes, both single instruments and arrays, designed to explore the radio spectrum were built. Though the very earliest radio telescopes were constructed by Jansky and Reber in the US in the 1930s, the real research activity began after the war with the building of a succession of radio telescopes in England at Manchester (Jodrell Bank) and Cambridge (Cavendish Laboratory), in Australia at Fleurs, Parkes, and Molonglo, in Holland at Westerbork. In the US radio observatories were built in the Owens Valley and Hat Creek in California, at Green Bank, West Virginia, near Marfa, Texas, on Kitt Peak in Arizona, in Michigan, Ohio, Massachusetts (two), Arecibo, Puerto Rico, and Murray Hill, New Jersey. In Bonn, Germany, Bologna, Italy, Ootacamund, India, and near Leningrad in the USSR other observatories were built. These facilities comprise a wide variety of single telescopes, interferometers, and arrays of telescopes which can detect radio sources at wavelengths ranging from a few meters to a few millimeters. Most recently plans have been made to build large millimeter wave facilities by Germany (Max Planck Society), Japan, France, and Great Britain.

In the short space of about twenty-five years radio astronomy has revolutionized our understanding of the universe. At the same time it has become very clear that with few exceptions radio observations and optical observations must go hand in hand, since the radio astronomers in general, when they detect a radio source, do not know what they are looking at, or where it is, unless it can be identified in optical wavelengths and studied with optical telescopes.

To extend studies of the universe to shorter wavelengths—into the ultraviolet and X-ray domains—we must get above the Earth's atmosphere. Some rocket and balloon observations were made in the twenty years after the war, but the observation times using these techniques are by necessity very short. The field was not really opened up until satellite observatories began to be flown. Preliminary investigations have been carried out in the ultraviolet using the Orbiting Astronomical Observatories and the Copernicus satellite and the International Ultraviolet Explorer (IUE). For X-ray astronomy the first major satellite instrument was Uhuru.

The development of X-ray astronomy is following a similar path to that of radio astronomy in that discoveries in this field also require observations in optical wavelengths to be fully exploited.

As far as other areas are concerned, gamma-ray astronomy is still in its infancy, and neutrino astronomy and gravitational wave astronomy still are in the very preliminary stages.

In the 1980s we have reached a new level of knowledge in radio, infrared, optical, ultraviolet, and X-ray astronomy. We have learned a great deal, particularly in the last thirty years, and our knowledge has kept in step with the increase in facilities and the improvement in instrumentation.

The articles in this monograph contain accounts of the history, planning, and completion of the latest generation of telescopes—a radio telescope, an optical/infrared telescope, an X-ray telescope, and the first large space telescope. They are all different as far as the research that can be done with them is concerned. Two are ground-based projects, and two are space telescopes. Three of them are totally funded by the US Government, while the fourth, the Multiple-Mirror Telescope, has been funded jointly by universities and the US Government (through the Smithsonian Institution). Indeed, nearly all of the money for the projects discussed in this volume has come from the US Government. In this respect they are similar to the majority of the major astronomical projects completed since the War. Very few of the post-war projects—the Lick 3-meter telescope, and the Las Campanas 2.5-meter telescope are two exceptions—have been built with private or state money. In contrast, in the first half of the 20th century, all of the large optical telescopes were built with private funds.

The change has brought much greater financial support for astronomy, and, at the same time, much greater dependence on one source of funding, which requires, at least in the US, that a large fraction of the scientists support the project. By the same token the role of the individual scientist, who by his own efforts raised the money, planned, and, to a large extent, dominated the project (G. E. Hale and O. Struve in the first half of the century are good examples) has diminished. Such individuals have been replaced by committees or groups of individuals. Moreover political factors enter, particularly for space projects.

The authors of these articles are each highly distinguished in their respective fields and have made major contributions to the design and construction of the telescopes about which they write. We have encouraged them to describe the many steps forward and the various setbacks that were encountered in bringing these projects to fruition. Of course, one of them, the space telescope, is not yet completed. We hope that these accounts will not only show how complex and difficult major projects have now become, but that they will also give help to those who are planning new projects which will probably only come to fruition toward the end of this century.

G. Burbidge
A. Hewitt
March 1981

Telescopes for the 1980s, pp. 1–61

THE VERY LARGE ARRAY

D. S. Heeschen

National Radio Astronomy Observatory,[1] Charlottesville, Virginia 22901

1. INTRODUCTION

After the powerful radio source Cygnus A was identified with a distant peculiar galaxy in 1954, two possibilities rather quickly became apparent. First, the very intense radio radiation, and its distribution in two lobes on either side of the parent galaxy, indicated the presence of extremely energetic phenomena that would challenge our understanding of the physical behavior of galaxies. Second, the great intensity of the radiation suggested that similar objects might be detectable at much greater distances than had hitherto been accessible to observations; thus these "radio galaxies" might prove useful probes of the distant regions of the universe. These two considerations have dominated much of the effort and instrumental development in radio astronomy since that time.

By the early 1960s it was clear that existing radio telescopes, especially in the US, were inadequate to carry extragalactic studies much further. Only the interferometer of two 90-foot dishes at the Owens Valley Observatory could compare favorably with European and Australian instruments for extragalactic work, and none of the existing telescopes anywhere in the world had the resolution or sensitivity that would obviously be needed to pursue extragalactic studies, or other radio astronomy problems either. The absence of any direct image-forming capability, such as is achieved so easily with optical telescopes and photographic plates, made radio observing particularly slow and difficult.

Stimulated by the obvious scientific challenges on the one hand and the glaring instrumental limitations on the other, many organizations and individuals turned their attention to the design of new radio telescopes. From these efforts emerged the great radio telescopes of the 1970s, such

[1]The National Radio Astronomy Observatory is operated by Associated Universities, Inc., under contract with the National Science Foundation.

8243-2902/81/0820-0001$02.00

as the Cambridge (Ryle 1972) and Westerbork (Högbom & Brouw 1974) arrays, and what will most surely be the preeminent radio telescope of the 1980s, the Very Large Array.

The Very Large Array (VLA) was motivated by a clearly perceived need, in the early 1960s, for an image-forming instrument of the greatest feasible resolution, sensitivity, and general versatility. While its designers were most strongly influenced by the opportunities and problems presented by extragalactic radio sources, the need for such an instrument was apparent in all areas of radio astronomy, and the VLA was, in fact, designed to be used for almost all kinds of radio astronomical studies.

The VLA is an array of twenty-seven 25-meter-diameter antennas, arranged along the arms of an equiangular wye (see Figure 1). It has the collecting area, sky coverage, and side-lobe characteristics of a fully steerable paraboloidal antenna 130 meters in diameter, and the angular resolution of a paraboloid 25,000 meters in diameter. It can be operated in any of four wavelength bands between 20 cm and 1.3 cm. Individual antennas can be moved along the arms of the wye, on railroad tracks, to provide different characteristics, such as resolution and sensitivity to extended

Figure 1 An aerial view of the central 1–2 km of the array, showing the wye-shaped track system and some of the antennas. The tracks extend out from the center a distance of 21 km. Individual antennas are displaced from the main track system on short spurs leading to concrete pads at each observing station. Stations unoccupied in this picture are used when the array is in a more compact configuration. In the least compact configuration most of the antennas would be outside this picture at distances up to 21 km from the center.

sources, for different problems. The array is located on the Plains of San Augustin, a high (elevation 2135 meters), remote, arid region about 95 miles southwest of Albuquerque, New Mexico.

The VLA has 10 to 100 times greater resolution and sensitivity than any other existing radio telescope, and its resolution is comparable to or greater than that attainable at optical and other wavelengths. Its speed, sky cover, ability to measure polarization, and ability to make high frequency resolution spectroscopic observations give it tremendous power and versatility for a wide variety of problems. As an image-forming device it is unsurpassed in the radio wavelength domain, and should remain so for a long time. Its influence on astronomy in the 1980s and beyond can be expected to be very great.

The initial discussions and feasibility studies that ultimately led to the VLA were begun at the National Radio Astronomy Observatory (NRAO) in 1961. Detailed design work began in 1964, and a first proposal for funding was submitted to the National Science Foundation (NSF) in 1967. The telescope was approved for construction in 1972 and construction began in 1973. Because of its nature the array can be built in stages, and funding was planned for a scheduled completion in 1981. At this writing (January 1981) the array is very nearly finished, and the scheduled completion in 1981 will be achieved. The partially completed array has been in limited scientific use since 1977.

Table 1 gives a rough chronology of various events associated with the VLA. Not all of the activities listed will be discussed. In particular, most of the committee activities related to reviews of radio astronomy and/or the VLA are omitted. All of the committees listed in Table 1 had some impact on the VLA, however. Those interested in this aspect of the VLA project can obtain more information from the references indicated in the table.

In Section 2 the development of the conceptual design and performance goals of the array is discussed, along with selected design problems and their solutions. The principle of aperture synthesis, which is basic to the array, is also briefly described. Site selection is discussed in this section as well. Section 3 gives brief descriptions of the principal elements of the array—antennas, rail system, electronics, waveguide, and computers. In these two sections the rationale for a particular engineering design or performance parameter is emphasized rather than the engineering detail itself. Section 4 describes some highlights and problems of the construction phase of the project. The performance and characteristics of the nearly completed array are described in Section 5. Finally, examples of work already done are given and some expectations for the future are described in Section 6.

Table 1 VLA chronology

Date	Event
1961–1963	Preliminary design work
1961	Report of NSF Advisory Panel on Large Radio Telescopes (NSF 1961)
1962	Phrase "Very Large Array" came into use
1964	"Whitford Committee" report (NAS 1964) Green Bank interferometer project begun VLA design group organized
1964–1967	Intensive design effort
1965	NRAO Users Committee formed First visit to Plains of San Augustin by NRAO personnel
1966	First VLA report distributed
1967	VLA proposals, Vols. I and II, issued (NRAO 1967) Plains of San Augustin proposed as site First "Dicke Committee" meeting (NSF 1967)
1967–1969	Further design work. Design group disbanded in 1969
1969	VLA proposal, Vol. III, issued (NRAO 1969) Second "Dicke Committee" meeting (NSF 1969)
1971	"Greenstein Committee" recommendations established (NAS 1972) VLA design work recommenced Stanford Research Institute undertook VLA feasibility study for NSF (SRI 1972) VLA proposal, Vol. IV, issued (NRAO 1971)
1972	Construction project management organized (J. H. Lancaster, project manager) VLA approved. First funding received
1973	Antenna contract signed
1974	Site work begun
1975	VLA project personnel move to New Mexico Two-antenna interferometer operating at site
1976	Ad Hoc Advisory Panel reviews VLA for NSF, Congress (NSF 1977)
1977	Scientific observing begun, with six antennas
1979	Twenty-eighth antenna completed
1981	VLA construction completed

2. PERFORMANCE GOALS AND CONCEPTUAL DESIGN

Design of the VLA began in the early 1960s with general discussions concerning the kinds of observations that were needed for extragalactic studies. Was enough already known about source characteristics to make a less than general purpose instrument acceptable? What sort of sky coverage, speed, and resolution were desirable? These questions turned out to be very controversial, and were the subject of extensive discussion both inside and outside NRAO for the next 10 to 12 years, right up to the beginning of VLA construction. The attitude taken to these questions determined the nature and cost of the instrument designed.

The NRAO design group tended, from the earliest discussions, to take an ambitious attitude toward the proposed instrument, partly because it was to be a "national facility" to be used by any qualified scientist with an appropriate program and, hence, ought to be flexible and versatile, but mainly because of scientific considerations. In fact, relatively little was known at that time about radio sources, galactic or extragalactic, but it was clear that they embraced a very wide range of properties and characteristics. Any particular existing radio telescope, on the other hand, could measure only a rather small range of properties. That is, the dynamic range of radio telescopes was extremely small compared to the intrinsic dynamic range of observables. This led to severe selection effects in almost all bodies of radio data. Further, all existing telescopes suffered, in varying degree, from effects of complicated antenna patterns or confusion due to inadequate resolution. Besides influencing the dynamic range, these effects made the interpretation of data more difficult. In the face of these considerations it seemed inappropriate to restrict the performance goals for the contemplated instrument until we were forced to do so for technical or economic reasons.

The first performance goal for the VLA was simply that it be a map-making instrument with the highest feasible sensitivity and resolution, at least comparable to that achieved with optical telescopes. From this initial idea the conceptual design and the performance goals evolved in parallel. A "highest feasible resolution" of the order of 10″ at 10-cm wavelength was adopted early as a performance goal. This, together with some general requirements of speed and flexibility, placed severe restrictions on the design of the instrument.

During the 1950s many ingenious and powerful radio telescopes had been developed. These included the various arrays built in Cambridge by M. Ryle, the "Mills Cross" and other systems in Australia, and the interferometer of two 90-foot dishes built at Caltech by J. G. Bolton. [A review of these telescopes is given by Christiansen (1963).] All of the techniques involved in these and other telescope systems were examined as to their applicability to the VLA. It was decided quite early in the design studies (about 1963) that the technique of earth-rotation aperture synthesis developed by Ryle was most suitable for the proposed instrument (see Section 2.3 for a description of aperture synthesis).

Thus, early studies, in 1961–1963, led to three major decisions: (*a*) the resolution was to be 10″ at 10 cm; (*b*) the telescope should be fast and versatile, a "national" instrument; and (*c*) aperture synthesis was the "best" approach.

Development of the VLA involved three distinct phases: (*a*) conceptual design, when the general nature of the instrument was established; (*b*) system design, when specific equipment to meet the concept was designed; and (*c*) construction. These three phases are discussed separately below, although they necessarily overlapped. Conceptual design was guided by performance goals, which in some instances were, in turn, modified as a result of conceptual or system design studies.

2.1 *Performance Goals*

The performance goals that emerged during this iterative design-study process are listed in Table 2 and described in more detail below.

Table 2 VLA performance goals in 1967

Resolution:	1 arcsec at 10-cm wavelength
Sensitivity:	0.1–1.0 mJy
Side lobes:	Similar to those of a parabolic antenna; first side lobe at −20 dB
Field of view:	1–10 arcmin
Sky coverage:	−20° to +90° declination
Speed:	$\leq 12^h$ to achieve above goals in a given region of sky
Polarization:	Full polarization capability
Frequency range:	$\geq$ 3 cm; multiple bands
Spectroscopy:	Yes
Versatility:	As versatile as reasonably possible

ANGULAR RESOLUTION There is no rational scientific basis for choosing a *particular* resolution for a telescope, because there are no known natural angular scale sizes. Greater resolution is always desirable, and indeed the desire for more resolution has been the driving force behind much of the development of radio astronomy instrumentation since the earliest days. The early goal of 10″ at 10 cm was soon found to be unnecessarily conservative. Scientific requirements certainly indicated the need for higher resolution than that, and the early design studies suggested that higher resolution would also be possible. A goal of one arcsecond at 10-cm wavelength was finally chosen because it appeared to be near the limit of what might be technically and financially feasible and because it was comparable with the capabilities of ground-based optical telescopes. There was a strong desire on the part of the VLA designers to be able to make radio "pictures," which could be directly compared with optical "pictures" so that the radio and optical structures of a given object could be compared on the same scale.

It was also established that the VLA system design should not preclude attainment of still higher resolution at some future time, by use of shorter wavelengths or by expanding the dimensions of the array.

SENSITIVITY The sensitivity of the VLA is limited by a combination of radiometer system noise, instrumental instabilities, bandwidth, atmospheric effects, collecting area, length of observing time, confusion due to side lobes and to unresolved sources in the main beam, and perhaps by other factors as well. Some of these parameters are under the control of the designer, some are set by the current state of the art in electronics, and some are beyond control. While the highest possible sensitivity is clearly the goal, it is also desirable that the design achieve some balance between the various parameters affecting cost and sensitivity. It would be unfortunate to pay a very large price for low-noise receivers, for example, only to find that the sensitivity is determined, at a much poorer level, by atmospheric fluctuations or confusion. A sensitivity goal of 0.1 to 1 mJy (1 mJy $= 10^{-29}$ Wm^{-2} Hz^{-1}) was finally established after considering the various limitations to sensitivity as well as the objective measuring of polarization of a significant number of sources to about 1% and achieving a dynamic range of 100 in studying the structure of sources.

SIDE LOBES During the early phase of the VLA design radio astronomers were having a great deal of difficulty interpreting source position and structure data obtained with antenna systems having complicated side-lobe and/or grating-lobe patterns. In contrast, data obtained with a simple parabolic antenna were relatively free of instrumental artifacts and relatively easy to interpret. As a result the VLA design group intuitively set as its goal an antenna pattern similar to that of a parabolic antenna. This would require a first side-lobe level of about -20 dB, and of course no grating lobes. Further constraints on side lobes arise from other considerations. To achieve a sensitivity of 0.1 mJy requires that rms side-lobe levels be at least 30 dB below the main beam to minimize confusion from sources outside the observed field of view. Distant side lobes must be still further suppressed to minimize confusion from the sun.

FIELD OF VIEW The field to be mapped in a single observation should be larger than the dimensions of most sources of interest, and should be sufficiently large to allow, for statistical studies, observation of a reasonable number of sources per day. Fields of view between 1 and 10 arcmin appeared to satisfy these criteria.

SKY COVERAGE In almost any technique suitable for large radio telescope systems there is a direct trade-off between cost and sky coverage. A narrow range of declinations can be covered more easily than can a large range. Further, in many types of radio arrays, performance, such as gain and resolution, is a function of declination. Most existing arrays only worked well at $\delta \gtrsim 30^\circ$ and could therefore observe less than half the total sky available to them. And, perversely, specific interesting objects seemed always to lie just outside the observable declination range. Because the VLA was intended to be a major national instrument for use by astronomers throughout the US and the world on a wide variety of problems, these compromises were considered unacceptable. Therefore, full sky coverage and performance parameters that did not vary strongly with declination were adopted as primary goals.

SPEED Speed is another parameter for which there is a direct trade-off with cost in many types of radio telescope systems. The speed with which an instrument can obtain data may have a strong influence on the power of the instrument, and on the kinds of problems for which it is used. As a national instrument for use by all qualified scientists, the VLA should clearly have some reasonable speed. But what is "reasonable"?

An interferometer consisting of one fixed and one movable element could achieve all of the performance goals so far described, except that of "reasonable" speed. With such an instrument, a region of sky could be mapped, with the desired resolution and sensitivity, in about five months, by observing with the movable antenna in a different location each day. It would require tremendous effort, but the initial capital costs for equipment would be low. By using some 200 fixed antennas the time could be reduced to a few minutes, but the cost would be prohibitive. Apparently reasonable speed lies somewhere between the extremes of five months and a few minutes.

As mentioned above, we decided fairly early in the design studies that the technique of aperture synthesis offered the best opportunity for realizing the desired performance goals. This approach uses some number of interferometers to collect the necessary data. It therefore offers a choice, as suggested in the previous paragraph, between either periodically moving antennas to get new interferometers or adding additional antennas. In such a scheme there is a natural breakpoint between cost and time, at a single "day's" (the time between rising and setting) observation. To save money by using fewer antennas than is needed for a one-day observing time it would be necessary to use the system in different antenna configurations over a succession of days. In practice, when scheduling considerations and the time required to move antennas are included, a reduction in the number of elements from that required for a 12-hour observing time probably increases to several weeks the time to complete a map. Since cost is roughly proportional to the number of antennas, and the rate of data acquisition to the square of the number, the savings are small compared with the increased time and the decreased scheduling and observing efficiency. On the other hand, to provide enough antennas to complete an observation in much less than 12 hours becomes very, very expensive. A goal of a 12-hour observing time was therefore adopted as the speed requirement.

POLARIZATION AND FREQUENCY RANGE Polarization and brightness as a function of position, frequency, and time are the only directly observable parameters of any celestial object (except for discrete spectral lines). It was important that the VLA, as a major national instrument, have both polarization capability and a broad wavelength range. These characteristics are heavily dependent on the state of the art in electronics and in antenna design and on the amount of money one is prepared to spend, and they were not further specified during the conceptual design.

SPECTROSCOPY Although the VLA was considered primarily a continuum instrument in the 1960s, the designers also recognized the importance of being able to use it for spectroscopic measurements as well. While the early design did not explicitly include spectroscopy as a priority requirement, it was important to design the array so as not to preclude its future use for spectroscopy.

VERSATILITY A catch-all goal, that the VLA be as versatile as reasonably possible, was adopted from the beginning of the program. Thus, for example, while the instrument would not be designed specifically for solar work, the design should not preclude its eventually being used for solar observations. In practice this meant that every detailed hardware design decision had to be examined from the standpoint of its overall effect on the performance by someone concerned with, and knowledgeable about, the scientific use of the total system. For example it is easy for an engineer, in designing a local oscillator system, to choose an option that is satisfactory from an engineering viewpoint, but that would seriously and unnecessarily compromise the scientific capability of the system. The concept of versatility also suggested that instrument parameters, such as wavelength, resolution, and sensitivity to extended sources, be adjustable to meet differing scientific needs.

These performance goals, or design parameters, guided the conceptual design of the VLA. Taken together, they imposed severe constraints on the design and strongly dictated the nature of the solutions. There were very few free choices or alternatives whose selection might be considered arbitrary. In particular, the requirements for angular resolution, sky coverage, side lobes, and speed pretty well determined the general nature of the instrument. One arcsecond resolution at 10-cm wavelength requires dimensions of the order of 35 km. This obviously rules out a filled aperture, or even a continuous aperture, and means that some kind of an array must be used. Full sky coverage requires that the array be two-dimensional; it is not possible to get a good, symmetric beam over a large declination range with a single line of antenna elements. The requirements of sky coverage, side-lobe levels, and versatility together establish the possible configuration. The speed and side-lobe requirements determine the number of antennas.

The adopted performance goals were by no means universally accepted as being necessary or desirable, within or without NRAO. They were the subject of argument and controversy for many years, right up until construction began. These debates, carried on in a rather large segment of the radio astronomy community, were quite useful. In the early stages of the design they helped establish the performance parameters. Continued debate after the parameters were set served to provide a more or less continuous review of the entire program.

Many people argued that full sky coverage was unnecessary, for example, or that the side-lobe requirement was too severe. Almost all of these arguments were directed toward reducing the cost of the instrument in exchange for a loss of performance. However, by the time the design had reached the stage where these performance goals were actually adopted, it was, in fact, very well-balanced, with performance, design, and hardware all optimized to the total system. A change in, say, sky coverage would affect other characteristics as well, and reduce the total array performance out of all proportion to the saving in cost. The one major exception to this appeared to be with respect to resolution; the dimensions of the array could be scaled down without severely affecting performance parameters other than resolution, and with significant savings in cost. This option was left open until almost the end of construction to allow for further cost reduction should it become necessary.

Other simple options for cutting costs, such as reducing bandwidth at the expense of sensitivity or reducing the number of frequency bands at the expense of overall versatility, were also left open. These options all had the desirable attribute of being recoverable in the future even if they were restricted initially. Modifications or reductions in the scope of the project that would fundamentally and irretrievably compromise the performance of the instrument were not acceptable.

This same philosophy was adopted during the construction phase, when very large and unanticipated inflation made it necessary to reduce the scope of the program, as discussed in Section 4. Again, those reductions were of such a nature that they were not irreversible.

One final "performance" goal had to be set. It is not possible to design an instrument such as the VLA without at least a rough idea of how much money may be available for its construction. Forty million 1964 dollars was somewhat arbitrarily chosen as an approximate budget figure to help guide the design. This amount was mentioned in the "Whitford Committee" report (National Academy of Sciences 1964) in conjunction with a recommendation that a large, high resolution array be built.

2.2 *Conceptual Design*

With performance goals established, the actual conceptual design proceeded in a straightforward manner. Design investigations were carried out in several areas: site selection, array configuration and number of antennas, antenna size, atmospheric effects, and hardware design. In addition, an interferometer was built at Green Bank to aid in the development of the array. Each of these activities is described briefly below.

SITE SELECTION The principal characteristics needed for the site were dictated by the proposed nature and performance of the array. An area at least 35 kilometers in diameter was required. It had to be relatively flat to facilitate moving antennas around on it. It should be, if possible, at a low latitude, because the performance of this type of array depends on the latitude of the instrument, and at a high elevation to minimize the effects of the atmosphere. It should be remote from large population centers, industrial or mining areas, and other areas of concentrated human activity, all of which are potential producers of radio emissions that can interfere with observations. And the site should be relatively sparsely populated, and not heavily utilized, to minimize acquisition costs and disruption to others. Finally, it should be easily accessible for visiting observers, and a pleasant place to live for the staff.

So a search was instituted for a large, flat, remote, accessible, unused area of the southwest. It turned out that these are very restrictive criteria. After an exhaustive search, primarily via topographic maps, only 34 sites were found that even marginally met the criteria. On inspection one was found to be an extremely rough lava bed, two were active oil fields, several were in Air Force bombing ranges, and others were eliminated for various reasons. Of the seven acceptable sites, one—the Plains of San Augustin—was clearly superior to the others.

The Plains of San Augustin are in New Mexico, about 95 miles southwest of Albuquerque. Ten thousand years ago the area was a lake. Today it is a large, remarkably flat grassland used only for grazing cattle. Sparsely populated, with no known mineral resources, arid, and at an elevation of 2135 meters, it is almost ideally suited for a large radio array.

The VLA itself occupies only one square mile at the center, purchased outright, and three leased strips of land, each about 180 meters wide and 21 km long. The construction of the array has noticeably changed the scenery, but has had almost no other impact on the use of the land. Cattle continue to graze as before. While a few of the ranchers have objected to the VLA on aesthetic grounds or as a matter of principle, most ranchers and others in the region have been extremely friendly and cooperative.

CONFIGURATION AND NUMBER OF ANTENNAS A variety of array configurations, with varying numbers of antennas, was investigated by computer simulation of their performance to find the configuration that would give the best average performance over the full range of declination, $-20°$ to $+90°$, with the smallest number of antennas, in an observing time of eight hours. The desire for versatility led to an additional requirement that it be possible to expand or contract the array, within the basic configuration, to give different performance characteristics for different problems. Circular, cross, tee, wye, and random configurations were among those investigated. Circular and random configurations do not lend themselves to expansion or contraction and were therefore eliminated, although both can give reasonable performance in other respects. Of the remaining possibilities, a wye configuration consisting of three arms separated by 120°, with one arm rotated away from north by about 5°, gave best performance. The arms must each be near 21 km in length to give the desired resolution.

Thirty-six antennas, 12 on each arm, were required to produce the desired synthesized antenna pattern in an eight-hour observation. The exact locations of antennas along the arms is not very critical, and there is no way to uniquely determine a "best" set of locations. The locations were ultimately chosen on the basis of a study by Y. L. Chow (Chow 1972) to both optimize performance and minimize the number of different locations needed for the four different configurations that were to be available.

After the array design was completed, but before the project was approved for construction, the side-lobe requirement was relaxed from -20 dB to -16 dB. This followed a parametric study of side-lobe levels versus number of antennas, which showed that the number of antennas needed for -16 dB side lobes was only 27. Greater side-lobe suppression required a large increase in the number of antennas, while lesser suppression allowed only a small decrease. Twenty-seven antennas and -16 dB suppression, therefore, appeared to be the better compromise between performance and cost. The additional suppression, to -20 dB, was just too costly.

ANTENNA SIZE The sensitivity of the VLA is determined by a number of factors, most of which must be set from other considerations. Sensitivity is proportional, for example, to the number of elements in the array. That number, however, as just discussed, is determined by the required synthesized antenna pattern. About the only "free" parameter that affects sensitivity is the diameter of the individual antennas. Array sensitivity is proportional to the square of the diameter of an individual antenna. Unfortunately the cost of an antenna is roughly proportional to the cube of its diameter. Thus the choice of diameter is, again, largely a trade-off between performance and cost. However, there are additional considerations. The beam pattern of the primary antenna serves to attenuate unwanted side lobes on the one hand and to restrict the field of view on the other. Too small an antenna makes suppression of side lobes more difficult. Too large an antenna may not allow sufficient field of view. In addition, individual antennas had to be large enough to allow sensitive calibration observations to be made, independently of the total array.

The above considerations do not themselves lead to a well-defined diameter, but to a moderately broad range of diameters. Within this range a 25-meter diameter was chosen as probably the largest that would fit the anticipated budget. In addition, groups in the US already had a lot of experience in the design and construction of antennas of that size, experience that would be advantageous in the construction of the VLA.

ATMOSPHERIC EFFECTS It was recognized that atmospheric phase fluctuations would be a principal factor in limiting the performance of the VLA. However, in the mid-1960s relatively little was known about the magnitude of such effects, and there was no experience available at obtaining coherent signals from atmospheric paths separated by the distances involved in the VLA. No one had ever made radio maps with one-arcsec resolution. Therefore, a number of programs were undertaken to investigate these potential problems. The Green Bank interferometer was used in an attempt to measure atmospheric phase fluctuations as a function of antenna separation. A fourth element was added to the interferometer at a distance of 35 km, in order to test the feasibility of doing coherent interferometry over distances and at wavelengths comparable to those of the VLA. Attempts were made to correlate measured water vapor content and interferometer phase errors with a view toward eventually developing an active correction system.

These studies showed that one-arcsecond mapping was feasible and provided a qualitative idea of the magnitude of atmospheric phase effects. The latter, together with some theoretical considerations, led to the adoption of an allowable instrumental phase error of 1° per gigahertz operating frequency. Thus, at a frequency of 5 GHz a maximum of 5° of instrumental phase error was considered acceptable. At that level instrumental and atmospheric phase errors were expected to be comparable.

HARDWARE DESIGN PROBLEMS As the VLA design began, two major elements of the electronic system were identified as being potentially most difficult. These were (*a*) the maintenance of phase stability and coherence in the local oscillator system over the large distances involved, and (*b*) the distribution of signals throughout the array. Design work was begun on these systems very early. Feasible designs were found, detailed, tested, and costed out before the VLA proposal was submitted. During the long interval between the submitting of the proposal and the beginning of construction, new and better solutions were developed, especially for the signal distribution system, so that the original designs were never implemented. They were extremely important, however, in demonstrating the overall feasibility and cost of the VLA.

The design of the antennas and of the track system was a matter of relatively straightforward engineering. No large risks or major innovations were involved. However, since a large fraction of the total cost of the VLA was tied up in these systems, a great deal of design effort was put into them, beginning early in the project, in order to identify the most economical designs.

GREEN BANK INTERFEROMETER A second, movable 26-meter-diameter antenna was built in Green Bank in 1964 for use with the already existing 26-meter antenna as a variable baseline interferometer. A third 26-meter antenna and a smaller, portable antenna were both added in 1966. As already mentioned, the latter was eventually carried to a distance of 35 km. This interferometer system was developed both as part of the VLA design effort and also as a powerful scientific instrument in its own right. It accomplished a variety of tasks for the VLA design.

The VLA antennas were to be movable, for reasons described later. Most movable antennas at that time were moved on tracks. To test a possible alternative scheme, the movable antennas of the Green Bank interferometer were placed on a large number of rubber-tired wheels and a roadbed was built on which the antennas could be pulled from one location to another. A few years' experience with that system showed conclusively that rails were the better alternative.

The interferometer was also used as a test facility for many of the electronic systems that were ultimately used in, or rejected for, the VLA. It was used to demonstrate the feasibility of making maps with one-arcsec resolution and to measure atmospheric phase effects. Data reduction and analysis techniques were developed and tested. Finally, and perhaps most important, the interferometer gave scientists and engineers direct experience in the design, operation, and use of much of the equipment and many of the techniques that were later incorporated in the VLA.

The final conceptual design emerged from the above considerations. (Table 3 lists the principal features of the VLA concept.) The VLA would consist of an array of 27 antennas, each 25 meters in diameter. The antennas would be positioned in an equiangular wye-shaped configuration, with each arm of the wye about 21 km in length. Every antenna would be individually connected to every other antenna, with appropriate delays, giving 351 separate interferometer pairs. A source or region of sky to be mapped would be observed for 8 to 12 hours and the output of each of the 351 interferometers accumulated. At the end of the observing period a Fourier inversion of the accumulated data would yield a radio map of the region observed.

Antennas move along the arms of the wye on standard railroad tracks and are moved off the main track to concrete foundations for observing. Four sets of observing stations are available, giving four different array configurations. The array resolution varies by a factor of about three between adjacent configurations, and there are corresponding differences in surface brightness sensitivity and other parameters.

Four frequency bands are available, and the observing band can be easily and quickly changed using computer control. All four parameters of polarized radiation can be measured simultaneously, and up to several hundred frequency channels and a variety of frequency resolutions are available for spectroscopic observations.

The array was originally designed as a 36-element array operating at 11-cm wavelength, with 21-cm and 3-cm operation as future add-on options. The change from 36 to 27 elements has already been described. Subsequent to submission of the original VLA proposal and prior to the commencement of construction, a period of about six years, a number of

Table 3 The VLA concept

Mapping instrument using aperture synthesis
27 antennas of 25-meter diameter
Equiangular wye configuration with arms 21 kilometers long
Antenna movable along wye on rail system
Four observing configurations, with different resolution
Four wavelength bands
Four polarization parameters measured
Spectral line capability
Full sky coverage

developments occurred that led to changes in the design. First, the ATS satellites were launched with powerful transmitters adjacent to the 11-cm radio astronomy band. Since the VLA was designed with a wide bandwidth, extending outside the protected radio astronomy band, it was decided to shift the primary operating wavelength to 6 cm. Then several electronic and antenna developments made a multi-band system feasible. These included development of an economical, four-frequency system front-end package, development of the waveguide signal distribution system, development of a rotatable assymetric Cassegrain secondary reflector for the antennas which would allow rapid switching from one wavelength band to another, and improvement in primary reflector design and antenna pointing techniques which allowed reliable operation to shorter wavelengths. All of these are described somewhat more fully in Section 3.

The conceptual and system design efforts described in this and the next section took place largely during the period 1964–1969. A VLA design group was formally established in the summer of 1964, with G. W. Swenson as chairman. Others in the group included B. G. Clark, D. S. Heeschen, D. E. Hogg, H. Hvatum, W. C. Tyler, C. M. Wade, and S. Weinreb. This group met approximately monthly for almost five years. Many others from NRAO and from the scientific community were also heavily involved.

Although the design group was never rigidly organized, in contrast with the later very structured project management and construction organization, there was a rough division of effort as follows: site studies, Wade and Swenson; array configuration, Hogg; array sensitivity and antenna size, Wade; electronics, Tyler and Weinreb; antennas, Hvatum; and data processing and computers, Clark.

2.3 *Aperture Synthesis*

Aperture synthesis is the name loosely applied to various techniques whereby observations with two or more small antennas can simulate, in some respects, observations made with a single, larger-aperture antenna. The basic idea goes back to the early days of radio astronomy (see, for example, McCready, Pawsey & Payne-Scott 1947), but the technique in its present form was first described by Ryle & Hewish (1960). The principle of aperture synthesis as used in the VLA has been described in detail elsewhere (Fomalont 1979) and only a brief qualitative description is given here.

If the signals from two antennas are appropriately combined, the result is one component of the Fourier transform of the brightness distribution of the region of sky toward which the antennas are pointed. The particular component is specified by the separation, d, and relative orientation, θ, of the two antennas. If, after one measurement at (d_1, θ_1), one antenna is moved to a new location specified by (d_2, θ_2), a second Fourier component can be measured. Thus, by repeated moves of one antenna, a large number of Fourier components can be measured. When a sufficient number of different Fourier components have been measured an inversion will produce a map of the region observed.

To map a region of sky 100-arcsec square with one-arcsec resolution would require measurement of more than 10^4 components, and therefore more than 10^4 moves of one antenna in the above scheme. This is clearly prohibitive. There are two ways to speed the process. One is simply to add more antennas. The number of Fourier components simultaneously measured with n antennas is $n(n - 1)/2$. Thus, 10 antennas can produce data 45 times as fast as 2, and 27 antennas 351 times as fast. The second way to speed the synthesis process is to take advantage of the fact that the apparent separation and orientation (d, θ) of an antenna pair, as seen from a celestial object, changes as the earth rotates. Therefore, a given fixed antenna pair can measure a large number of different Fourier components if the region of sky is tracked from rising to setting.

Both of these time-saving techniques are used in the VLA. Twenty-seven antennas produce 351 interferometers, which can track a source for 8–12 hours and measure more than 10^5 different Fourier components of the brightness distribution of the region observed. Fourier inversion of the data produces a map with resolution and sensitivity similar to that of an aperture very much larger than the diameter of the individual antennas.

3. SYSTEM DESIGN

In this section some general considerations of the system design are discussed. More detailed descriptions are available in the VLA Proposal (NRAO 1967, 1969), Thompson et al. (1980), and references therein.

Design of the VLA proceeded in four broad areas: 1. site and wye, 2. antennas and feeds, 3. electronics, and 4. computers. This breakdown was used in both the design and construction phases of the VLA project, to delineate areas of responsibility. Interfacing between these areas is generally simpler than interfacing between elements within a particular area.

3.1 *Antennas and Feeds*

The VLA required an antenna 25 meters in diameter, with a solid surface of sufficient precision to allow operation at wavelengths as short as one centimeter. The design of antennas of that size for operation at three centimeters wavelength was well understood, and we felt that extension to one centimeter would present no special problem, except perhaps in pointing. That proved to be the case. However, the requirement for 28 such antennas (27 in the array plus 1 spare) offered the opportunity for economies in construction, which had to be anticipated in the design.

The NRAO engineering staff, led by W. G. Horne, developed a design that was then used to establish performance specifications for the final design and construction contract. It was also used to compare and judge designs submitted by bidders. Design study contracts were given to a number of firms to assist in this development. The design finally adopted consisted of an altitude-azimuth mounted, fully steerable Cassegrain antenna. The main reflector deviates slightly from a paraboloid to maximize the efficiency of the Cassegrain system (Williams 1965). The subreflector is a modified asymmetric hyperboloid. The asymmetry causes a small displacement of the Cassegrain focal point, away from the geometrical axis of the primary reflector. As the subreflector is rotated about the axis, the focal point describes a circle in a plane just in front of the vertex of the primary reflector. By placing feeds for different frequencies around this

a

Figure 2 (*a*) The general structure of the antennas is apparent in this ground level view down the east arm of the array. (*b*) A schematic diagram of an antenna, with the main components labeled. Arrows show the signal path, from sky to control building. →

circle, the operating frequency can be changed by simply rotating the subreflector. Figure 2*a* shows several VLA antennas; Figure 2*b* is a schematic diagram of an antenna, with the principal components labeled.

The antennas were designed to operate at full precision in winds to 15 mph, and with slightly reduced precision (principally in pointing accuracy) in winds up to 40 mph. They can survive winds of at least 125 mph, even if coated with 1 cm of ice or 20 lbs/ft^2 of snow. The latter conditions are rare. High winds are not uncommon at the site, however, and wind speeds of 90 mph have been recorded on several occasions.

During the course of the structural and mechanical design of the antennas, considerable attention was given to the relationships between scientific performance requirements, mechanical or structural specifications,

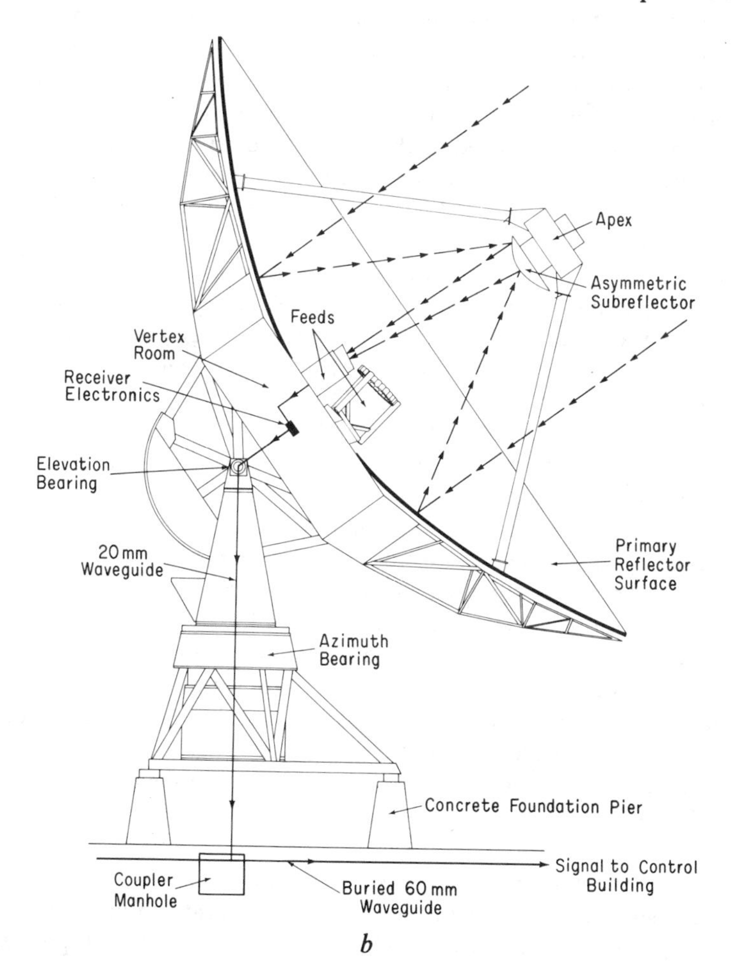

b

and cost. For example, from the standpoint of the scientific performance of an interferometer array it is desirable that the phase centers of the antennas be structurally in the same position on each antenna, and that they not change with antenna orientation. This translates into a requirement, among others, that the two axes of the antenna intersect. But if too close a tolerance is placed on the intersection requirement, the manufacturer may be forced to adopt measurement, adjustment, and control procedures that are prohibitively expensive.

Considerations of this kind, involving compromises between scientific performance, engineering or manufacturing design and construction, and cost, were encountered in many areas of the VLA design, as they are in the design of any scientific instrument. In the case of the VLA they were handled by insisting that all aspects of the design be reviewed jointly by a combined group of scientist/users, design engineers, and builders, however time-consuming and unnecessary such reviews might seem to be on occasion to various members of the review group.

The antennas are designed so they can be picked up and moved with a "transporter" (described in a later section). When an antenna is being used for observations it is bolted to a concrete foundation consisting of three pedestals.

Each antenna is equipped with feed horns for four wavelength bands. The feeds have dual outputs, which can give either opposite circular or orthogonal linear polarization. Design of the feeds, especially for the 18–21-cm band, proved to be one of the more challenging problems (Weinreb et al. 1977, Gustincic & Napier 1977). The 18–21-cm feed is a corrugated horn with a hybrid lens of dielectric and waveguide elements. The 6-cm feed is also a corrugated horn, with dielectric lens, while the 2- and 1.3-cm feeds are multi-mode horns.

A system was also designed for simultaneous operation in the 6-cm and 2-cm bands. It consists of a reflector mounted over the 2-cm feed which is transparent to 2-cm radiation, but reflects 6-cm radiation into another reflector that, in turn, reflects it into the 6-cm feed. System temperatures are increased slightly and aperture efficiencies decreased slightly at both wavelengths but the system could be very useful for various kinds of investigations. It fell victim to inflation, however, and has not yet been implemented. The cost, while not excessive, was significant and the system was judged to be of lower priority than other features on the array.

3.2 *Wye*

Making the antennas movable has several advantages: (*a*) it gives greater scientific flexibility to the array by allowing different configurations and, therefore, different performance characteristics for different problems;

(*b*) it allows the antennas to be assembled at a single location, with a consequent saving in construction costs; (*c*) it allows the antennas to be rotated back to a central maintenance facility on a regular schedule, considerably easing maintenance.

The track system that provides this flexibility is shown, for the central part of the array, in Figure 3. It consists of two standard-gauge railroad tracks, spaced 5.5 m apart, running the full length of each arm of the array. Spur tracks, perpendicular to the main line, run to the observing stations, which are offset 30 m from the center line of the main tracks. A total of about 130 track-kilometers is required. The design adheres as closely as possible to standard railroad practice and materials, to minimize costs. Since both speeds and weights are low by railroad standards the system presents no particular design problems, but much effort went into insuring economy of construction and of maintenance.

A specially designed transporter is used to move antennas. It has four

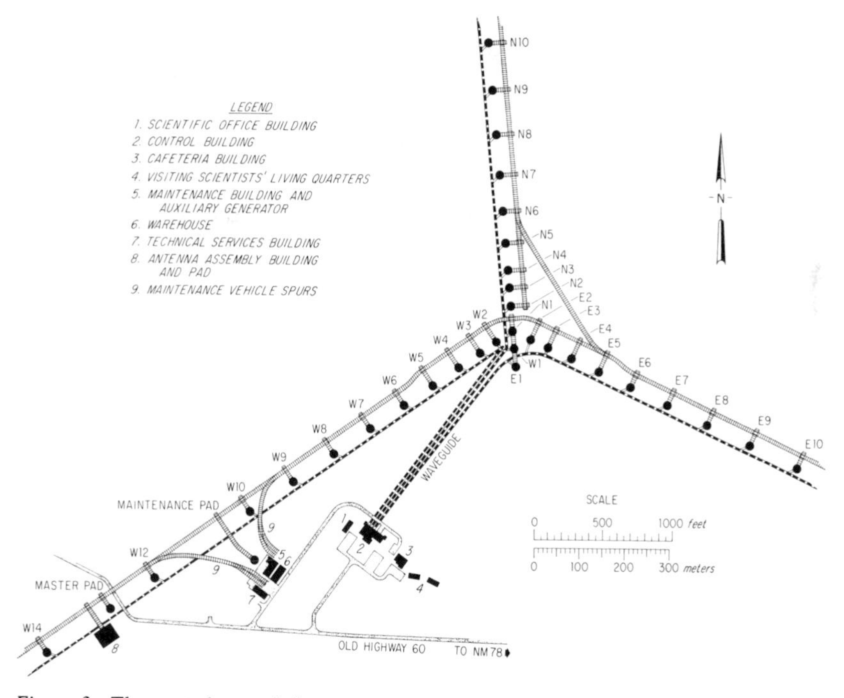

Figure 3 The central part of the array, showing rail system, antenna stations, waveguide, and buildings. Antenna stations are indicated by filled circles (Thompson et al. 1980). There are 24 stations on each arm. In the most compact configuration the 27 innermost stations would be occupied with antennas. In the most extended configuration only E8, N8, and W8 in the drawing would be occupied; all other antennas would be located outside the limits of the figure.

six-wheeled trucks, two on each track. These trucks can be hydraulically raised and lowered, individually or separately, and they can also be rotated through 90°. This enables the transporter to lift an antenna off the three pedestals and to transfer it from a spur track to the main track at right angles to the spur. Diesel-hydraulic power is used to move the transporter. A diesel-electric generator is mounted on the transporter to provide power to critical components of the electronics system during an antenna move. The transporter weighs 90 tons and can move a 215-ton antenna at a speed of 8 km/hr. With two transporters, a change of configuration may require one and one half to five days, depending on which configurations are involved. During a change in configuration only a few antennas are unavailable for observing at any given time.

3.3 *Receiver System*

A simplified block diagram of the receiver system is shown in Figure 4. It is basically a conventional interferometer system, with only a few features or problems that are peculiar to the VLA. The signal from an antenna is amplified in the front end, converted to an intermediate frequency, modu-

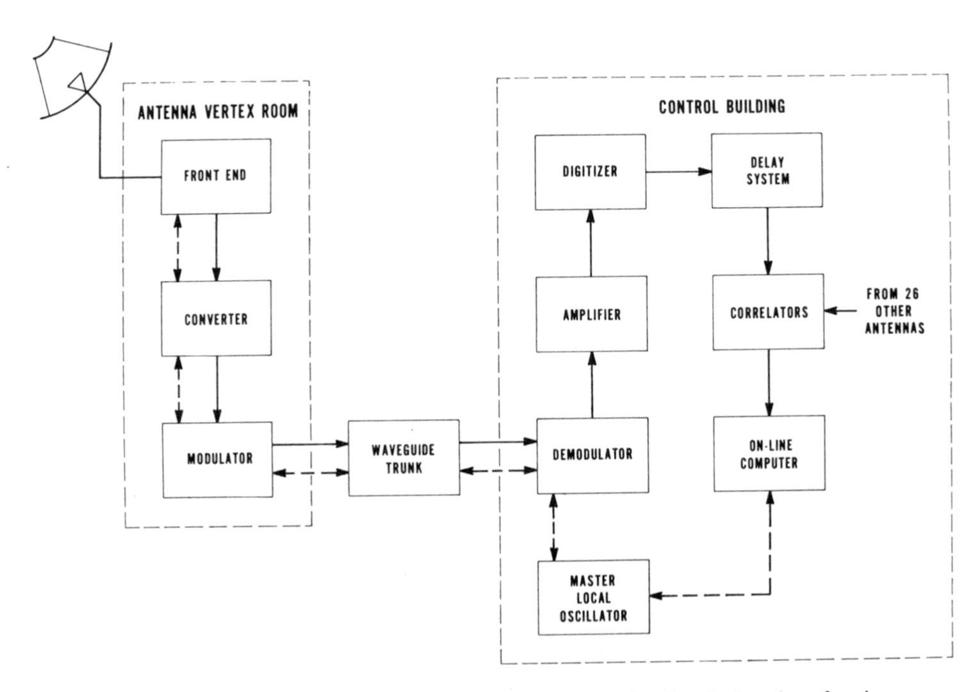

Figure 4 Simplified block diagram of the receiving system, showing the location of major subsystems.

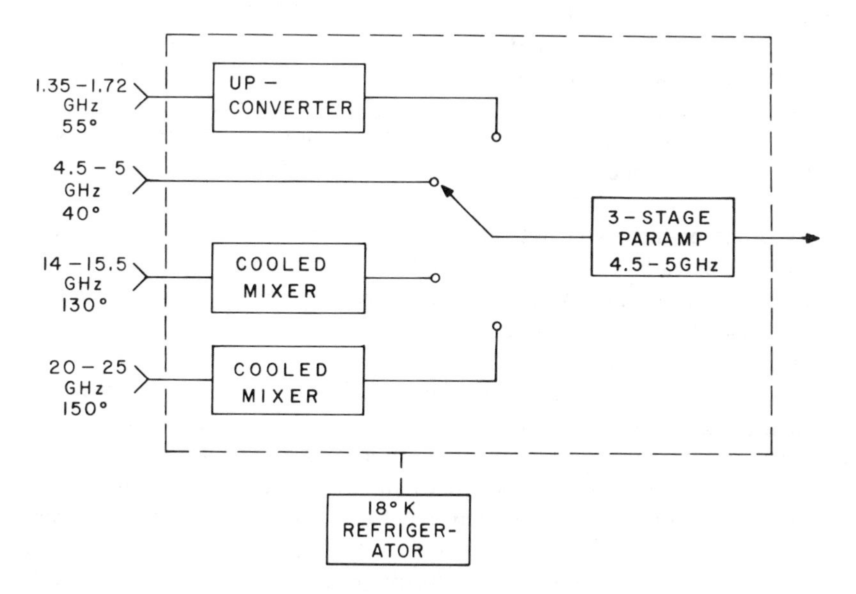

Figure 5 Block diagram of the principal components of a receiver front end.

lated on a carrier frequency, and transmitted back to a central control building. There it is demodulated, amplified again, digitized, and passed through a delay system that compensates for the different air and transmission line path lengths of different antennas. The delayed digital signal is then split into 26 components and correlated with a signal from each of the other antennas.

Most of the special problems and unusual features arise because of the large distances and large number of elements involved in the array, and because of its frequency flexibility and wide bandwidth. Problems of signal distribution and phase stability are common to all interferometer systems, for example, but are considerably enhanced with 351 interferometers and distances of 21 km.

The multi-frequency front end was originally developed for the 43-meter telescope at Green Bank. The VLA front end, shown diagrammatically in Figure 5, consists of the following elements: a 6-cm parametric amplifier, used as the first stage of the 6-cm system; an 18–21-cm upconverter, which uses the 6-cm paramp as a second stage; and 2-cm and 1.3-cm mixers, which also use the 6-cm paramp as a second stage.

Two complete sets of these units, along with appropriate switches and calibration signals, are packaged in a single refrigerator, which maintains the front ends at 18 K. The refrigerator is mounted in an equipment room at the vertex of an antenna. The two independent channels allow processing of both polarization outputs from the feeds. Combination of, for example, the right hand (R) and left hand (L) circular polarization components from two antennas then gives the four combinations (RR, LL, RL, LR), which in turn define the four polarization parameters of the received radiation.

Any one of the wavelength bands 18–21 cm, 6 cm, 2 cm, and 1.3 cm may be selected by computer control. Changing from one band to another requires a few seconds, primarily to allow rotation of the secondary reflector, as was described previously.

The cooled four-frequency front-end package illustrates some of the complex compromises that must be adopted in the design of an instrument like the VLA. First, cooling the front ends, amplifiers, and frequency converters yields greater sensitivity, a major goal of the VLA. But cooled systems cost more than uncooled ones and are more difficult to maintain. In particular, the cryogenic system itself is a principal source of trouble. Second, putting the front ends for four frequency bands into a single package adds considerably to the flexibility and versatility of the VLA and probably also allows modest cost savings in initial construction. But the contents of the refrigerated box are varied and complex. If anything at all goes wrong, the entire box has to be shut down for repair. This adds to the difficulty of maintenance and puts additional demands on reliability of components, particularly since it is a cooled system with all the attendant problems of temperature recycling, accessing, cooling times, etc. Third, maximum sensitivity and high reliability seldom go together. It is usually necessary to sacrifice some performance in order to gain reliability. And finally, high performance and high reliability usually carry a high price tag. In the 18–21-cm and 6-cm bands the VLA front ends are close to present sensitivity limits. At 2 cm and 1.3 cm they are about a factor of five less sensitive than the best presently available for reasons of both economy and reliability.

The local oscillator system must deliver tunable signals, at several frequencies, to each antenna for use with the mixers and parametric amplifiers there. These signals must be synchronous in phase and should be phase stable to a precision of about 1° of phase per gigahertz of observing frequency, over a period of a few hours. The effect of longer-period phase instabilities can be eliminated by occasional observations of calibration sources. All of the local oscillator signals are derived from a single master oscillator located in the central control room. They are tunable in discrete steps so that, for spectroscopic observations, the 50-MHz observing band can be located anywhere within the total band of the receiver/feed combination being used. Local oscillator signals are transmitted to the antennas via the waveguide system described in the next section. Local oscillator reference signals are also transmitted, via the waveguide, from each antenna back to the central control room. Phase variations in the round-trip path are measured and are used to apply a correction to the measured Fourier component.

The stability, coherence, purity, and reliability of the local oscillator system are critical to proper performance of the array. Design of this system was considered to be one of the more difficult problems of the VLA electronics design. Two different systems were designed and tested before the final design was approved.

Another critical and difficult element of the VLA design was the signal transmission system. Three different kinds of signals must be handled: (*a*) stable local oscillator signals must be sent to and from the central building to each antenna, as described above; (*b*) four 50-MHz-wide intermediate frequency (IF) signals, carrying the output of the receiver, must be sent from each antenna back to the control building; (*c*) monitor and control signals for the antennas and the electronics must be transmitted back and forth between the antennas and the control building. For an array of 27 antennas at distances of up to 21 km from the control building this is a formidable task.

The initial VLA design in the mid-1960s utilized several 41-mm and 22-mm-diameter buried cables for transmission of these signals. Microwave free-space transmission was considered but rejected because of interference to the array and multi-path problems due to antennas in the link path near the array center. Optical carrier systems were not sufficiently developed at that time and, in addition, they do not satisfy a reciprocal path requirement for a round-trip local oscillator phase correction system. That is, in order to attribute one half of a measured round-trip phase error to the outgoing path the return path must have identical phase errors. This cannot be realized with an optical system because optical modulators do not function as demodulators, whereas the same microwave mixer can be utilized for both modulation and demodulation.

The coaxial cable system was feasible, but cumbersome and restrictive. Fortunately, in the early 1970s a much better data transmission medium was developed by US and Japanese telephone companies. This was a circular waveguide using the TE_{01} mode of transmission. That particular mode, almost a scientific curiosity for decades, has attenuation of the order of 1 dB per km for a bandwidth of 30 GHz, in a 60-mm-diameter pipe. This is to be compared with a loss of 30 dB per km and a bandwidth of 2 GHz for the 41-mm cable of the coaxial cable system.

The system finally adopted for the VLA consists of one circular waveguide, 60 mm in diameter, running along each of the three arms of the wye. This waveguide carries local oscillator signals, IF signals, and monitor and control signals back and forth between antennas and the central control building. Signals to be transmitted are amplitude modulated onto carrier bands in the frequency range 27 to 53 GHz. Each antenna on an arm of the wye is assigned a different 2.4-GHz-wide frequency band. At each antenna station a waveguide coupler couples appropriate signals into and out of the waveguide. The maximum attenuation in the waveguide system from an antenna vertex room to the central control room is 56 dB. No repeater amplifiers are required, and the reciprocal path requirement for local oscillator phase correction is satisfied. To minimize thermal expansion effects and help maintain phase stability of the local oscillator signals, the waveguide is buried a minimum of one meter under ground. It is also pressurized with dry nitrogen to help reduce attenuation by oxygen absorption and to prevent internal corrosion. External corrosion is inhibited by a plastic sheath and cathodic protection system.

The signal transmission system is one example of how the VLA was a beneficiary of a very rapidly developing technology. The technology is, in fact, still developing rapidly, and if the VLA were designed today it might well utilize another method instead of circular waveguide in its signal transmission system. Other examples, where the VLA benefited from developing technology, are in the areas of electronics and computers. Staying close to the forefront of the technology is not without risk, however. Incorporating a new, unproven device or technique into a large complex system can be costly if the device or technique does not work as expected. So the risks and potential gains must be weighed, and be in some sort of rough balance. In addition, there is a tendency to constantly redesign to take advantage of the developing technology, and thereby delay construction. At some point a design simply has to be adopted and frozen if costs and schedules are to be met and the instrument to be built.

The delay line and correlator systems are other elements of the electronic system that changed considerably between initial and final design. A plane wave from space impinges on different antennas of the array at different times, depending on the location of the antenna and on the point of origin of the incoming signal. In addition, the time required to transmit a signal received at an antenna back to the central control building depends on the location of the antenna. To compensate for these time differences the signal from each antenna must be delayed by an appropriate amount so that all signals arrive at a common point in the proper phase.

The total delay required is 140 microseconds. Further, the delay must be variable and accurate to about two nanoseconds.

Several delay line systems were developed and tested on the Green Bank interferometer. A hybrid system consisting of a combination of cable, lumped constant, and acoustic delays was found feasible. A continuously variable all-acoustic delay was also investigated, and looked promising. Again, however, the delay of the program in the late 1960s worked to its advantage. Digital techniques and circuitry were developing very rapidly, and, in particular, their cost was dropping dramatically. By the time the VLA design had to be frozen, in the early 1970s, a completely digital delay system was feasible, and relatively inexpensive, and it was adopted.

In the final design, the signals from each antenna are digitized when they reach the control building, passed through a variable digital delay line, split into 26 components, and digitally correlated (multiplied) with a signal from each of the other 26 antennas. For continuum observation the cross-correlations are determined only for zero time delay. For spectroscopic observations the correlations are obtained for a series of time delays and then Fourier transformed to obtain the correlation as a function of frequency.

Two other interesting features of the digital delay and correlation system are worth noting. First, the number of identical units was large enough to justify the design and production of two special integrated circuits, a dual 3-level multiplier and a 12-bit integrating counter. These devices were not only economical for the VLA, but several other radio astronomy groups were able to use them in correlation systems they were building. Second, the system has a built-in automatic test and replacement capability. A test signal is periodically injected and processed. Correlator failures are detected and delay line failures are detected and automatically replaced by spare units.

The electronics system is also responsible for the generation and transmission of control and monitor signals. The antennas, subreflectors, and various elements of the electronics are under computer command via a digital control system. In addition, a variety of data concerning the performance of the antennas and electronics is monitored and available in the control room for calibration and error or malfunction diagnosis. Some of the monitor data are used for automatic calibration and control. The system monitors more than 200 parameters for each antenna, although not all of these data are presently used. The monitor system is potentially a very powerful tool for calibration and quality control of the observational data and for diagnosis of equipment faults. Understanding and implementing the system is a major, continuing job at the VLA.

3.4 *Computer System*

The task of a VLA computing system was succinctly described in the VLA Proposal, Vol. II, page 20–2 (NRAO 1967) as follows:

> The philosophy that has been evolved for the computer system for the VLA is that in normal operation the operator should have to do as little as possible. The computer must be informed of the position in the sky that it is wished to observe and the length of time that the observation should continue. The computer will then point the antennas, set the receiver gains, acquire, condense, and record the receiver outputs, monitor many tens of receiver checkpoints, keep a log of the system behavior, and inform the operator if any part of the system is not behaving according to the computer's internal specifications. After the completion of the observation, the computer will sort the observed data points onto the μ-ν plane (the μ-ν plane is the Fourier transform of the sky plane), calculate and apply various calibration corrections, combine observations, apply a weighting specified by the observer, perform the Fourier inversion, and output a map of the region of sky under study. This will be done asynchronously with the computations necessary for observations, but at a rate such that these computations will not fall behind. No backlog should be allowed to form.

Implementation of the philosophy has proven extraordinarily difficult, and remains the principal challenge for the VLA designers and builders. It was recognized from the beginning of the project that the total computer task described above was a large one. During the late 1960s and early 1970s several things transpired to make the job especially challenging.

First, computer hardware technology, perhaps more than any other, was evolving at a rapid rate. This was fortunate for the VLA because without that rapid development VLA performance could have been severely restricted simply by limited computer capability. But the rapidly developing technology also made it difficult, in a long term project with limited financial resources, to establish a well-integrated, advanced design that could anticipate and take full advantage of the new developments. Second, the sophistication and power of the data processing and display techniques used by astronomers on interferometer and aperture synthesis data has increased tremendously in the past few years. Third, the use of unedited or computer-edited data has not been as extensive as was originally contemplated. Fourth, there has been a very large expansion in the VLA performance expectations over those originally specified. Spectral line capability was originally expected to be modest. It now encompasses up to 256 frequency channels. The maximum resolution has gone to 0.1 arcsec, from 1 arcsec, by extending the wavelength coverage to 1 cm. The numerical field of view (number of picture elements in a map, or number of data points) was first expected to be about 100×100. Now 1000×1000, or more, is routinely called for. Further, the map-making mode of operation of the VLA was expected to generally involve a minimum of a few hours of observing time per region mapped. But the VLA is so sensitive that "snapshots" involving only a few minutes' observing per source or region are very powerful for many problems. To make full use of the potential of these "snapshots" requires almost as much data processing as would a full eight-hour observation. All of these developments demonstrate the great power and flexibility of the VLA. However, they also greatly increase the computing load.

Finally, there is one further consideration that has complicated the computer system design. The amount of data generated by the VLA for even one map is sufficiently large that not many computer installations available to astronomers are able to handle it.

There are several consequences of the above developments to the VLA computer design. First, the computing job is even larger than originally anticipated, though this is a happy development since it is a direct result of the VLA's being more powerful than originally expected. Second, the computer data processing task is less well defined than it first was, for it now includes a much wider variety of data reduction options and the whole gamut of "post-processing" analysis techniques and programs. The task of data processing, at this point, is almost an open-ended problem that defies logical definition. Third, the "best" processing techniques are not easily identified because of rapid developments in the computer field as well as our lack of experience in dealing with the large amount of data generated by an instrument like the VLA.

For all of these reasons it was decided that the data processing part of the computer system could not really be designed before some experience was obtained with the VLA. It would have to grow and evolve with the instrument. That is what has happened and is still happening. Designing and acquiring the hardware part of the system on a faster schedule would have been possible simply by providing for much more computing capacity than was originally believed to be necessary. This approach would have been prohibitively expensive.

The computer design that has emerged divides the computational tasks between two basic systems. The on-line computer system handles all of the on-line, real-time computing jobs including array operations, electronics control and monitoring, and data acquisition and a first stage of calibration. This system employs five identical mini-computers. As a control system it is programmed for minimum interaction with the observer.

Among its other tasks, the on-line computer collects the observational data and passes them on to the second system, the off-line computer. Off-line data reduction and analysis, ranging from data editing and calibration to map making via Fourier transformation to further processing and analysis of maps, is done in this system. The software allows a very wide range of processing to be conducted with extensive observer interaction. Development of both the software and hardware of the off-line system is still under way.

The analysis of astronomical data can be a lengthy process which, for a given observation, may proceed in fits and starts as the investigator gets new ideas or additional, complementary data. The designers never planned for all analysis of VLA data to take place at the VLA site. It would be unfortunate to constrain observers in this way. A VLA observer may take away with him, on magnetic tape, calibrated VLA data or images for further processing elsewhere. Unfortunately, there are very few computing facilities accessible to astronomers that can handle these VLA data. In fact, this is a general problem developing for investigators using other astronomical instruments as well. Some solution to the problems of data processing must be found soon.

3.5 *Intermission*

The design effort discussed thus far culminated in the issuance of Volumes 1 and 2 of the proposal to build the VLA (NRAO 1967). There followed several years of review and debate about the merits of the instrument. The two "Dicke" Committees of the NSF (NSF 1967, 1969) and the Astronomy Survey Committee of the National Academy of Sciences were involved in review of the VLA proposal. The latter committee, chaired by Jessie L. Greenstein, ultimately recommended that the VLA be given first priority among all astronomy instrumentation projects for the 1970s (National Academy of Sciences 1972). That recommendation was probably the final crucial step in gaining approval for the VLA.

Some design work was continued during 1967 and 1968, and a third volume of the proposal, describing results of that work, was issued in January 1969 (NRAO 1969). The VLA design group was then disbanded and all direct VLA design work was terminated until such time as it appeared the instrument might be approved. The intermission lasted until the spring of 1971.

4. CONSTRUCTION

Construction of the VLA was approved by Congress and the National Science Foundation (NSF), and first construction funds were made available in fiscal year (FY) 1973, that is, in November 1972. Design work on electronics and antennas was recommenced in the spring of 1971, under the direction of H. Hvatum, in anticipation of that funding.

It was agreed (between NSF and the Office of Management and Budgets) that the project should be funded at about $10 million a year. This rate of funding led to a longer construction period, and consequent higher costs, than was originally planned, and it further delayed actual scientific use of the instrument. It did have some advantages, however, in that a longer period was available for detailed design and testing of some elements, which probably resulted in a better instrument overall. So, although there was unhappiness at the time with the schedule imposed by the funding rate, in the long run it probably did no harm and perhaps did some good.

The total cost of the VLA, including that due to the stretched-out schedule, was estimated in 1972 to be $76 million. This figure included a $6 million contingency and an allowance of 6% per year for inflation. Perhaps somewhat surprisingly, it was consistent with the $40 million estimate of the early 1960s. Because the design of the system was well advanced, and because a large fraction of the money was for relatively straightforward items such as antennas and railroad track, the cost estimate was basically sound. NRAO was determined to build the VLA on schedule and within budget.

NRAO decided to act as its own prime contractor for the VLA rather than bid the entire job to a single contractor as a turnkey operation. The latter approach was favored by several potential contractors. However, to keep within the time and cost schedules, NRAO would have to maintain tight control over the job, minimize design changes and "extras" in subcontracts, and make sure that detailed design and specifications were fully developed before awarding construction contracts. The tight control necessary by NRAO and the diverse nature of the project, ranging from conventional railroad construction to state-of-the-art electronic and computer technology, seemed to preclude a turnkey approach.

A VLA management and construction organization was set up in the fall of 1972. This group, under the direction of J. H. Lancaster, was responsible for the detailed design work and setting of specifications and all aspects of the construction. H. Hvatum retained responsibility for the overall basic technical design. Major construction contracts were ultimately given out in only a few cases, as in the construction of antennas and transporter, antenna feeds, rail system, and site buildings and facilities. However, these contracts account for about 45% of the total cost of the array. All other work was done by the VLA staff. At the peak of the construction in 1979 there were about 130 people on the VLA construction staff, exclusive of contractor personnel.

Shortly after the project got under way it faced a major financial crisis as a result of the Mid-East war, the oil embargos, devaluation of the dollar, and the consequent acceleration of inflation. To try to keep within the $76 million budget a number of changes were made in the program. The simultaneous dual-frequency system was eliminated. The number of transporters was reduced from three to two. These were the only changes that directly affected the scientific usefulness of the array. The site building program was drastically modified, with simpler prefabricated buildings replacing the more permanent structures originally planned. This was true for all but the control building and the cafeteria. The prefabricated buildings are adequate but will not last as long as the buildings that were originally planned. An airstrip was also eliminated.

A number of other elements of the project were postponed pending clarification of the budget situation. These included elimination of one half of the IF system (which affects the array sensitivity by $\sqrt{2}$) and the last 4 km of track (which would reduce the array resolution by 4/21). Eventually, economies were found in other areas of the construction program, especially in the cost of materials for the rail system, and most of the postponed items were reinstated in the program.

4.1 *Site and Wye*

Site and wye work included all building, utility, and railroad construction, and installation of the waveguide. The building, utility, and railroad construction was done by contractors, while NRAO staff personnel were responsible for installation of the waveguide. Figure 3 shows the central part of the site, including buildings and waveguide runs.

A number of curious, totally unanticipated minor problems cropped up in connection with the site and wye work that had nothing to do with actual construction. The first problem arose in acquiring access to the site. The north arm of the wye extended 2 km into one of a very few irrigated patches of land in New Mexico outside of the Rio Grande valley. The owner of the property was unhappy at the idea of a railroad running into the land and was clearly planning to extract a high price for it. The question was resolved simply by cutting 2 km off the north arm, so that it stops just short of the irrigated land.

New Mexico law requires that all proposed construction areas be surveyed for the existence of sites of special archaeological interest and that any such sites found be investigated before construction work commences. One site of interest was found along the southwest arm of the array, and it was determined that it should be excavated. But, while the archaeologists thought the site was important and needed investigation before being disturbed by construction, no one was willing to pay for it. Neither the New Mexico nor the NSF archaeology programs, nor any university program within the state, was prepared to fund the necessary "dig." After a lengthy period of bickering, NRAO and radio astronomy had to finance a bit of archaeological research.

A team of archaeologists from New Mexico State University did the field work during the winter and spring of 1978. Remnants of three cultures were found at the site: the Folsom Culture of 10,000 years ago; the Cochise Culture of 3,000 years ago; and the Mogollon Culture of about 1,500 years ago. Thousands of artifacts ranging from points, utilized flakes, and bone fragments, to pottery shards, manos, and lithic tools were found and catalogued (Beckett 1980).

The NRAO had originally intended to utilize used rail and railroad ties for the railroad construction, and, in fact, a supplier had been found and prices tentatively agreed upon. But in 1973 the price of used rail soared far beyond the budgeted price, and threatened to jeopardize the entire program. A desperate search for alternate sources of rail led to the discovery of large quantities of abandoned track on military bases around the country. With the aid of the NSF some of this track was declared surplus and made available to NRAO for the cost of taking it off the bases and shipping it to New Mexico. The availability of this surplus rail was instrumental in keeping the VLA cost within budget in the face of much larger than anticipated inflation. In fact, enough money was saved to allow restoration of most of the postponed elements of the program.

Burial of the waveguide was a critical task. To avoid excess attenuation and multi-mode propagation in the waveguide, it must have a minimum radius of curvature of one km. Thus, it had to be installed flat and straight and in such a way that it would remain flat and straight for the life of the VLA. Several different schemes for burying the waveguide were developed and tested at Green Bank and at the VLA site. The most promising of these was then used to install about 3 km of waveguide along the southwest arm connecting the first two antennas. This installation was accomplished two years before the schedule called for further waveguide burial, allowing time for additional testing. Those tests suggested that the waveguide was settling unevenly at an unacceptably fast rate. The burial procedure was therefore modified. The new procedure provided better initial performance and also eliminated the deterioration with time. Fortunately, the original run of waveguide stopped settling well before its performance became unacceptable.

Because the waveguide burial was critical to the entire array performance and because it required careful measurement and control, it was done entirely under the direct supervision of NRAO personnel. This was one of several instances where NRAO could obtain better performance and quality control, for a given cost, by doing the job itself.

4.2 *Antennas and Transporter*

The design and construction of 28 antennas was awarded as a single contract to E-Systems, Inc., after an intensive and very competitive bidding and selection process. This, and the first transporter, were the only instances where the same contractor was responsible both for design and construction. This method of procurement was chosen in order to take full advantage of the potential cost economies inherent in the construction of 28 identical units. E-Systems, in the end, produced 28 antennas of high quality that fully met specifications, at a reasonable price.

The contract itself was complicated by the fact that in order to realize the savings in mass producing the antennas the contractor needed contractual assurances that all 28 would in fact be built. NRAO on the other hand had to operate under the constraint of annual funding from Congress that precluded a contractual commitment for the full amount of the contract. The final contract was a fixed price agreement with various options built in to protect both parties.

The 1973 price increases caused some early difficulties. The antenna contractor was suddenly faced with projected material costs much larger than originally estimated. After protracted negotiation the contract was modified in a variety of minor ways. The basic elements of the contract, including performance specifications, cost of the antennas, and fixed-price nature of the contract, remained the same.

The antenna contract required, first, that a detailed design be produced and approved by NRAO. NRAO had already developed a detailed design of its own against which that of the contractor was compared. After the design was approved two prototype antennas were built and erected at the site. These antennas were thoroughly tested by NRAO engineers before the contractor proceeded further. The remaining 26 antennas were then produced at an average rate of about one antenna per seven weeks for the next several years. Because of the extensive design, and the prototyping and testing that occurred prior to and early in the contract, very few changes or extras were required or allowed, and the original fixed price survived almost intact.

After an antenna was received from the contractor it was tested to see that it met specifications. It was then moved to an outfitting pad where the installation of cabling, subreflectors, electronics, cryogenic equipment, and miscellaneous lights, ladders, and other gear was carried out by NRAO personnel. Finally, it was moved to a regular observing station along one of the arms of the array and "first fringes" were obtained by using it with another antenna as an interferometer. A bottle of champagne was broken over the first antenna to come off this assembly line, and several bottles were consumed by the staff when the last "first fringes" were obtained.

The antennas were fabricated off-site and then assembled at the site in a special assembly building, 34 meters tall and containing special cranes and other equipment. Now that the antenna construction is complete, the building will be used for antenna maintenance. One antenna will be rotated back for routine maintenance and repair every seven weeks. Thus, each antenna will be serviced once approximately every $3\frac{1}{2}$ years.

The antenna assembly building was the first VLA structure at the site, except for a few trailers. It looked very conspicuous and out of place, a totally foreign object in the midst of the natural beauty of the Plains of San Augustin. Now, nestled among 27 gleaming antennas, it is less conspicuous. The antennas themselves somehow do not look out of place, at least not to a radio astronomer.

E-Systems also designed and constructed the first transporter. After several years of experience the design was modified, the first transporter upgraded, and a second transporter built to the new design.

4.3 *Electronics*

Most of the electronics system was designed and built by NRAO personnel. S. Weinreb provided the detailed electronics block diagram, which specified the interfaces and design requirements for the various modules and subsystems. The subsystems were then assigned to a group of NRAO engineers (front ends, local oscillator, IF transmission, digital delay, and correlators) and to several outside contractors (monitor and control, waveguide signal distribution, and antenna feeds). Several large building blocks, such as power supplies, paramps, and cryogenic refrigerators, were purchased as finished products from manufacturers, but for the most part the system was assembled and tested by NRAO personnel. In general, the same group responsible for detailed design of a particular subsystem was also responsible for manufacturing it. Considerable redesign of the electronics was done as a result of experience with the first units built. Subsequent units were then built to the new design, and the earlier units retrofitted to the new design as time allowed. This is typical of any radio astronomy electronics system, and it is expected that VLA electronics will undergo a more or less continuous process of evolution and upgrading. However, in a construction project with a fixed budget and time schedule such a process can be dangerous. A major challenge for the VLA project managers was to allow, and indeed encourage, the right amount of redesign and retrofit, but not let it get out of hand and endanger the budget or the schedule.

One aspect of the VLA hardware in general, and the electronics in particular, that received increasing attention as the construction project progressed was reliability of components. As soon as a reasonable amount of hardware was built and operating, statistics of failure rates were accumulated. It was quickly found, for example, that the compressors used in the electronic cryogenic systems had a very high failure rate. Another supplier was found who built more reliable units. Various other high failure rate components were also identified and were replaced with more reliable ones.

4.4 *Computer System*

The computer system was designed, assembled, and programmed by NRAO personnel under the direction of B. G. Clark. Those activities are still very much in progress. The on-line computer used to control the array was designed and procured early in the construction program. Software design and development of this computer has proceeded on schedule. The first elements of the off-line computer, to be used for data processing, were also designed and acquired early in the project. Further development of the system, and of the software for it, has been dependent on experience gained with the partially completed array. As a result, the capability of the system has lagged behind the demands placed on it (see earlier discussion in Section 3). This situation will probably persist for some time. Data processing is certainly the principal irritant and bottleneck encountered in using the VLA. It strongly affects the speed with which any individual or group can process data. However, it should not affect what or how much can be done with the array.

4.5 *Construction Summary*

In spite of real problems, petty irritations, miscellaneous distractions, and excessive inflation, VLA construction has proceeded pretty much on schedule and within budget. Construction should be completed early in 1981 for a total cost of just over $78 million. This may be compared with the 1972 projections. At that time, when construction began, a completion date in mid-1980 and a cost of $76 million were estimated. Scientifically, the VLA meets or exceeds all of its performance goals.

4.6 *Transition to Operations*

The nature of the VLA makes it both possible and desirable to begin some scientific observing well before the instrument is completed. Not only does this give valuable scientific results with the instrument earlier than would otherwise be the case, but it also makes possible debugging and necessary redesign and other adjustments at an early stage, and allows development of calibration and observing techniques, data processing procedures, and array operations and maintenance procedures.

Observing, and all the developments that go with it, began in a small way as soon as there were two antennas in 1976. More officially, scientific use of the partially completed VLA was opened to both guest investigators and scientific staff in 1977. Initially six antennas were available. The number grew steadily as more antennas were completed and instrumented. Now all 27 antennas are in regular use.

There has been a planned, gradual development of an operating staff to allow a smooth transition from construction to operations. The transition plan was meant to accomplish a number of goals. First, the expertise gained during construction should, in so far as possible, be transferred to operations. Second, as many people as possible should be transferred from construction to operations, partly for the reason mentioned above, partly to minimize the hardships occasioned by layoffs, and partly because the possibility of a permanent position is an important factor in recruiting high quality construction staff. Third, the transition should be accomplished with minimum cost and with minimum disruption to the construction program.

By and large the transition from construction to operations has proceeded smoothly. Some problems have arisen because the total number of people needed for construction, plus operation and maintenance in the later stages of the project, is considerably greater than that needed for either construction or operation alone. To avoid having to hire, and then soon release, too many people, the operations were developed a bit more slowly than would otherwise have been desirable. The problem, however, has not proven to be too serious, and, more important, the relation between construction and operation has been kept in reasonable balance. In the process, a great deal of interesting new scientific work has been done with the partially completed VLA. Some of that work is described in the final section.

5. PERFORMANCE AND EXPECTATIONS

It is dangerous to speculate about what may be accomplished with a new instrument. Such speculation is, of course, necessary during the design phase when performance goals must be set, and it is also necessary in convincing others that the proposed instrument should, in fact, be built. Thus are produced the standard "shopping lists" of expected accomplishments, which tend to look very much alike irrespective of which instrument is being discussed. Once an instrument is built and operating, however, it can speak for itself, and what it actually has accomplished becomes more important than any of the earlier speculation. Then, too, the most exciting contributions are often those that were unanticipated.

The partially complete VLA has now been in limited use, for scientific investigations, for about four years. Although its capabilities during this period have been, on average, perhaps only one tenth those expected of the full array, the work done thus far does give an indication of the power and scope of the instrument. In the following sections the performance of the array, as built and tested to date, is summarized and some early observations presented. They, mixed with a little imagination and speculation, may give some indication why so many astronomers are excited about using the VLA for their investigations.

5.1 *Performance Characteristics*

The expected performance of the VLA, based on experience obtained with the partially completed array, is summarized in Tables 4 through 7. The sensitivity listed in Table 4 is for detection of a point source of radiation with signal/noise equal five. Dynamic range is a complicated function of system performance and atmospheric stability. The figure in Table 4 is conservative. Under favorable conditions dynamic ranges $> 300:1$ have already been achieved with the incomplete array.

Four wavelength bands are available. Changing from one band to another is accomplished in less than one minute. This is particularly useful

Table 4 VLA general performance parameters

Resolution	$0''.13$ to $350''$
Sensitivity[a]	up to 0.085 mJy (5σ)
Dynamic range[a]	$> 100:1$
Field of view	$1.2'$ to $30'$
Side lobes[a]	-16 dB maximum
Wavelength range	1.3 to 21 cm in 4 bands
Sky coverage	$-25° \leq \delta \leq 90°$

[a]For 8–12 hour observing time.

Table 5 Antenna and electronic parameters

Wavelength	Aperture efficiency (%)	System temperature (K)	rms noise[a] (mJy)
18–21 cm	50	50	0.02
6	65	50	0.02
2	54	240	0.10
1.3	46	290	0.14

[a]For 27 antennas, 50-MHz bandwidth, 8-hour integration time.

because it allows the continuum spectrum of a source to be determined with relative ease. The short wavelength limit, 1.3 cm, is set by the characteristics of the antennas. Twenty-one centimeters is probably the longest wavelength that can be used in the Cassegrain configuration. At longer wavelengths the physical dimensions of the primary feed needed to illuminate the Cassegrain reflector become awkwardly large. Within the wavelength range 1.3 cm to 21 cm additional observing bands could be added rather easily. Operation at wavelengths longer than 21 cm is also possible in the future with feeds at the primary focus. However, such an arrangement would not permit as rapid changes of observing band as is possible with the present Cassegrain system. Table 5 lists some antenna and electronic parameters for each band.

Four configurations of antennas, A, B, C, and D, are available (see Table 6). They give different resolutions and sensitivities to extended sources and therefore allow some tailoring of the instrument for the problem.

Table 6 VLA beamwidths

Wavelength (cm)	Primary beamwidth	Synthesized beamwidth[a]			
		A	B	C	D
18–21	26′–30′	2.0″	6.6″	22″	71″
6	8.6′	0.6″	2.0″	6.5″	21″
2	2.9′	0.2″	0.7″	2.2″	7.1″
1.3	1.9′	0.13″	0.4″	1.4″	4.6″
delay beamwidth[b]		1.2′	3.9′	13′	42′

[a]For 27 antennas, 15-dB taper, 8-hour observing time.
[b]For 50-MHz bandwidth.

The field of view of the array may be limited by the primary beamwidth of an individual antenna, shown in Table 6, or by the "delay" beamwidth, also listed in Table 6, which is a measure of distortion and attenuation resulting from loss of coherence away from the center of the field at large system bandwidths. It may also be limited by the particular sampling of Fourier components measured or by the specific processing techniques used in analyzing the observations. The lowest value given in Table 4 is for a 50-MHz bandwidth and is independent of wavelength. The large value is for 21-cm wavelength and is set by the beamwidth of the individual antennas. Still larger fields, up to 2–3 times the primary beamwidth, can be mapped if needed, but with the consequent attenuation due to the primary beam. Fields of $80' \times 80'$ have been mapped at 21 cm, for example, when searching for confusing sources.

For spectral line studies various combinations of frequency resolution and total bandwidth are available, ranging from 16 channels with 3.125-MHz resolution for a bandwidth of 50 MHz to 256 channels with 381-Hz resolution and a total bandwidth of 97 kHz. A few of the possible combinations for the D array are shown in Table 7.

To obtain a map with maximum resolution, sensitivity, and dynamic range requires eight to twelve hours of observing. However, the array also has very high sensitivity even with much shorter observing times. For example, in fields where high side-lobe levels and low dynamic range can be tolerated, a sensitivity of about 1 mJy at 6-cm wavelength can be reached in just 5 minutes of observing time. This makes possible a wide variety of statistical studies, investigations of variability, and exploratory searches for new sources of radiation or new types of phenomena. This "snapshot" mode of observing has already proven fruitful with the incomplete array, and it is expected that the full VLA will be extensively used in this manner.

Table 7 D-array spectral line parameters

Frequency resolution	Number channels	Bandwidth	Temperature sensitivity (K) at various wavelengths[a]			
			21–18 cm	6 cm	2 cm	1.3 cm
763 Hz	512	391 kHz	5.0	12	80	113
12.2 kHz	512	6.25 MHz	1.2	0.77	5.1	7.2
48.8 kHz	256	12.5 MHz	0.63	0.38	2.6	3.6

[a]For 27 antennas, 8-hour observing time.

The detailed performance characteristics of the VLA are somewhat complex, especially those relating to dynamic range, field of view, sensitivity, and spectral line measurements. Many of them are under direct control of the observer, who can adjust parameters to give the best overall performance for a given problem. Performance characteristics are discussed in more detail by Thompson et al. (1980) and in a VLA users' manual prepared by R. M. Hjellming (1978).

5.2 *Solar System and Galactic Studies*

The VLA will be used for a wide variety of solar system and galactic studies, from the determination of the vernal equinox using observations of asteroids to the measurement of interstellar magnetic fields using observations of Zeeman splitting of the 21-cm line. Only a few examples will be mentioned here.

THE SUN The VLA is expected to have a major impact on investigations of solar active regions. It has been known for many years that microwave radiation is associated with solar flares and active regions. Studies of that radiation can provide information about high energy electrons generated in flares and about the magnetic fields associated with solar active regions. Radio measurements may provide the key to understanding the physical processes occurring in these regions. Because the regions themselves are small and the phenomena associated with them are both small in scale and transient in time, very high time and spatial resolutions are required for these studies.

With the VLA used in a "snapshot" mode, maps of active regions may be made with spatial resolution of 0.1–1 arcsecond and time resolution of 10 seconds. Such maps should be extremely useful for investigating active regions.

Some solar observations have already been made with the partially completed VLA. For example, Marsh & Hurford (1980) have obtained maps of a few microwave bursts at 2-cm and 1.3-cm wavelength, with spatial resolution of about 1 arcsecond and time resolution of 10 seconds. They found that in each case the microwave emission at first came predominantly from a compact source located between the Hα kernels of the flare. As a burst developed the source of radio emission became larger and elongated along the magnetic field lines joining the Hα kernels.

The flexibility of the VLA, either through parallel subarrays at different frequencies or quick switching of frequencies, has been most useful in creating "three-dimensional" pictures of flares at different heights in the atmosphere. The images in different frequencies are quite different and help investigators study the optical depth and magnetic structure of the burst.

Solar astronomers are finding evidence for rapid, small-scale magnetic variations in flares; they hope these variations can be understood from VLA measurements of the change in polarization in bursts like these.

In another program, Kundu & Velusamy (1980) observed a solar active region with the VLA at 6-cm wavelength with a resolution of 3.5 arc-seconds. They found that the radiation formed a loop structure connecting two sunspots of opposite polarity. Several compact sources of emission were observed within the loop structure. They suggest that the diffuse emission in the loop is optically thin thermal emission from a hot (10^6 K) gas and that the compact sources are due to gyro-resonance.

These, and similar programs already undertaken, have demonstrated the potential power of the VLA for studies of solar active regions. The full array will allow still greater resolution and sensitivity to be brought to bear on these problems and may be expected to make a substantial contribution to our understanding of the sun.

PLANETS The atmospheres of the major planets, and of Venus and Titan, are opaque at optical wavelengths, thus their surfaces and lower atmospheres can only be observed at radio wavelengths. But only the VLA has sufficient resolution and sensitivity to detect the smaller objects and to obtain detailed maps of larger objects. The VLA can detect and resolve the thermal emission of all of the planets (except perhaps Pluto) and of several satellites, including Titan and the Galilean satellites of Jupiter. It will, therefore, be used to study the lower atmospheres or surfaces of these objects. For example, high-resolution observations of limb-darkening effects together with the wavelength dependence of surface temperature will be used to determine the chemical nature and vertical distribution of microwave absorbers in the atmospheres of Jupiter, Saturn, Uranus, and Neptune.

The VLA will also be used to investigate the structure of nonthermal emission features, such as those exhibited by Jupiter. Figure 6 shows Jupiter at four meridian longitudes, as observed with the VLA at 20-cm wavelength. The thermal disk radiation is clearly visible, as is a more complex distribution of nonthermal radiation. The observations were made by J. A. Roberts, G. C. Berge, and R. C. Bignell (in preparation).

ASTROMETRY Radio astrometry is a relatively new and promising field. The techniques of interferometry as used in the VLA and other radio interferometers allow the direct determination of declinations, independent of any optical reference system, to a precision set by the resolution

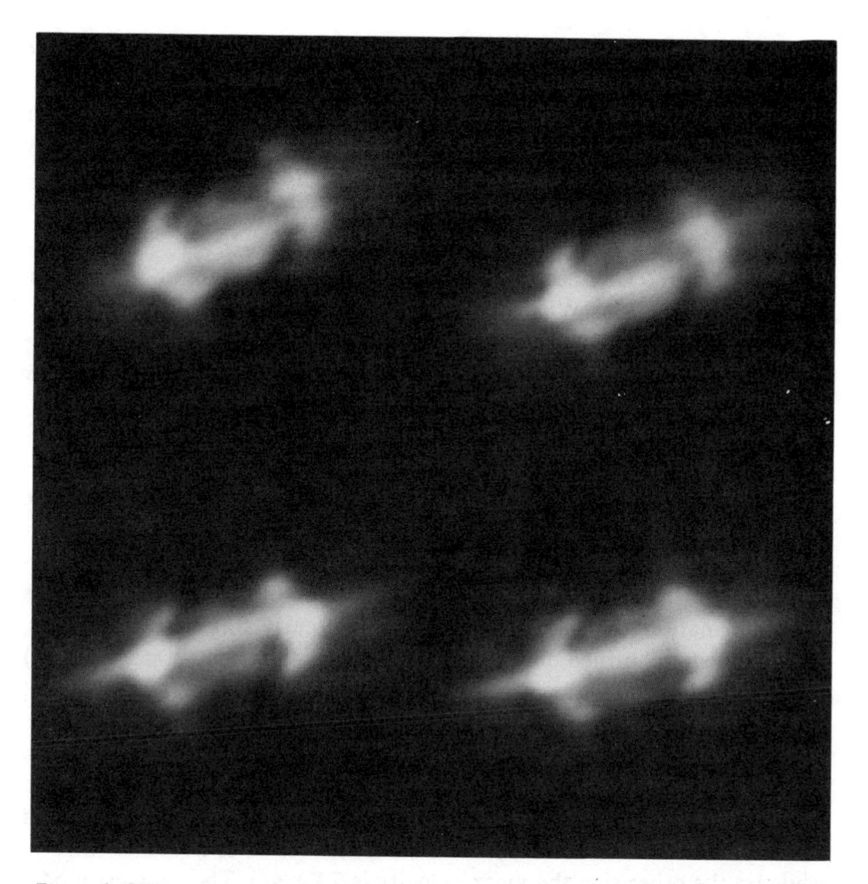

Figure 6 VLA radiographs of Jupiter at 20-cm wavelength. Clockwise from upper left the central meridian longitudes are 199°, 214°, 304°, and 319°. The separation of the brightest spots on either side of the disk in each radiograph is about 50″. The half-power beamwidths are 2.2″×5.5″ (G. L. Berge, R. C. Bignell, J. A. Roberts, in preparation).

and stability of the instrument. In the case of the VLA, this is about 0.05 arcseconds for a single measurement. Differential right ascensions can be determined with similar precision. At present it is necessary to adopt a zero point in right ascension from optical measurements. However, the VLA has sufficient sensitivity to observe the brighter asteroids and from them determine directly the vernal equinox. Thus, it will be possible to establish a reference frame with the VLA that is completely independent of any optical reference frame, is uniform over the whole sky, and is relatively free of regional systematic errors. Further, a truly "inertial" reference system can be established by choosing distant extragalactic sources to define the system.

Repeated measurements over a long period of time can be used to refine certain fundamental geophysical constants such as the precession constants and earth tide parameters. Many of the measurements needed for this purpose are also required for calibration of the array. Thus, array calibration and some aspects of astrometry will be incorporated into a single long-term program.

SPECTROSCOPY The spectral line system will be the last to be fully implemented on the VLA. When it is completed in 1981 it should be an extremely useful instrument for studies of neutral hydrogen and many molecules in our own and other galaxies. Lines of at least eleven molecules lie within the VLA wavelength bands.

With the spectral line system, and especially the software for data analysis, not yet completed, only relatively simples types of spectral line investigations have been undertaken thus far. Most of the programs have utilized the high sensitivity of the array to determine accurate positions and radial velocities of galactic OH and formaldehyde masers.

Studies of the spatial and velocity distribution of OH and H_2O masers give information about the kinematics of the early stages of proto-stars, before the stars "turned on." S. Hansen and K. Johnston (in preparation) have studied the spatial and velocity distribution of OH masers in Orion, for instance. Their VLA map is shown in Figure 7. It shows the positions and velocities of a large number of OH masers in the region of Orion A. The OH masers are symmetrically distributed with respect to an infrared source and an SiO maser at the center of the distribution. The mean velocity of the OH masers on one side of the center is about 19 km s^{-1}, while on the other side the mean velocity is 7 km s^{-1}. The simplest model to account for distribution is one of a flattened disk gravitationally bound to a central object of mass 80 $M_{\odot}$. The SiO maser, indicated by the triangle in Figure 7, may originate in this central object, which is probably a massive star in the early stages of star formation. The study of OH, formaldehyde, and H_2O masers and of other evidence of star formation such as compact ammonia clouds will be a major area of VLA spectral line observations.

Another area where the VLA and its spectral line system may be of special value is in the study of H I and OH absorption lines, both in our own and in other galaxies. Absorption line observations can take full advantage of the high spatial and frequency resolution of the array. Studies of OH absorption against the nuclei of galaxies may give information about noncircular motions in these galaxies. Studies of H I absorption of sources shining through the outer reaches of galaxies may give informa-

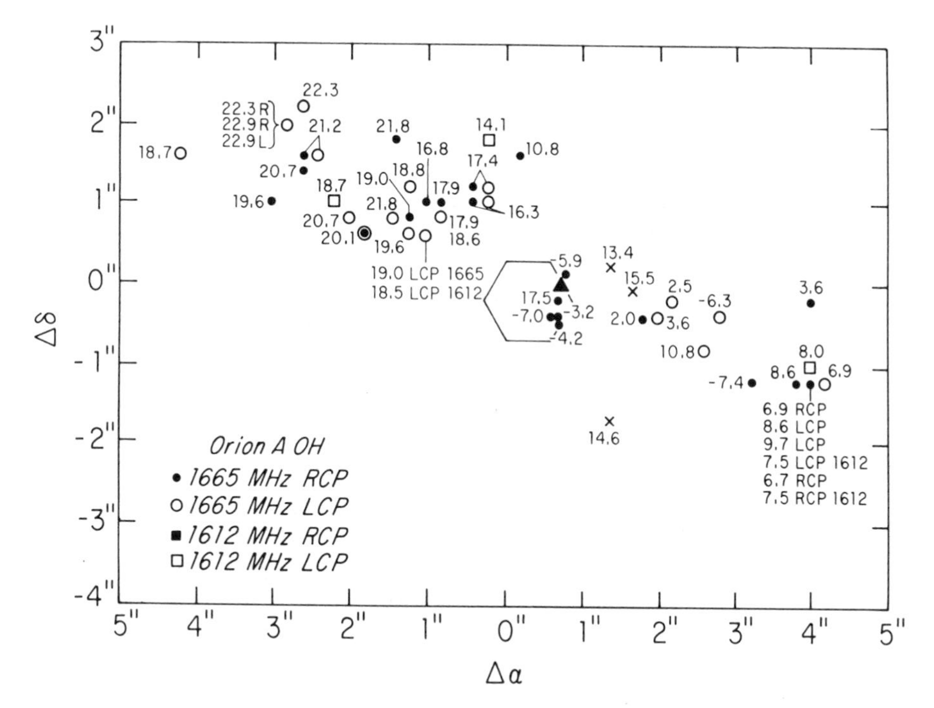

Figure 7 Positions and radial velocities of OH masers in Orion A, measured with the VLA. Each circle and square represents one or more OH masers. The number identifying a feature is its radial velocity (km s^{-1}). The hexagon shows the size and position of the infrared source IRC2. The triangle indicates an SiO maser. The map center is $\alpha = 5^h\ 32^m\ 45.05^s$, $\delta = -5°\ 24'\ 23''$ (1950). The absolute uncertainty of the maser positions is $\pm$ 1″; the relative positions are accurate to better than $\pm$ 0.2″ (S. Hansen, K. Johnston, in preparation).

tion about motions and matter in these otherwise inaccessible regions of the galaxies.

Radio spectroscopy with spatial resolution less than about 30 arcseconds is an essentially unexplored field. Until now such observations have been limited to absorption line interferometry of the 21-cm line of hydrogen, and to low-sensitivity very long baseline interferometry of the OH and H_2O masers. The VLA brings new spatial resolution and sensitivity to this field, as well as new wavelength coverage. Because so little is known in the areas that VLA spectroscopy will be able to explore, it is difficult to predict what may result, but it is going to be interesting to do the exploring.

STARS AND H II REGIONS When the VLA design was begun, radio emission from stars other than the sun was unknown. Today many types of stars have been detected as radio emitters. With a few exceptions, radio emission from stars is unresolvable even with the 0.1–1 arcsecond resolution of the VLA. Nevertheless, the array is being used extensively for stellar studies. Its great sensitivity allows more distant stars of a given luminosity to be observed and thus increases the sample available for study. It also allows detection of nearby stars of lower radio luminosity and thereby broadens the range of observable phenomena. Finally, its speed and sensitivity make possible a wide range of investigations of radio variability.

SS 433 The 6-cm structure of the peculiar star SS 433 is shown in Figure 8. The observations were made April 5, 1980, by R. M. Hjellming

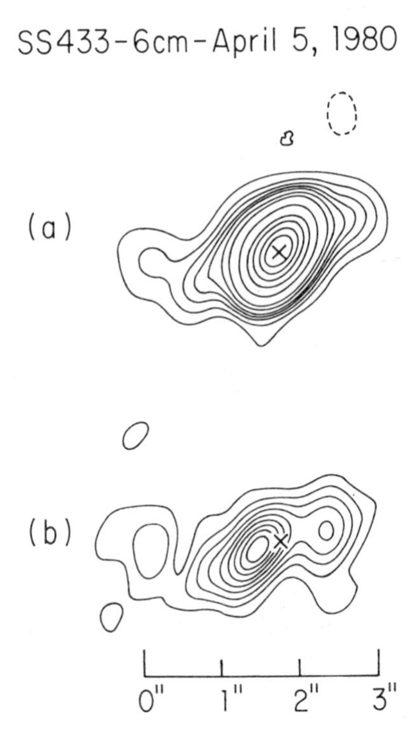

Figure 8 VLA maps of SS 433. (*a*) Total intensity contours. Contour levels are −1, 1, 2, 3, 4, 5, 7, 10, 20, 35, 50, 65, and 80% of the peak flux of 340 mJy. (*b*) Linearly polarized intensity. Contour levels are 10, 20, 30, 40, 50, 60, 70, 80, and 90% of the peak flux of 4.2 mJy. The half-power beam size in both maps is about 0.5″×0.8″, as indicated by the third brightest contour in (*a*). The optical position of SS 433 is indicated by the crosses (R. M. Hjellming, K. J. Johnston, in preparation).

and K. J. Johnston (private communication). They now have VLA maps of this star at six epochs and find that the radio structure changes drastically from month to month. The radiation in the "jets" is about 15% polarized. Hjellming and Johnston suggest that inflow of material along the poles of a precessing accretion disk produces the particles and magnetic fields responsible for the radio emission.

Mass loss The phenomena of stellar winds and mass loss are important to the evolution of stars and to the interaction between stars and the interstellar medium. The VLA may play an important role in studying these phenomena. The value of the VLA to studies of mass loss from early-type stars has been pointed out by Barlow (1979). Assuming a 3σ detection limit for the full array of 0.05 mJy at 6-cm wavelength, the VLA can detect a B supergiant star like P Cyg at a distance of 24 kpc, a Wolf-Rayet star such as γ Vel at 13 kpc, and a main-sequence O8 star at 700 pc. Only the radio flux, the terminal velocity of the stellar wind, and the distance to the star are needed to determine a mass-loss rate. Some work of this nature has already been reported. For example, Abbott et al. (1980) used up to 17 antennas in November 1978 and July 1979 to observe 15 O, B, and A stars. They detected six of them and derived mass-loss rates of 2×10^{-6} to 25×10^{-6} $M_{\odot}$ yr^{-1}.

Some late-type stars also exhibit mass loss, and VLA programs are under way or planned to study mass loss in supergiant and peculiar M stars, T Tauri stars, and other late-type objects.

NGC 2359 Figure 9 shows radio and optical emission of the ring nebula NGC 2359 (Schneps et al. 1981). The authors call this a stellar wind bubble caused by a Wolf-Rayet star. They did not detect radio emission from the star itself and put an upper limit of 7×10^{-5} $M_{\odot}$ yr^{-1} to its mass-loss rate. There is good correspondence of optical and radio emission over most of the shell. Optical obscuration probably is responsible for the apparent absence of optical emission in the northeast part of the shell. Schneps et al. used these radio and optical observations, and observations of CO, to investigate the evolution of stellar wind shells.

S128 The H II region S128, shown in Figure 10, has been studied with the VLA by Ho et al. (1980). Besides the 20-cm observations shown they also have 6-cm observations, and are using the VLA data, along with other radio and optical data, to study star formation in the region. The compact northern source shown on the map is resolved and is thermal. It is not seen optically, presumably because of obscuration. The authors suggest there are three distinct regions of star formation in this area, the compact source, the extended source, and an H_2O source just north of the

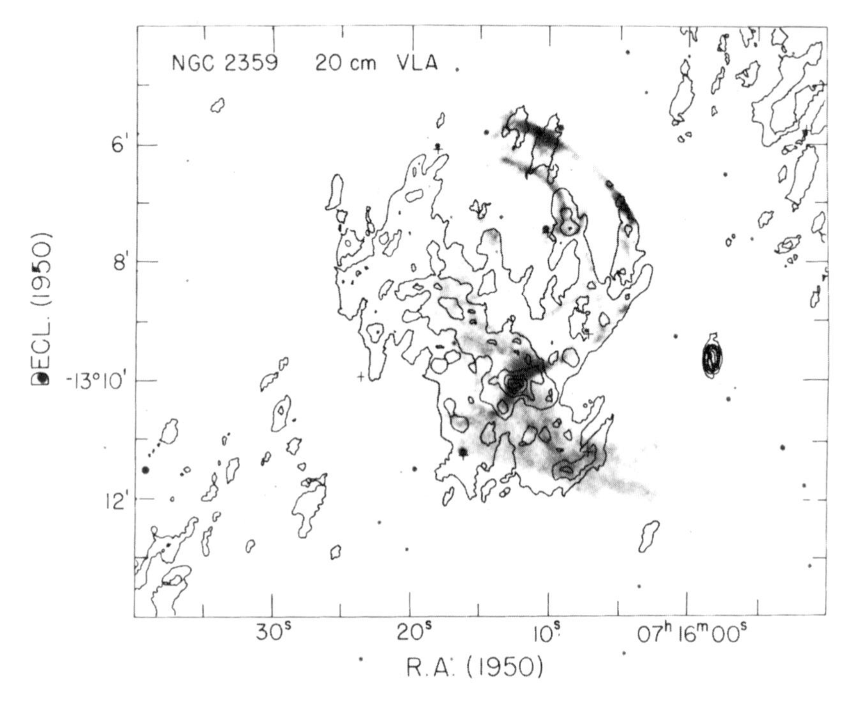

Figure 9 Ring nebula NGC 2359. A 20-cm VLA map overlaid on an [O III] photograph. Contour levels are 1, 2, 3, 4, 5, and 6 mJy/beam area. The beam shape is shown by the point source just west of the nebula. A large (10″×20″) beam has been used to achieve good surface brightness sensitivity at low declination with only 11 antennas (Schneps et al. 1981).

compact source. The ability to obtain maps at two wavelengths, with comparable resolution, is important to this and many studies.

5.3 *Extragalactic Studies*

The problems posed by radio galaxies provided the original impetus for design and construction of the VLA, and, while much has been learned about them in the intervening years, much is also still unknown. Observation of these objects has occupied a major part of the VLA time since it became available and this will probably continue for some time. It is clear that VLA observations of the detailed structure and polarization of radio galaxies, especially such features as jets and tails, are going to contribute significantly to our understanding of the physics of these objects and of the nature of galactic and intergalactic plasmas and magnetic fields.

The VLA is also being used heavily to study the radio properties of other extragalactic objects, including quasars, Seyfert galaxies, and "normal" spiral and elliptical galaxies. The high sensitivity and resolution it

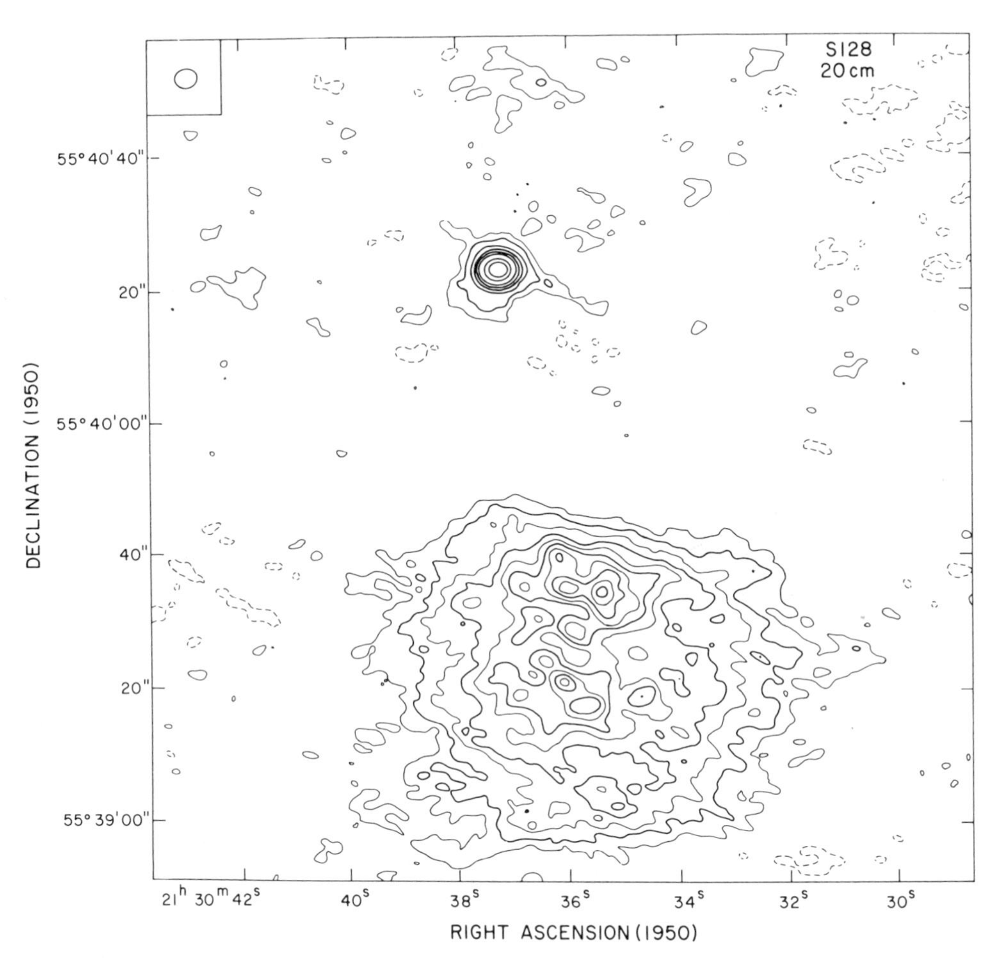

Figure 10 The H II region S128 at 20-cm wavelength. The peak flux density of the small bright northern source is 33 mJy/beam area. Contour levels are −5, 5, 10, 20, 30, 50, 70, and 90% of the peak flux. The half-power beam shape is shown in the box, upper left (Ho et al. 1980).

provides is beginning to turn up previously unknown phenomena in many of these objects.

One difficulty for many kinds of extragalactic studies is the absence of any radio distance indicator. Data from the VLA may stimulate and facilitate searches for new distance indicators.

RADIO GALAXIES VLA observations are already providing a mass of new data on the jets, tails, and other structural features of radio galaxies. Some examples are shown in Figures 11–16.

M87 M87 is perhaps the most famous and intensively studied object with optical and radio jets. The 2-cm radio map, Figure 11, was made by Owen et al. (1980) in March 1979, using 12 antennas. This is the only object in which a direct comparison is available between optical and radio jets emanating from the nucleus of a galaxy. Owen et al. have made a detailed comparison which shows that the optical and radio brightness distributions are remarkable similar all along the jet.

NGC 1265 NGC 1265 is the prototype and most extensively studied head-tail radio galaxy, first studied in detail by Ryle & Windram (1968). This head-tail morphology is now generally believed to be the result of interaction between a moving radio galaxy and the surrounding medium. The VLA radiograph, Figure 12, was obtained by Owen et al. (1978) in November 1977 using 7 antennas. It shows only the inner part of a much more extensive source. The entire region of the map is contained within a low surface brightness optical halo of the galaxy.

1638 + 538 This source, Figure 13, appears similar to, but some five times larger than, NGC 1265. The map by Burns & Owen (1980) was

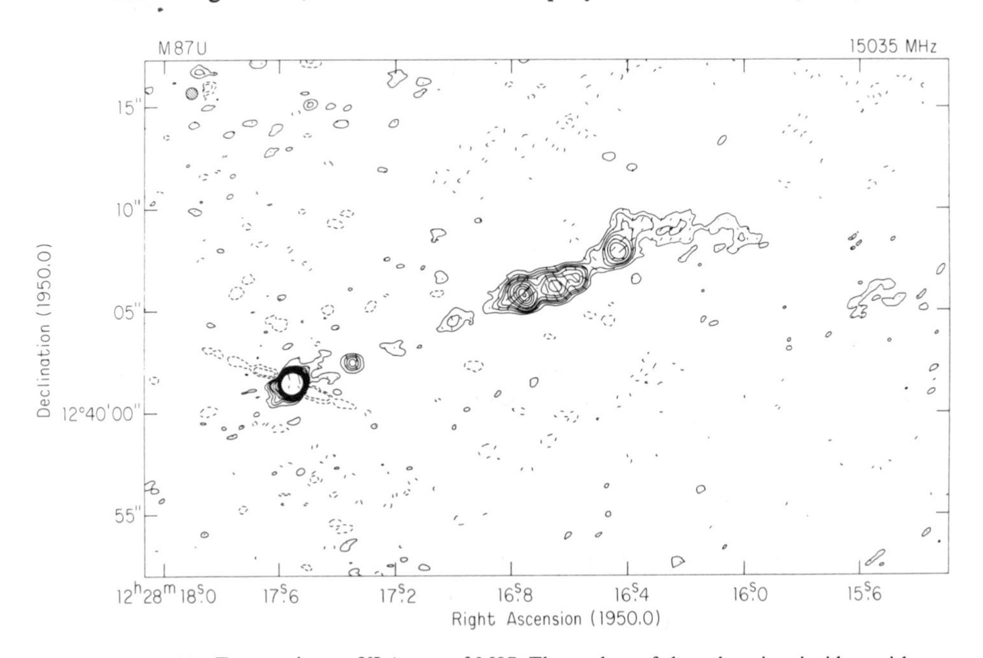

Figure 11 Two-centimeter VLA map of M87. The nucleus of the galaxy is coincident with the strong source at lower left. The half-power beam, shown in the upper left corner, is 0.6″ in diameter. Contour levels are (35, 30, 25, 20, 16, 12, 8, 6, 4, 3, 2, 1) × 10.17 mJy/beam area (Owen et al. 1980).

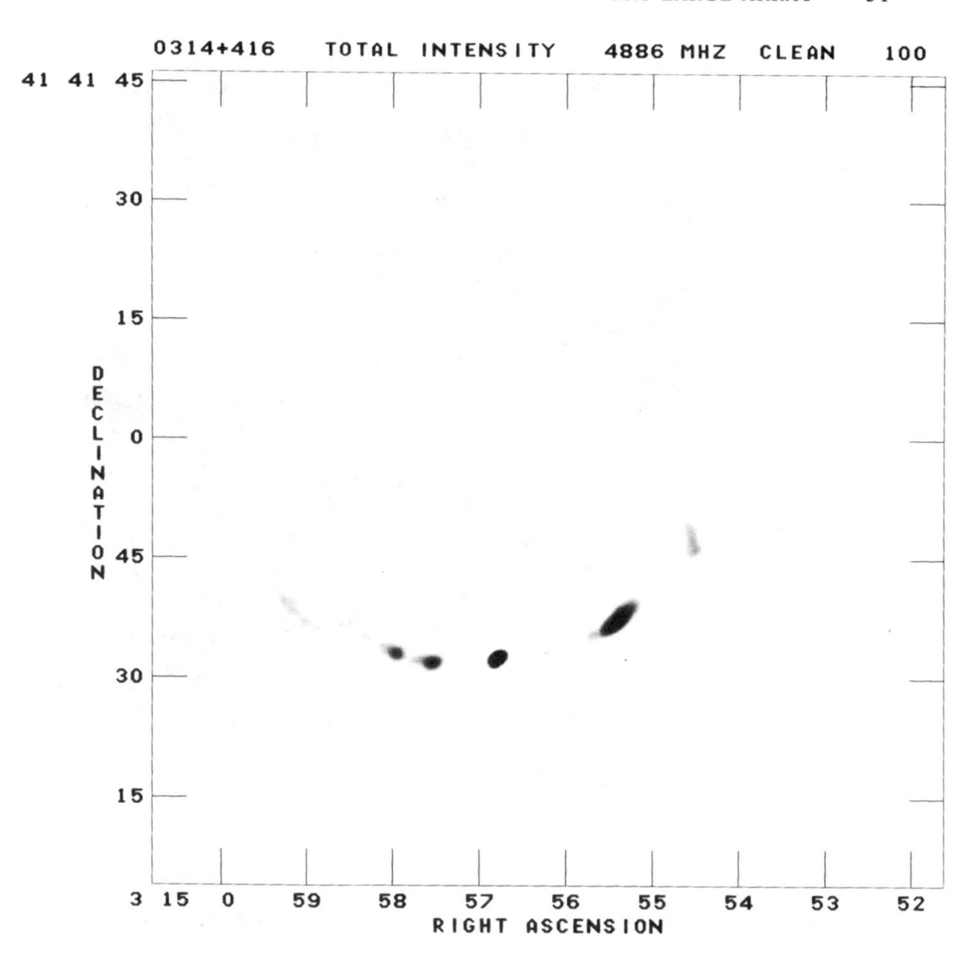

Figure 12 VLA radiograph of NGC 1265 at 6-cm wavelength. Intensities in the radiograph have been clipped so that only levels $\leq$ 5 mJy/beam area are displayed. The half-power beam size is about 1.4″×1.0″ (Owen et al. 1978).

made in November 1978 using 11 antennas. It shows two long, curved jets emanating from the galaxy nucleus and ending in more diffuse "blobs" or tails. As in the case of NGC 1265, the jets lie largely within the optical image of the galaxy while the tails are outside the optical image.

1919 + 479 This map, Figure 14, was made August 1979 with 16 antennas. It shows very complex source structure. There is a single, long curved jet with at least two distinct knots. The "tails" on either side appear relatively symmetric (Burns 1980).

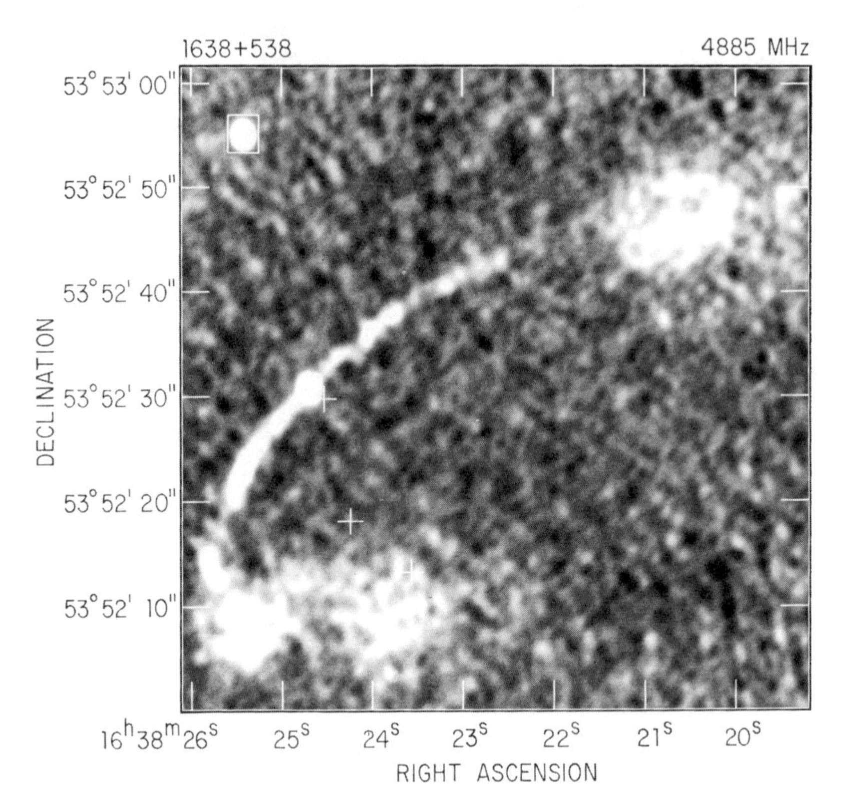

Figure 13 Six-centimeter VLA radiograph of 1638+538. Peak flux density is 11.8 mJy/beam area. The half-power beam shape is indicated in the box upper left. The radio source is identified with the brightest galaxy in a rich cluster. The galaxy position is shown by the cross near the central compact component of the source (Burns & Owen 1980).

NGC 6051 NGC 6051 is a cD galaxy in a poor cluster. The radio appearance of the galaxy, shown in Figure 15, is similar to that of many tailed radio galaxies in rich clusters. But in this object most of the distorted structure of the tails lies within the optical image of the galaxy rather than in the cluster medium (Burns et al. 1980). Kriss et al. (1980) have made observations of NGC 6051 with Einstein Observatory which indicate that the X-ray and radio morphologies are roughly comparable.

3C 315 3C 315, Figure 16, shows a very different structure. The observations were made by E. B. Fomalont, A. H. Bridle, J. A. Högbom, and A. G. Willis (E. B. Fomalont, private communication). The radio source is associated with a close pair of galaxies. Rapid precession of radio jets may account for its morphology.

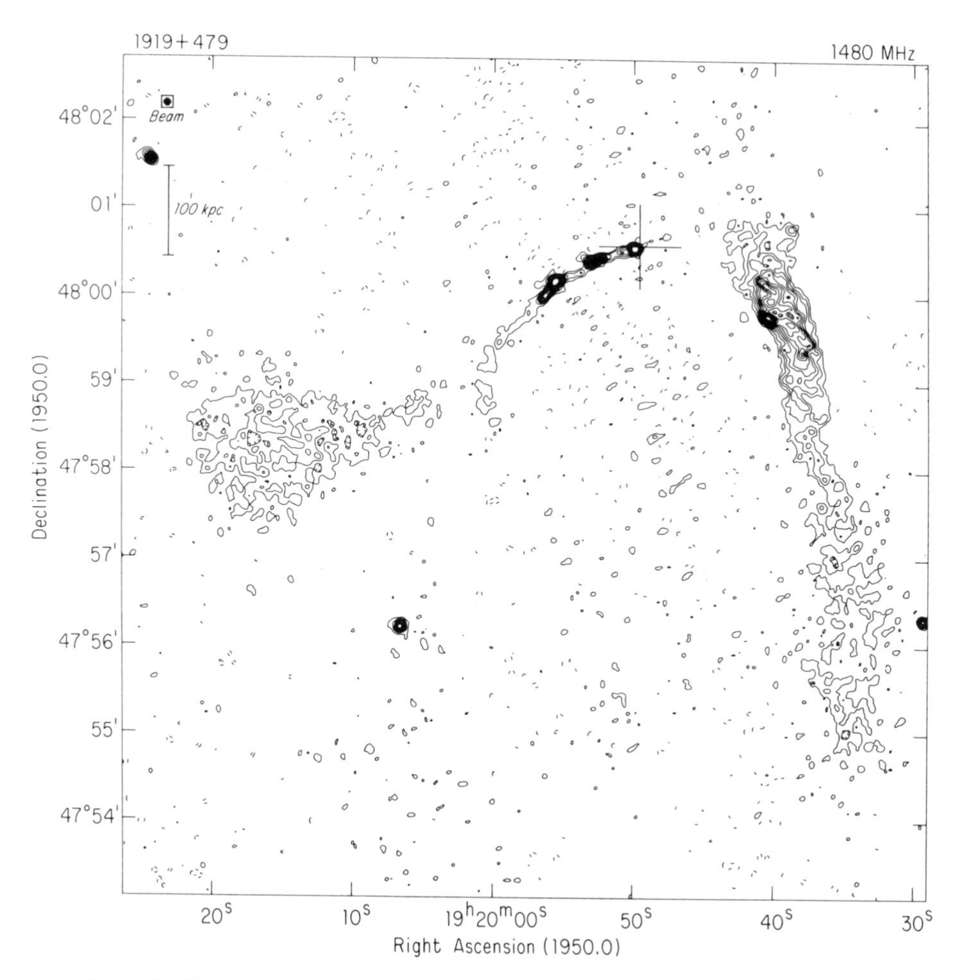

Figure 14 Twenty-centimeter map of 1919 + 479. The contour interval is 0.4 mJy/beam area and the peak flux is 13.4 mJy/beam area. The cross indicates the position of a cD galaxy identified with the radio source (Burns 1980).

It is clear from the above examples that VLA observations are providing a wealth of detailed, quantitative information for use in developing and testing specific theories to explain the general phenomena observed in radio galaxies. It seems likely that considerable progress will be made in the next few years in our understanding of the physical processes that determine the nature of these objects. Fomalont (1980) has given a good review of extended radio sources, based largely on recent VLA observations.

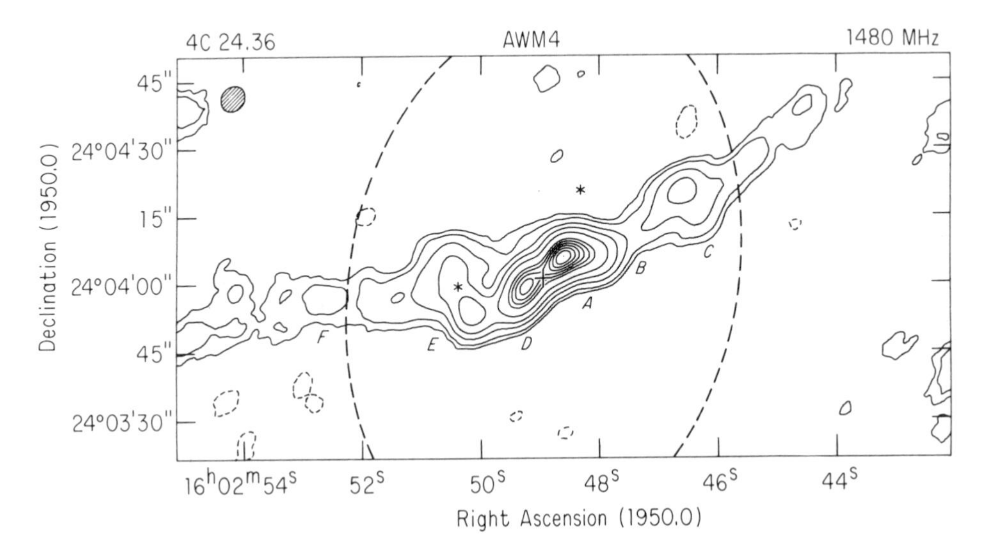

Figure 15 VLA map of 4C 24.26 at 20 cm. Contour levels are −1.8, −0.9, 0.9, 1.8, 4.5, 9.0, 18, 36, 45, 54, 63, 72, 81, and 90 mJy/beam area. The cross marks the center of the cD galaxy NGC 6051 and the dashed ellipse shows its approximate optical extent (Burns et al. 1980).

Most of the work to date has been on nearby radio galaxies, where evolutionary and cosmological effects are unimportant. As the problems associated with them become better understood, and as the VLA becomes more powerful, investigators will turn more of their attention to increasingly distant radio galaxies to see what can be learned about their evolution and about the evolution of the universe.

Another area that will receive increasing attention from the VLA is that of the relationship between radio galaxies, quasars, and other extragalactic sources. Some work of this nature is already in progress, of course, and some of it is discussed below. But again, as the VLA comes into full use with its full capability, more work will be undertaken on low-luminosity phenomena of relatively nearby objects.

QUASARS There is presently considerable interest in the fundamental differences, if any, between radio "loud" and radio "quiet" quasars. Several studies have been made or are in progress with the VLA to investigate this question. Condon et al. (1980) used the VLA in June 1978, with 9 antennas, to search for 6-cm emission from a sample of 22 bright optically detected quasars. They detected 9 of them, to a sensitivity limit of 0.5 mJy. The radio and optical flux limits of their survey are such that any quasar not detected by them must have a spectral-flux distribution that increases from radio to optical wavelengths. They found no evidence for two intrin-

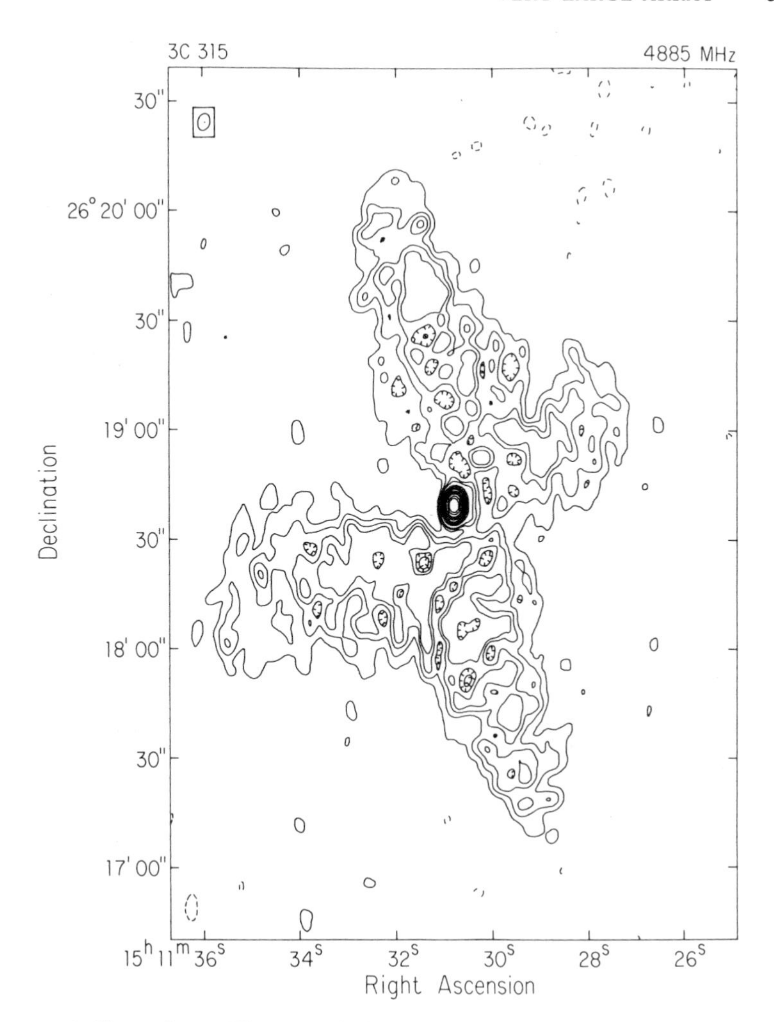

Figure 16 Six-centimeter VLA map of 3C 315. The peak flux is 125 mJy/beam area and contour levels are −1, 1, 2, 3, 4, 8, 12, 16, 20, 30, 40, 50, 70, and 90 mJy/beam area (E. B. Fomalont, A. H. Bridle, J. A. Högbom, A. G. Willis, in preparation).

sically different kinds of quasars, radio "quiet" and radio "loud," though such a difference has sometimes been suggested. In comparing their results with other work that included optically fainter quasars, Condon et al. concluded that the ratio of radio to optical luminosity of quasars must depend on optical properties and/or the distances of the quasars.

OTHER GALAXIES The VLA will be used extensively to study the radio properties of other extragalactic objects such as Seyfert galaxies, elliptical and spiral galaxies, and BL Lac objects, with the two-fold goal of understanding the processes occurring in the objects and of determining the relationships between various classes of objects.

NGC 1052 Figure 17 shows the 20-cm emission of NGC 1052 (J. M. Wrobel and D. S. Heeschen, in preparation), a nearby elliptical galaxy. The 20-cm morphology, a core with weak components on either side, is similar to that of many radio galaxies. The central core is unresolved and has a self-absorbed synchrotron spectrum. The weaker components have typical steep nonthermal spectra. The radio luminosity of NGC 1052 is much less than that of a typical radio galaxy and is intermediate between that of radio galaxies and of "normal" elliptical galaxies.

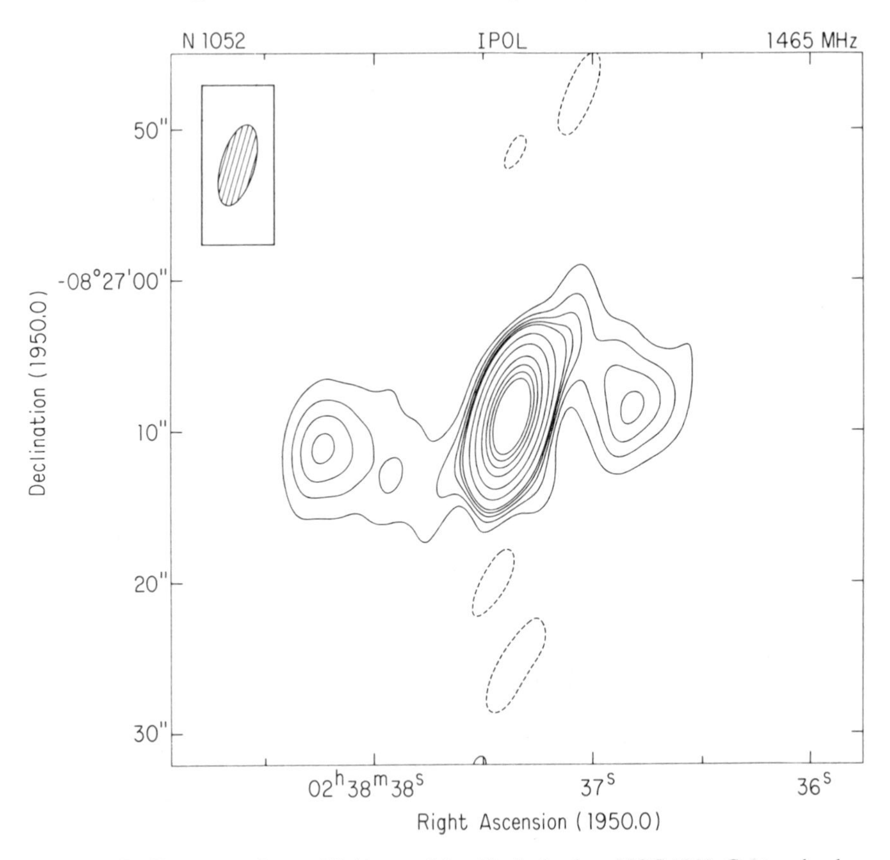

Figure 17 Twenty-centimeter VLA map of the elliptical galaxy NGC 1052. Contour levels are −0.55, 0.55, 1.1, 1.65, 2.2, 2.75, 5.5, 11, 22, 33, 44, and 55% of the peak flux of 734 mJy/beam area (J. M. Wrobel and D. S. Heeschen, in preparation).

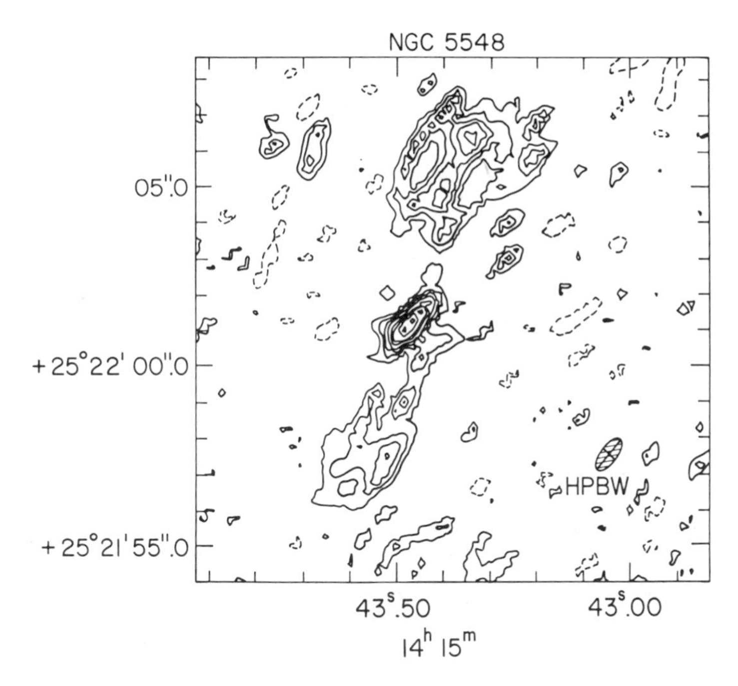

Figure 18 VLA map of the Seyfert galaxy NGC 5548 at 6-cm wavelength. The compact central source is coincident with the optical continuum nucleus. Contour levels are −10, 10, 15, 20, 25, 30, 40, 50, 70, and 90% of the peak flux of 1.38 mJy/beam area (A. S. Wilson, J. S. Ulvestad, R. A. Sramek, in preparation).

NGC 5548 NGC 5548 is a spiral galaxy with a Seyfert I nucleus. The VLA map, Figure 18, shows three 6-cm components, the central of which is coincident with the optical nucleus of the galaxy. This is one of a very few Seyfert galaxies that shows multiple radio components. The central component is unresolved by the 2.5×1.5 arcsecond beam. Again, the structure is similar to that of many radio galaxies, but the luminosity is much less (A. S. Wilson, J. S. Ulvestad, R. A. Sramek, in preparation).

NGC 3310 The VLA map, Figure 19, of this peculiar spiral galaxy shows a close correspondence with the Hα distribution of the galaxy (Balick & Heckman 1980). The radio flux of the southwest source, about 10 mJy, is consistent with its being thermal emission from the large H II region shown in the Hα map. Some of the other radio emission may be nonthermal synchrotron emission from supernova remnants in regions of recent star formation.

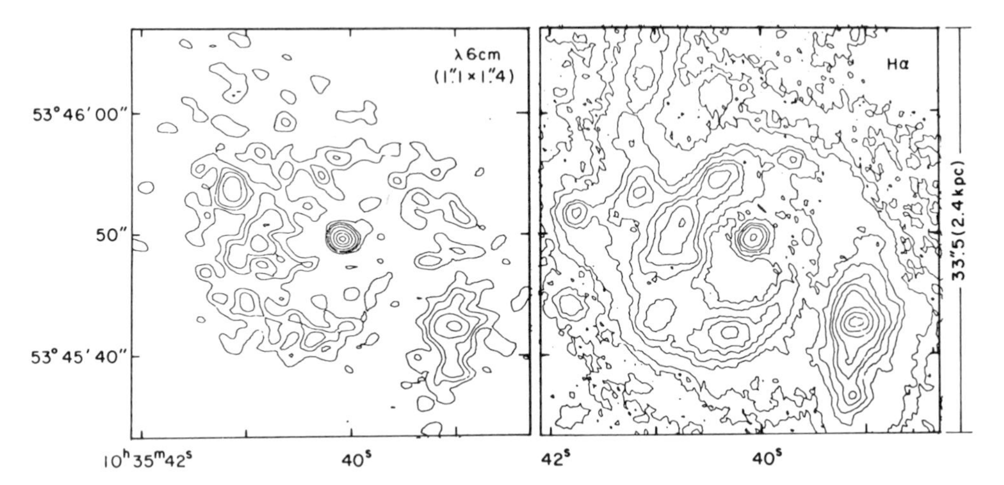

Figure 19 Six-centimeter and Hα contours of the spiral galaxy NGC 3310. Contour intervals of the VLA map, on the left, are 10, 20, 30, 50, 70, and 90% of the peak flux of 24.5 mJy/beam area. The right-hand map shows Hα intensity contours, to the same angular scale and about the same angular resolution as the VLA map (Balick & Heckman 1980).

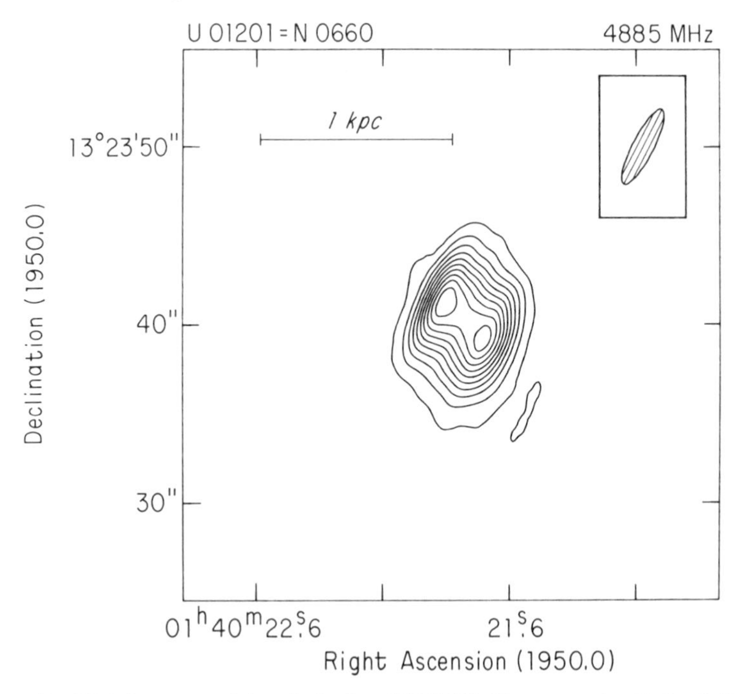

Figure 20 VLA 6-cm map of the spiral galaxy NGC 660. The contour level is 2 mJy/beam area. The half-power beam shape is indicated in the box, upper right (Condon 1980). The elliptical beam shape shown here, and in several of the other figures of low-declination objects, results from the fact that when the observations were made there were antennas on only one arm of the wye. The completed array will have a circular beam.

NGC 660 This is an example, Figure 20, of a double-appearing source in a spiral galaxy. It might be a symmetric double, a core with a jet, or a ring seen edge-on (Condon 1980). The radio spectrum is nonthermal, and Condon suggests that the emission may be the result of recent star formation in the inner part of the galaxy.

Markarian 8 Figure 21 shows the radio emission, at 6 cm, from the clumpy, irregular galaxy Markarian 8 (D. S. Heeschen, J. Heidmann, and Q. F. Yin, in preparation). As in the case of NGC 3310, there is excellent correspondence between the radio and optical emission. Whether the radio emission is thermal or nonthermal is not yet known, but VLA observa-

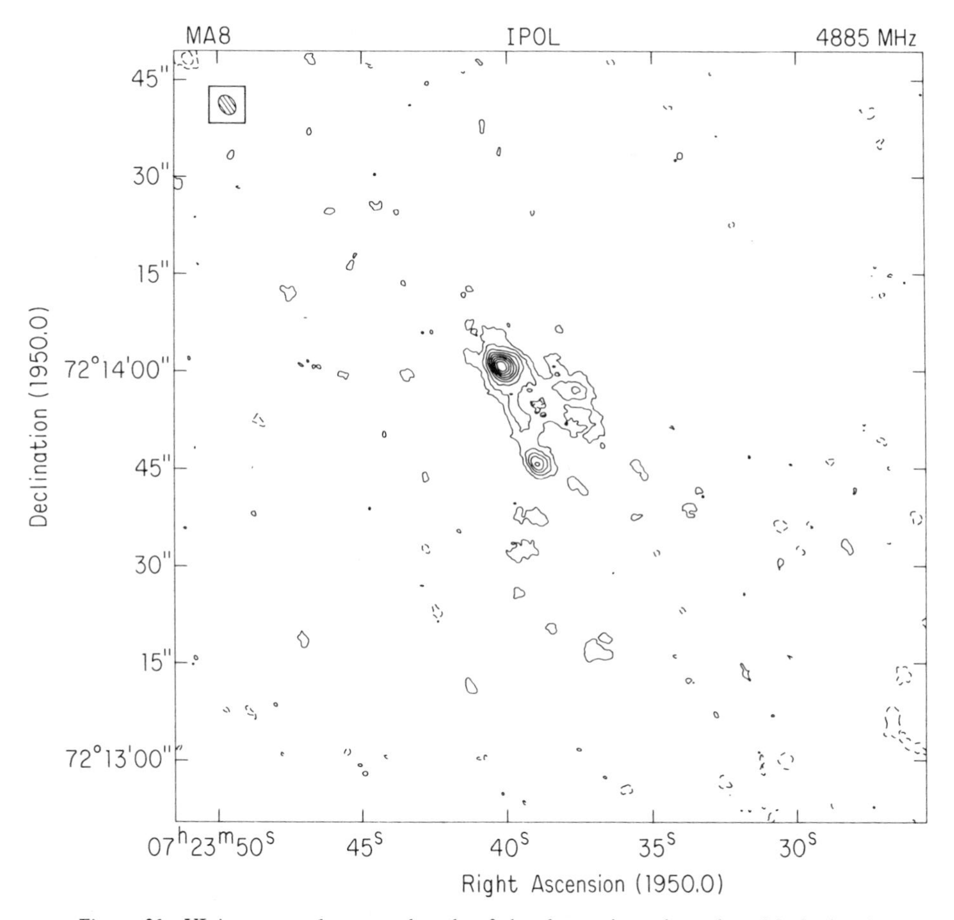

Figure 21 VLA map, at 6-cm wavelength, of the clumpy irregular galaxy Markarian 8. The contour interval is 10% of the peak flux of 1.5 mJy/beam area (D. S. Heeschen, J. Heidmann, Q. F. Yin, in preparation).

tions at 20 and 2 cm, with resolution comparable to that at 6 cm, will resolve that question.

The radio emission of objects such as those shown in Figures 17–21 is just now beginning to be studied in any detail. In general, the emission is weak and the emitting regions are small. Only the VLA has sufficient resolution and sensitivity to obtain good quantitative data on any significant number of them. VLA observations of these objects, when combined with UV, X-ray, optical, and IR data, may bring new understanding to problems of star formation and evolution in galaxies.

STATISTICAL STUDIES AND COSMOLOGY Except for the studies of optically selected quasars mentioned above, the VLA has not yet been used for statistical and other investigations that would fully exploit its highest sensitivity and resolution. Such studies, for example of source distributions and dimensions, will certainly eventually be made to levels of 0.1 mJy and 1 arcsecond. They will surely yield data of cosmological interest about the structure, distribution, luminosity, and evolution of extragalactic objects. Perhaps more than any other kind of work, these investigations require the full power of the VLA. Accumulation of sufficient bodies of data will be time-consuming, even with the VLA's speed. But observations with 0.1-mJy sensitivity can probe a vast new volume of the universe, until now essentially inaccessible. It would be presumptuous to predict what may be found.

6. CONCLUSION

The VLA will be a powerful tool for astronomy throughout the 1980s and beyond. It is a national facility, available to any qualified scientist with an appropriate program, without regard to the scientist's institutional affiliation. The instrument has already been used by more than 200 scientists from 50 institutions. It is at present oversubscribed by a factor of about five, so only a small fraction of programs proposed can actually be accomplished. Further information about use of the VLA can be obtained by writing the Director, NRAO, Charlottesville, Virginia 22901.

Design and construction of the VLA required the dedicated talents and efforts of many individuals. The NRAO scientific, engineering, technical, and business staffs have contributed heavily, from the beginning in 1961 to the present, as have many others in the scientific community. I am sorry it is not possible to recognize each individual by name.

It is a pleasure to thank the following people for providing VLA data and other material for this article: B. Balick, C. Bignell, J. Burns, J. Condon, E. Fomalont, A. Haschick, R. Hjellming, W. Jaffe, K. Johnston, F. Owen, M. Schneps, A. Thompson, C. Wade, S. Weinreb, A. Wilson, and H. Zirin.

References

Abbott, D. C., Bieging, J. H., Churchwell, E., Cassinelli, J. P. 1980. *Astrophys. J.* 238:196–202

Balick, B., Heckman, T. 1980. *Astron. Astrophys.* In press

Barlow, M. J. 1979. *Proc. IAU Symp.* 83: 119–30

Beckett, P. H. 1980. *Department of Sociology and Anthropology, Rep. 357.* Las Cruces: New Mexico State Univ. 307 pp.

Burns, J. O. 1980. *Mon. Not. R. Astron. Soc.* In press

Burns, J. O., Owen, F. N. 1980. *Astron. J.* 85:204–14

Burns, J. O., White, R. A., Hough, D. H. 1980. *Astron. J.* 86:1–15

Chow, Y. L. 1972. *IEEE Trans. Antennas and Propagation* 20:30–35

Christiansen, W. N. 1963. *Ann. Rev. Astron. Astrophys.* 1:1–18

Condon, J. J. 1980. *Astrophys. J.* 242:894–902

Condon, J. J., O'Dell, S. L., Puschell, J. J., Stein, W. A. 1980. *Nature* 283:357–58

Fomalont, E. B. 1979. In *Image Formation from Coherence Functions in Astronomy, IAU Colloq.* 49:3–17

Fomalont, E. B. 1980. *Proc. IAU Symp. 94.* In press

Gustincic, J. J., Napier, P. J. 1977. *Proc. IEEE Symp. on Antennas and Propagation,* p. 364. Stanford

Hjellming, R. M. 1978. *An Introduction to the Very Large Array.* Socorro: National Radio Astronomy Observatory. 369 pp.

Ho, P. T. P., Haschick, A. D., Israel, F. P. 1981. *Astrophys. J.* 243:526–38

Högbom, J. A., Brouw, W. N. 1974. *Astron. Astrophys.* 33:289–301

Kriss, G. A., Canizares, C. R., McClintock, J. E., Feigelson, E. D. 1980. *Astrophys. J.* 235:L61–L65

Kundu, M. R., Velusamy, T. 1980. *Astrophys. J.* 240:L63–L68

Marsh, K. A., Hurford, G. J. 1980. *Astrophys. J.* 240:L111–L114

McCready, L. L., Pawsey, J. L., Payne-Scott, R. 1947. *Proc. R. Soc. London Ser. A.* 190:357–75

National Academy of Sciences. 1964. *Ground-Based Astronomy, A Ten Year Program.* Washington, DC: National Academy of Sciences—National Research Council. 105 pp.

National Academy of Sciences. 1972. *Astronomy and Astrophysics for the 1970's,* Vol. 1. Washington, DC: National Academy of Sciences. 136 pp.

National Radio Astronomy Observatory. 1967. *The VLA. A Proposal for a Very Large Array Radio Telescope.* Vols. I, II. Green Bank: National Radio Astronomy Observatory. 361 pp.

National Radio Astronomy Observatory. 1969. *THE VLA. A Proposal for a Very Large Array Radio Telescope.* Vol. III. Green Bank: National Radio Astronomy Observatory. 117 pp.

National Radio Astronomy Observatory. 1971. *The VLA. A Proposal for a Very Large Array Radio Telescope.* Vol. IV. Green Bank: National Radio Astronomy Observatory. 184 pp.

National Science Foundation. 1961. *Astrophys. J.* 134:917–39

National Science Foundation. 1967. *Report of the ad hoc Advisory Panel for Large Radio Astronomy Facilities.* Washington, DC: National Science Foundation. 19 pp.

National Science Foundation. 1969. *Report of the Second Meeting of the ad hoc Advisory Panel for Large Radio Astronomy Facilities.* Washington, DC: National Science Foundation. 23 pp.

National Science Foundation. 1977. *Report of the ad hoc Advisory Panel for the Very Large Array (VLA).* Washington, DC: National Science Foundation. 25 pp.

Owen, F. N., Burns, J. O., Rudnick, L. 1978. *Astrophys. J.* 226:L119–123

Owen, F. N., Hardee, P. E., Bignell, R. C. 1980. *Astrophys. J.* 239:L11–L15

Ryle, M., Hewish, A. 1960. *Mon. Not. R. Astron. Soc.* 120:220–30

Ryle, M., Windram, M. D. 1968. *Mon. Not. R. Astron. Soc.* 138:1–21

Ryle, M. 1972. *Nature* 239:435–38

Schneps, M. H., Haschick, A. D., Wright, E. L., Barrett, A. H. 1981. *Astrophys. J.* 243:184–96

Stanford Research Institute. 1972. *VLA Feasibility Study,* ed. R. I. Presnell, Parts 1–3. Menlo Park, Calif: Stanford Research Institution. 156 pp.

Thompson, A. R., Clark, B. G., Wade, C. M., Napier, P. J. 1980. *Astrophys. J. Suppl.* 44:151–67

Weinreb, S., Balister, M., Maas, S., Napier, P. J. 1977. *IEEE Trans. Microwave Theory and Tech.* 25:243–48

Williams, W. F. 1965. *Microwave J.* 8(7): 79–87

Telescopes for the 1980s, pp. 63–128

THE MULTIPLE MIRROR TELESCOPE

Jacques M. Beckers and Bobby L. Ulich

Multiple Mirror Telescope Observatory,[1] Tucson, Arizona 85721

Robert R. Shannon

Optical Sciences Center, Tucson, Arizona 85721

Nathaniel P. Carleton, John C. Geary, and David W. Latham

Smithsonian Astrophysical Observatory, Cambridge, Massachusetts 02138

J. Roger P. Angel, William F. Hoffmann, Frank J. Low, Ray J. Weymann, and Neville J. Woolf

Steward Observatory, Tucson, Arizona 85721

1. INTRODUCTION

On May 9, 1979, the Multiple Mirror Telescope, or MMT, was dedicated. The construction and tuning of the MMT is now complete except for the automatic coalignment of the six telescopes which remains to be fully implemented. The six telescopes can be aligned very well manually by the observer so that a major program of astronomical observing has been going on at the MMT since the dedication date.

[1]The Multiple Mirror Telescope Observatory is a joint venture of the Smithsonian Institution and the University of Arizona.

8243-2902/81/0820-0063$02.00

The MMT was conceived in the late 1960s when scientists at the Smithsonian Astrophysical Observatory and at the University of Arizona decided it would be possible to use six existing 1.8-meter "egg crate" mirrors to construct six parallel Cassegrain telescopes whose images were to be combined into one, by so-called beamcombining optics. In this way they could effectively create a telescope with a combined collecting area equivalent to that of a 4.5-meter-diameter single mirror, thus making it, in that sense, the third largest optical telescope in the world. The need for such a large collecting area motivated the final decision to construct the MMT. The telescope is designed to be optimal for both optical and infrared astronomy of faint and low-contrast objects. Its large edge-to-edge distance in the primary mirror assembly (6.9 m) creates, in addition, the opportunity to do very high resolution imaging, both at infrared and optical wavelengths using, for example, Michelson and speckle interferometry.

The MMT is located in Arizona on the top of Mount Hopkins (110° 53′ 04″.4 longitude, 31° 41′ 19″.6 latitude) at an elevation of 2600 meters, a site selected because of its general location in the astronomically active Tucson area, its good seeing and photometric properties, its reasonably high altitude making it a dry and relatively dark site, and its large number of clear nights. Figure 1 gives a general overview of the Mount Hopkins

Figure 1 The multiple mirror telescope (MMT) is located on the flattened top of the peak of Mt. Hopkins. This view looking west shows the relatively small flat area surrounding the telescope and the steep drop-off of the mountain beyond. Building height is 16.8 meters, width 19.5 meters, and depth 13.4 meters.

Figure 2 The six main MMT mirrors and the central seventh guide/alignment telescope mirror are supported on the alt-azimuth mount by the open tubular optical support structure. The building rotates with the telescope.

site. The telescope is located on the relatively small, flat top of the rather sharply peaked mountain. This places it effectively high above the general surroundings. In addition to the design of the building and telescope itself, the location is, we believe, the source of the very good image quality obtained with the MMT. With our limited observing experience to date the average image quality is about one arcsecond (full width at half maximum of seeing disk) with images of 0.5 arcsecond being recorded rather frequently.

Figure 2 shows a closeup of the telescope itself. It is supported by an altitude-over-azimuth mount, which, under computer control, can track objects at both sidereal and nonsidereal rates. The telescope axes use ball-bearing supports rather than the conventional pressurized hydrostatic-bearing support. They are driven by a torque motor-pinion gear arrangement rather than by the traditional worm gear. The alt-azimuth mount cradles the optical support structure (OSS) which contains the six 1.8-m Cassegrain telescopes, and the beamcombining optics, as well as the 76-cm Cassegrain guide/alignment telescope, and which was designed to provide the required stiffness at a minimum cost in weight. The telescope and its mount is located within a building that rotates with the telescope azimuth. This building, in addition to housing the telescopes, contains the telescope control room, the data collection and analysis room, workshops, a conference room, and offices.

The optical paths in the MMT are shown in Figure 3. Each Cassegrain telescope has a $f/2.72$ 1.8-m-diameter primary mirror. The secondary mirror changes the beam to a $f/31.6$ focal ratio. The beam is then transferred via a flat tertiary mirror and beamcombiner to the MMT's quasi-Cassegrain focus located on the central axis of the OSS. The image scale there equals 3.6 arcseconds/mm. The steepness of the facets of the beamcombiner determines the "final f-ratio" of the MMT, which is defined by the envelope cone of the six converging beams. The angle of this cone and the size of the unvignetted field of the MMT are related. For the existing $f/9$ beamcombiner, this field equals 52 arcseconds. Vignetting sets in slowly, however, so that the MMT has a "useful field" ($<$ 50% vignetting) of 4 arcmin. A "faster" beamcombiner will increase the field of view at the cost of a lower filling factor of the incident light cone and of a somewhat larger image deterioration due to the larger image plane inclinations. Because of the polarization effects caused by the tilted tertiary and beamcombiner mirrors, the MMT as it is now is not suitable for polarimetry, nor can it be used well for interferometry using other than opposite mirror pairs.

The MMT presents, in many ways, a major departure from traditional telescope technology. This departure, risky as it may have originally appeared, has resulted in a relatively low-cost facility and a new technology that will undoubtedly change the ways telescopes of the future will be built. The most significant departures from conventional optical telescope technology are (*a*) the use of *light-weight "egg crate" mirrors,* which resulted in substantial weight savings of the telescope, (*b*) the use of an *alt-azimuth mount,* which simplifies the gravitational effects on the structure, thus again resulting in a greater economy. Alt-azimuth mounts have, of course, been used elsewhere, as in the USSR 6-meter telescope and in radio telescopes. The MMT mount represents, however, a major advance in precision and speed of tracking, (*c*) the use of a *ball-bearing support* rather than hydrostatic bearings resulting in cost savings and less maintenance, (*d*) the use of *spur gear drives* rather than worm gears, and (*e*) the use of *multiple coaligned light collectors* rather than a single monolithic mirror. The MMT has proved that all of these new concepts are viable, although many engineering refinements are still to be made, as is the case in all new telescopes. Observing runs began on the MMT in 1979 and already the telescope has proved itself to be a very good astronomical facility, producing major scientific returns for its sponsoring institutions.

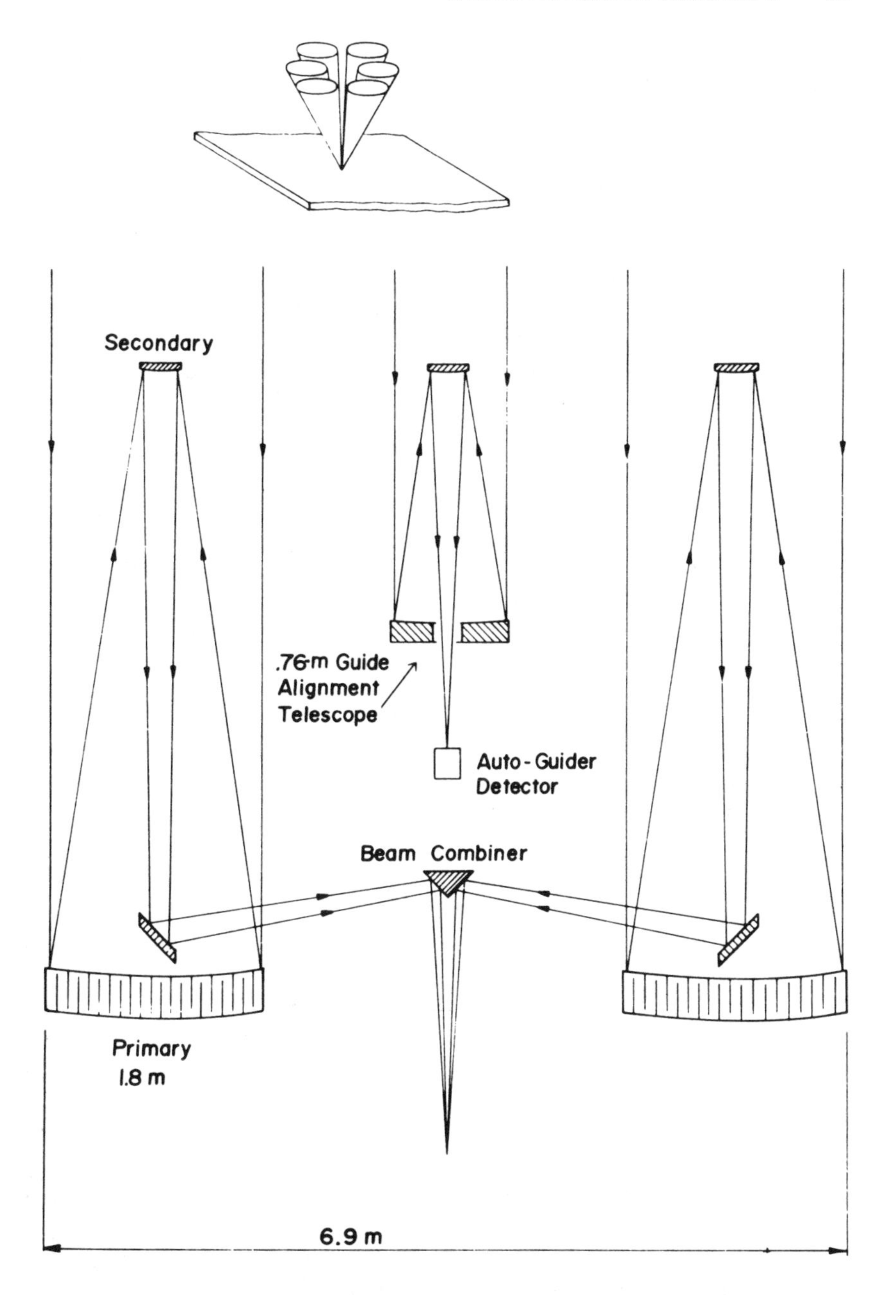

Figure 3 Optical paths for an opposite main mirror pair and for the central guide/alignment telescope. On top the angular filling is shown for an image at the quasi-Cassegrain focus.

The MMT was constructed jointly by the Smithsonian Institution and the University of Arizona on the basis of a Memorandum of Understanding signed on December 23, 1971. Manpower and funding were provided by both institutions to complete the telescope. The NSF provided grants for specific tasks related to the construction of the telescope and its auxiliary instrumentation. After the dedication of the MMT in May 1979, the two sponsoring institutions, on July 1, 1979, entered into a new Memorandum of Understanding for the operation of the MMT, creating the Multiple Mirror Telescope Observatory (MMTO). The MMTO is charged with the responsibility of operating and maintaining the MMT facility, as well as with the completion of a number of remaining engineering tasks. It has an engineering, technical, operations, and administrative staff of about twenty, and an annual budget of about one million dollars which is divided equally between the two sponsoring institutions with the University of Arizona deriving part of its contribution from an NSF grant. The major unfinished engineering task relates to the automatic coalignment of the six telescopes. Presently the telescopes can be successfully coaligned manually to sub-arcsecond accuracy. The automatic laser coalignment system can maintain this to 1 arcsecond rms. The latter is, however, not good enough, especially because of the exceptionally good seeing often encountered on Mt. Hopkins, which requires maintaining the coalignment to a fraction of the 0.5-arcsecond seeing disk. Continuous updating of the coalignment on star images themselves, as well as improved laser coalignment, are both being pursued.

This and other engineering efforts proceed at the same time that a substantial astronomical observing program is in progress. Up to 75% of the nights available in a month have been used for astronomical observing. The MMT is of sufficient complexity to require a staff of telescope operators to run the telescope and to assist the observing scientist in the acquisition of data. It is particularly suited for deep sky spectroscopy and imaging, for high resolution spectroscopy, for infrared observations, for precision photometry of faint objects, and for optical and infrared interferometry. Section 7 of this paper, written by R. Weymann, elaborates on the science done and to be done on the MMT. Other sections (with individual authors indicated) will deal with the history of the MMT [Sections 2 (Beckers), 3 (Woolf), and 4 (Weymann)], with the design and performance of the MMT [Section 5: 5.1–5.3 (Carleton), 5.4 (Shannon, Beckers), 5.5 (Hoffmann), 5.6 and 5.7 (Low), 5.8 (Ulich), 5.9 (Carlton, Beckers)], with MMT instrumentation [Section 6: 6.1 (Beckers), 6.2 (Angel, Latham), 6.3 (Low), 6.4 (Geary)], and with concepts for future multiple objective telescopes [Section 8 (Woolf)].

2. EARLY MULTIPLE OBJECTIVE TELESCOPES

The history of astronomical telescopes is to a large extent a history of the manufacture of ever larger and larger telescope objectives. The size at which lenses and mirrors could be manufactured has been the major factor in determining the size of astronomical telescopes in the past. Because astronomers have always wanted telescopes with larger and larger collecting area to probe weaker and weaker objects, the idea of combining the light from a number of smaller astronomical objectives into one, with the goal of creating an image with the brightness of an effectively larger telescope, must have occurred to many in the past. The astronomical literature contains a number of references to concepts and implementations of such multiple objective telescopes. Some of these have already been reviewed by Ingalls (1951), Jacchia (1978), and Green (1979).

Table 1 summarizes the literature describing devices that were actually constructed and tested. Some of these consist of arrays of single mirror segments, which combine to form effectively a single objective in the sense that a single image is formed (Type A). Others, like the MMT, form initially separate images, which are combined by means of relay optics or electronics into one final image (Type B). Table 1 excludes multiple aperture devices aimed at interferometric studies only. By far the greatest early effort was that by Guido Horn-d'Arturo who, as the director of the Bologna Observatory in Italy, pursued the construction of the so-called "tessellated mirror" from 1932 to 1953 with major interruptions caused by the Second World War. His was a zenith-pointing, fixed telescope 1.8 meters in diameter with limited celestial tracking provided by the motion of the photographic plate. It was Horn-d'Arturo's concept to place a number of these zenith-pointing telescopes, each covering 80 arcminutes in latitude, 150 km apart in geographical latitude to study a larger area of the sky. He also viewed his tessellated (or segmented) mirror as a prototype for much larger telescopes 13 and 18 meters in diameter. Although these larger mirrors remained stationary, pointing at the zenith, he proposed increasing sky coverage by using complex pointing prime focal plane optics (Horn-d'Arturo 1955).

Table 2 summarizes the literature on concepts for multiple objective telescopic devices. Although recent designs for multiple objective telescopes have received much attention, it is interesting to examine some of the pre-MMT literature. The paper by Synge (1930) is especially fascinating because it appears to be the earliest description of an MMT-like device (Type B). His description of ideas for the optics and the support arrangement for the optics in fact resembles the MMT so closely that the

Table 1 Other multiple objective telescopes

Reference	Type[a]	Objective type	Diameter element (cm)	Number of objectives	Total collecting area (cm^2)	Comments
Oxmantown (1828)	A	Mirror	~ 10	2	180	Goal was not large size but elimination of spherical aberration.
Horn-d'Arturo (1935)	A	Mirror	10	20	2160	Zenith pointed. Track by moving photographic plate.
Vaisala (1949)	A	Mirror	32	7	5600	
Horn-d'Arturo (1950, 1953, 1955, 1956a,b)	A	Mirror	20	61	22000	Zenith pointed. Track by moving photographic plate. 1955 article has sketch for 13- and 18-meter mirrors by L. T. Johnson
Hanbury Brown (1964)	A	Mirror		252	300000	Angular resolution only 6 arcmin.
Burke et al. (1968)	A	Mirror	60	248	780000	Angular resolution only 0.5 deg.
Faller (1978)	B	Lens	19	80	23000	~ 3-arcsec resolution and field of view.
Basov et al. (1979)	A	Mirror	40	7	11000	
Grainger (1979)	B	Mirror	38	7	8000	
MMT (Carleton & Hoffmann 1978)	B	Mirror	180	6	150000	

[a]Type A = segmented mirror telescope with prime images coincident.
Type B = multiple telescope device with images combined by relay optics.

MMT, although conceived independently, could be viewed as the realization of the Synge concept. The paper also points out the weight (and hence cost) advantage of such a telescope ("the weight of such a telescope would tend to increase as the square of the equivalent aperture, while that of an ordinary telescope would increase as the cube"). The Fastie (1961, 1967) articles are also of interest because they describe an optical configuration actually now used to couple the MMT to the MMT spectrograph (see discussion in Section 6.2). By aligning the star images of the multiple mirror device along the spectrograph slit, and by aligning the principal axes of all incident light cones by means of an array of small refracting prisms, the large aperture (4.5-m) telescope can be coupled to a spectrograph which has the higher spectroscopic resolution of a spectrograph designed for a 1.8-meter telescope, without loss of light-gathering power.

Table 2 Concepts for multiple objective telescopes

Reference	Type	Objective type	Diameter element (cm)	Number of objectives	Total collecting area (cm^2)	Comments
Fotescu & Heilbrun (1909)	B	Lens				*UK Patent No. 21456.*
Synge (1930)	B	Mirror				Concept of optics and support structure is very close to MMT design. Points out the cost advantage of such a telescope since weight is proportional to square rather than cube of diameter as for single telescopes.
Fastie (1961, 1967)	A/B	Mirror		7		Points out the advantage of a multiple mirror device in enhancing the throughput of a spectrograph employing the same optical system used for the MMT spectrograph.
Thomas (1965)	B	Mirror				*US Patent No. 3507547.*
Meinel (1970)	B	Mirror				Study of phased MMT telescope.
Mertz (1970, 1980)	A	Mirror				Aricebo-like disk.
Hall (1978), Barr (1980)	A/B	Mirror			4,900,000	Various concepts for 25-m next generation telescope (NGT).
Richardson & Grundmann (1978) Grundmann & Richardson (1980)	B	Mirror	300	69	4,900,000	Also refers to previous study concepts involving two 72-in, 178 75-in, 25 1.5-m telescopes, and 6 3-m telescopes.
Learner (1978)	B	Mirror				Describes three arrays of small separate telescopes similar to NGT "Singles Array."
Meinel (1978)	A/B	Mirror				Examines fixed telescope fed by siderostats.
Meinel (1979)	B	Mirror				Discusses various MMT-type concepts.

Table 2 (*continued*)

Reference	Type	Objective type	Diameter element (cm)	Number of objectives	Total collecting area (cm^2)	Comments
Steshenko (1979)	A	Mirror			4,900,000	USSR plan for 25-m telescope.
Nelson (1980)	A	Mirror	140	60	780,000	
Meinel & Meinel (1980)	A	Mirror		36	780,000	
Woolf & Angel (1980)	B	Mirror	500	8	1,570,000	

Most of the design references listed in Table 2 are very recent. Multiple objective structures are considered in almost all concepts for future large telescopes, not only because the size of mirrors that can be constructed is limited, but also because the weight and cost of multiple aperture telescopes is lower. We may expect the length of Tables 1 and 2 to grow rapidly in the future.

3. EARLY HISTORY OF THE MMT PROJECT

After the pioneer work of Synge and Horn-d'Arturo on the two types of multiple objective telescope, there was a lull while technology developed to the point that precision telescopes of this kind became practicable. It was, indeed, an attempt to develop a large telescope of Type A by P. Connes that led to the reinvention of the MMT.

In the late 1960s, Frank Low at the Lunar and Planetary Laboratory (LPL) of the University of Arizona had developed techniques for the detection of faint objects at 10-μm wavelengths using a 1.5-meter telescope. He believed then that he was successful in part because he used a moderate-size aperture, and he wondered how to increase sensitivity still further. While visiting the Connes telescope in 1967, he decided that, although there would be thermal problems associated with the approach, the general process of adding mirrors was appropriate, and one could operate a number of 1.5-meter telescopes with a common focus. Then, in 1968 when he had returned to the US, he discussed this possibility with Aden Meinel at the Optical Sciences Center of the University of Arizona.

Meinel was concerned about increasing the light grasp and resolution of surveillance telescopes in space. After considering various ways of combining light, Meinel calculated the optical details of a Type B system. He also developed the methods of obtaining a phased field for such a telescope using prisms (Meinel 1970).

Low reported on the MMT concept at a meeting in Pasadena on July 28, 1969. Six months later, he also reported on it to the infrared panel of the National Academy of Sciences studying the future of US astronomy in the 1970s. During a part of this period, Robert Noyes of the Smithsonian Astrophysical Observatory (SAO) was visiting the Lunar and Planetary Laboratory.

When the Smithsonian Astrophysical Observatory set up an observatory on Mt. Hopkins in the mid-1960s, it left the summit area of the mountain clear to permit the construction of a giant telescope. After the design phase and much of the construction of the 1.5-m telescope was complete, attention was turned to the problem of making a giant telescope for the summit at a moderate cost. Fred Whipple gathered together a team consisting of Nat Carleton, Larry Mertz, and Tom Hoffman. They explored a number of schemes, each with some disadvantages. The most favored of these options was an Aricebo-like bowl made up of mosaic elements (Mertz 1970). In a phone call to Whipple, Gerard Kuiper at the Lunar and Planetary Laboratory mentioned the developing MMT concept. Meinel informed Whipple that he could obtain some surplus lightweight mirror blanks from the US Air Force, and so a number of explorations to develop a joint project were initiated.

The Smithsonian team embarked on a detailed study of the MMT concept to search for hidden flaws. Ray Weymann of Steward Observatory persuaded the National Academy of Sciences Optical Panel, of which he was a member, to support this effort. And the first mechanical studies were made of how to support the optical components. These revealed that the original mechanical concept of the telescope, while possibly appropriate for space, had too much flexure, and, indeed, that there was a problem in keeping the foci together. Weymann and Mike Reed then came up with the laser alignment scheme.

The entire joint project, however, nearly foundered over the choice of site, with some favoring Mt. Hopkins, and others favoring Mt. Lemmon. Mt. Lemmon, being higher, might be a better infrared site, while Mt. Hopkins, having a lower sky brightness, was a better site for visible wavelengths. Because of Meinel's support Mt. Hopkins was chosen, though even after a road was started up it, the exact location of the future telescope was uncertain.

4. R. J. WEYMANN: PERSONAL RECOLLECTIONS

In the following personal recollections of some of the events and decisions concerning the construction of the MMT during the years that I was closely associated with the project, I cannot vouch for precise detail or exact chronology—many of the documents no longer exist and many of the comments were mercifully unrecorded. Nevertheless, I believe in essence the account is accurate.

For some time, the group at the Lunar and Planetary Laboratory (mainly G. P. Kuiper and F. J. Low, but also including H. L. Johnson) had discussed with A. B. Meinel, then at the Optical Sciences Center of the University of Arizona, the construction of a large MMT-type telescope to be located on Mount Lemmon. The availability of several lightweight 1.8-m blanks made this a realistic possibility. At that time the LPL group conceived of the instrument primarily as an infrared telescope. Kuiper was especially interested in its use for IR spectroscopy with the Michelson spectrometers then being developed.

My first "official" introduction to the program came after a conference on comets in April 1970 at which time Frank Low, Gerard Kuiper, and Fred Whipple described to me their concept of the project. Whipple, as Director of Smithsonian Astrophysical Observatory, had himself, along with several colleagues at SAO, been interested in approaches to a low-cost large aperture telescope and responded favorably to an invitation to join in a collaborative effort between SAO and LPL. As Director of the Steward Observatory, my reaction was that such a device offered equal promise as a telescope for optical spectroscopy and for small-field optical astronomy in general. Moreover, if the University of Arizona were serious about creating one of the leading Astronomy Departments in the country, then participation in the MMT project offered an opportunity not likely to occur again. Fortunately, the University Administration also took this point of view, and Steward Observatory soon became involved in the project.

At that time the National Academy of Sciences Survey Committee, chaired by J. L. Greenstein, was being convened, and the Optical Panel, of which I was a member, was drawing up a list of priorities. At an early meeting in Pasadena a southern copy of the Hale 5-m telescope was designated the top priority optical project. This left me disturbed. For one thing (let us be frank) it was a case of the rich getting richer, since it was envisioned that the instrument would be administered by the Pasadena group. This would squelch healthy competition, which is often felt to be highly desirable for progress in science, especially by those without ready access to first class facilities. More fundamentally, however, it seemed a less-than-visionary project to propose for the coming decade. Bruce Gregory, acting as executive secretary for the committee, put it succinctly: "How will a duplicate of an instrument, designed and conceived in the 1930s, fare against the plans being advocated by the radio astronomers on the scale of the VLA?" With this in mind, the Panel was persuaded to hear a presentation by Aden Meinel on the MMT concept (and also one by Dick Miller on optical interferometry). The final upshot of the deliberations was that the Optical Panel and full Committee report included recommendations for construction of an MMT prototype and studies of its possible application to a large (10 m–15 m) instrument.

Meinel made an effective and enthusiastic presentation, but, unfortunately, left the clear impression in the minds of NSF representatives present that Defense Department funding was all but secured and that little help would be required from NSF, an impression that plagued us for several years. Meanwhile, several key decisions concerning the telescope were being made and the division of responsibilities roughed out. One of the early decisions was to fabricate $f/2.7$ primaries rather than a much slower system. Proponents of the $f/2.7$ design argued (correctly, I believe) that the advantage of a very compact almost "square" structure outweighed the disadvantages of more difficult figuring and the unavoidable requirement to slump the existing blanks, and the risks that process entailed.

One of the most controversial issues involved the selection of the site: Kuiper and Low of LPL strongly advocated the Mt. Lemmon site: it was developed and accessible, making construction cheaper, and was slightly higher than Mt. Hopkins, which had some slight advantage for infrared observations. SAO, of course, preferred Mt. Hopkins, because they already had a facility there. I, too, preferred Mt. Hopkins, partly because I was concerned about the effect of sky brightness on optical astronomy, but also because the realities of funding seemed to require major support through SAO and this seemed unlikely to materialize if Mt. Lemmon was chosen as the site, a view that Meinel also shared. The decision for Mt. Hopkins was, thus, a very disappointing one for the LPL group and caused their enthusiasm for the project to markedly diminish for several years.

The division of responsibility for the telescope design and construction was never a major issue: SAO personnel, especially Nat Carleton and Tom Hoffman, had considerable interest and expertise in the area of the building and the mount, whereas the University of Arizona, because of the presence of the Optical Sciences Center, took responsibility for the optics. Mike Reed, a former graduate student at Arizona with a background in electronics, was recruited by Meinel to lead the work on the alignment system. The only grey area, initially, was responsibility for the optical support structure or "tube," but funding and engineering realities both dictated that this should fall within SAO's domain. A formal Memorandum of Agreement, committing both institutions to a major effort in building the telescope and setting out the basic guidelines involving equal access to the telescope, was signed on December 23, 1971. Arizona lacked a project leader to oversee the University of Arizona work on the project and, after considering several possibilities, W. F. Hoffmann, then at NASA, was selected. Hoffmann accepted the position and joined the project in early 1973. He and Nat Carleton assumed the day-to-day task of overseeing and coordinating the University of Arizona–SAO effort.

One further important design decision was made subsequent to this: in considering various structural constraints and opportunities afforded by the compact structure and limited "swing" of the tube associated with its alt-azimuth mount (an equatorial mount was never really seriously considered), Tom Hoffman and Nat Carleton suggested bringing the building structure up very close to the telescope, abandoning totally the concept of the traditional dome. They proposed enclosing the telescope with a small box-like "barn," entirely rotating from ground level up. This idea was at first fiercely resisted, partly because of concern over restricting access to the telescope but also for reasons that now seem fairly irrelevant. While it was realized that the very open structure would minimize inside-outside air temperature differences, there was also concern that the box-like structure, coupled with the telescope proximity to the ground, might give rise to bad seeing. Happily, one of the pleasant surprises to date at the MMT has been the frequency of excellent seeing.

Funding for the telescope proved a continuing problem. Despite the Greenstein Report, there was very little sympathy at NSF for funding Arizona's efforts on the MMT. Partly this arose from the delicate question of multi-agency funding, but partly it arose from some longstanding misunderstandings, difficulties, and personality conflicts between SAO/SI and NSF. A "catch-22" situation developed: it was said to be necessary to first demonstrate with a small prototype the soundness of the concept, but then this approach was criticized on the grounds that what was really required was a comprehensive proposal for a full-scale MMT! In 1971 SAO presented its plans to the House and Senate Interior Subcommittee and, with the support of the Arizona delegation, obtained initial funding in fiscal year (FY) 1972 for their portion of the project.

Arizona then turned to the State Legislature for support. At that time a proposal to request enabling legislation for Pima County to enact a light-pollution ordinance was also being considered and I had had conversations with Senator Douglas Holsclaw about this legislation. Late one evening he called me very apologetically to say that the legislation had failed to clear the Senate Majority caucus, and he added that, as a general maxim, special legislation whose full implication and background wasn't clearly understood usually failed, even if it didn't cost anything. This was alarming, since the implications of the MMT appropriation might not have been all that clear either and it certainly did cost something. In response to my question, Senator Holsclaw admitted that the MMT appropriation was indeed not doing very well either, and agreed that additional information might be helpful. I collected all the material I had and took it to the post office at 1 a.m.! The MMT appropriation did pass (along with the light-pollution legislation).

This event marked the first time that I felt the MMT would become a reality, or at least the University of Arizona's participation in it, though funding continued to be, and remains, very tight. Optical Sciences had several cost overruns, and Reed had to cut many corners. In this connection, I would like to record the contribution that Greg Sanger made to the project. He was given day-to-day responsibility for the Optical Sciences operation while he was still a graduate student working on a dissertation; his responsibilities were later taken over by R. Parker. Despite all of the grumbling about the progress of the optics and the budget problems, the fact is that the MMT optics turned out to be very good, to the credit of both of them, as well as Dr. R. Shannon who had overall responsibility for the project at OSC.

A major criticism of the entire project was that it should not have been begun until funding was fully identified. Realistically, given the particular set of circumstances surrounding the MMT, to wait until full funding was located would have been tantamount to abandoning the project. A second criticism was that the project lacked a single leader with clear authority to make decisions, but, instead, developed a large and complicated administration. The platitude was (and is) that you "can't build a telescope by committee." In fact, the collaborative nature of the program, the geographical separation of the two groups, and the way the responsibilities were divided made the administrative machinery that evolved more or less inevitable, and certainly necessary.

This is not to say that errors of judgment and design philosophy were not made. In retrospect, I think we failed to draw a clear enough distinction between the secondary actuator system and the laser stabilization scheme for maintaining automatic alignment. Possibly too little attention was given to using stars for alignment, partly because too much weight was given to being able to offset guide during full moon, which, given the small field and the state of TV guiding at that time, seemed to require an internal alignment system. Some decisions such as this might have benefited from extensive outside peer review—in fact, such a review by several eminent astronomers was held on November 20, 1975, but earlier and more extensive reviews would perhaps have been useful.

In any event, the "committee criticism" has always seemed to me somewhat beside the point. In a project as complex as the MMT, the essential ingredient is a group of dedicated and competent individuals who can have their ideas carefully evaluated and their progress coordinated by a larger group whose members enjoy a certain amount of mutual trust and goodwill. The real danger of the committee structure, as I see it, is that it can become a battleground for debates revolving, not around the merits of the technical issues at hand, but around the self-esteem of groups and individuals within them.

I do not know what concepts embodied in the MMT will eventually be incorporated into the large ground-based telescope envisioned for the future in the Greenstein Report and subsequently in the Field Report, but I am convinced that the construction of the MMT has been a significant and worthwhile effort that will benefit not only the SAO and Arizona users of the instrument but the whole astronomical community as well.

5. DESIGN AND PERFORMANCE OF THE MMT

5.1 *Optical Support Structure*

The basic optical layout of the MMT, based upon the six 1.8-meter primary mirrors, was the driving concept in the design of its supporting structures, except for a few important early thoughts that flowed the other way. One of these was the general realization that the six-shooter layout should make the support designers' job easier, since it provided the opportunity to locate major structural members in between the six barrels. Another was the decision, stemming from Meinel's thinking on conventional telescopes, that the primary focal lengths should be 2.7 times their diameters, rather than some larger value that might have permitted grinding and polishing the blanks in their original flat configuration. The argument was that the savings from a more compact telescope structure, and *a fortiori* those from a smaller dome structure, must outweigh the additional costs of optical figuring. This decision was made without a detailed cost comparison, as was also the decision to use an alt-azimuth mount. This latter decision was based not so much on cost comparison (the savings are obviously substantial) as it was a decision that the problems of field rotation would not be so onerous with the alt-azimuth mount as to require the extra cost of an equatorial mount. After these early considerations, the structural designers essentially accepted the problem of filling in support around more than 100 optical components of the main telescopes and of the guide-alignment system, as well as 2000-kg instruments plus associated guider optics and the like. Hoffman (1978, 1980) describes the engineering concepts included in the optical support structure, mount, and building.

For simplicity, we wished to confine our active-optics system to correction of the tilt and focus of the secondary mirror in each optical train. It would obviously be preferable to correct the translation and tilt of all primaries and secondaries so that their axes always coincided with an ideal hexagonal figure. The simpler scheme should be adequate, however, so long as two conditions are met: 1. the misalignments of the six individual Cassegrain telescopes (due to gravitational and thermal flexure) must be small enough to maintain image aberrations below a tolerable limit, and 2. the directions of the incoming beams, as seen at the focal plane, must not vary so much that their interaction with the optical systems of instruments would suffer errors such as vignetting.

The relation of these conditions to specific structural deflections is a complicated vector problem, but we may summarize it crudely. The growth of the images due to misalignment will be less than 0.2 arcsec if the relative displacement of primary and secondary mirrors is held below about 0.7 mm, and the angle between the primary and secondary axes is held below about 2.5 arcmin. These displacements and tilts can cause image motions in the final focal plane of up to 25 and 300 arcsec, respectively. The directions of the beams will not shift by more than 1% of their angular diameters if the primary mirrors are, in addition, held to about 2 mm of their nominal positions.

These criteria controlled the configuration and stiffness of the support structure. They also imposed the important consideration that the structural members should not differ too much in thermal response time, so that expected temperature changes of a few degrees Celsius per hour would not warp the structure.

Another important constraint on the structure was its interaction with two servomechanical systems—those of the main-axis drives, and those of the secondary-mirror controls. For these systems to operate stably at given bandwidths, the resonant frequencies of structural modes with which the servos might interact must be kept well above the highest desired frequency of response. For the main drives, stiffness against wind buffeting and other disturbing torques was calculated to require resonances above 5 Hz or so for the main structural modes. For the secondaries, the possibility of correcting for the residuals of these major disturbances required resonances of the secondary supports above 50 Hz. The resonant frequencies of a structure are, in fact, closely related to its deflections under load, and so it turned out to be a pleasant coincidence that these two different constraints could be satisfied about equally well by a given structural design.

It was decided fairly early to use a truss structure (i.e. one with long, thin members that act principally in simple compression and extension). A structure of extended cylinders, plates, etc., was considered less suited for picking up the many localized loads of mirror cells, instruments, and the like. The materials considered were steel, aluminum, and fibre composites, with steel being chosen as the most cost-effective.

Starting from this point, the detailed design of the optical support structure demanded considerable ingenuity (particularly with an added constraint of using only standard-size steel members wherever possible), but it was duly accomplished with even some room to spare. One potentially useful feature of the design was to locate the attachment points of the primary cells so that they were not on the main load-bearing paths to the elevation-drive axles. This makes it possible to adjust the cross sections of specific mirror-hanger members so as to correct any deflections that might exceed tolerances, and to do so with relatively little effect upon other critical deflections. We have not had occasion, however, to use this feature, because the predicted performance of the structure has been so well realized in practice. We have not measured many of the critical deflections directly, but have let well enough alone, since the optical performance of the telescope is up to specifications.

5.2 *The Telescope Mount*

The optical support structure, weighing 45 metric tons with all of its cargo, is in turn supported by a massive steel yoke. The alt-azimuth geometry permits it to rotate in elevation on two simple preloaded spherical roller-bearing assemblies. The yoke is built of internally braced box sections of 25-mm plate with central reinforcement by a 75-mm wall cylinder, supported on an angular-contact thrust ball-bearing race, which is 2.5 m in diameter and contains 130 50-mm balls. This bearing was much less costly than a hydrostatic bearing system, and preliminary tests of the yoke showed it to rotate very smoothly. Under a full load of 120 tons it operates with an average coefficient of friction of 3×10^{-4}, and variations are only a small fraction of the average. The yoke and bearing have a compliance of 0.1 arcsecond per 3000 Nm moment about axes perpendicular to the rotation axis, to resist temporary imbalance and wind forces. The torsional stiffness about the azimuth axis is sufficient for the lowest fundamental frequency oscillation to be about 5 Hz (but see below).

Both axes of the telescope are driven by essentially identical systems employing dc torque motors and straight spur gears of moderate quality. Each axis has two motor-and-gearbox combinations that oppose one another during tracking to eliminate backlash, and the motor speeds are governed by tachometer feedback in a stable servo rate loop. The precision of the system for tracking is derived from on-shaft 24-bit encoders of the Inductosyn type. These encoders have an absolute accuracy of 1 arcsec and locally have the full 24-bit precision (0.08 arcsec). The operation of the telescope is ordinarily completely controlled by the Nova 800 minicomputer, which is dedicated to this purpose. It converts object coordinates to the current epoch, calculates corrections for refraction and aberration of light (and for telescope flexure and misalignment), and commands appropriate position sequences for slewing and tracking in alt-azimuth coordinates (see Section 5.8).

The analytical predictions of the behavior of the yoke and drives have been well borne out, in general. One omission in those predictions was discovered, however, in the factory assembly and testing. The mount has a mode of oscillation corresponding to a rocking on its azimuth bearing, since the balls of this bearing are confined in a race whose cross section has a radius somewhat larger than that of the balls themselves. This motion was considered in the general calculation of resistance to external forces, but its interaction with the drive system was not properly included. What generates this interaction is the placement of the two azimuth drive gearboxes close to one another in one side of the yoke (chiefly for convenience). Thus, when they exert a torque upon the mount, they also exert a net force as well, and this force can excite the rocking mode. Placement of the drives on opposite ends of a diameter would have prevented this interaction, but would have provided a constant tilting of the mount due to the torque bias that prevents backlash (though this latter effect can be compensated in the pointing-correction matrix). At present this rocking mode contributes to the low resonance frequency of the mount (2.2 Hz) and thus limits the stiffness in the azimuth drives.

5.3 *The MMT Building*

We expected from the start that the operation of the MMT and its associated instrumentation would be remotely controlled, for the most part, with the observer viewing the focal plane via a low-light-level television system. At the same time, we recognized that the MMT was itself a complex and novel piece of equipment, and that it would generate a large variety of novel accessory instruments. Therefore, in planning the housing for the MMT, we strove to provide both easy access to the telescope itself, and a large amount of laboratory space immediately adjacent to it, for the electronics used during observations and for the calibration and checkout of optical instrumentation.

The alt-azimuth geometry offers unique advantages over an equatorial mount for the design of the telescope housing, since the rotation of the housing follows the telescope in one coordinate. Therefore, instead of a shell dome on top of a building whose working space is necessarily below the level of the telescope, we designed an enclosure in which the telescope is completely embedded, and which is slaved to it in azimuth rotation.

There were important spatial, structural, mechanical, and thermal considerations in the design of the building to house the MMT. (Figure 4 shows a schematic of the building.) Spatially, we wished first of all to provide an observing floor at the level of the top of the yoke, and then to work up and down from that level to provide additional space. We needed to be very efficient in the use of space, both to make access convenient and

Figure 4 Cutaway view of the MMT and its building. Height of building equals 16.8 meters.

to make the best use of a very restricted area on the mountaintop site. Structually, we needed to provide support for the working space and for a shutter mechanism to expose the telescope. Mechanically, we wished to ensure smooth rotation of the building, and accurate following of the telescope. The building should permit observing in winds up to 72 km/hr, and should survive winds up to 225 km/hr. Thermally, we had to prevent heat from the enclosed spaces or from the sun-warmed exterior from creating temperature gradients in the optical paths.

The resulting building, shown in Figures 1, 2, and 4, is unusual by astronomical standards. It is rectangular (width 19.5 meters, depth 13.4 meters, height 16.8 meters) because it is less expensive to enclose space in such a shape, and the penalty in terms of wind torques and buffeting is calculated to be small. The lower part functions as a deep structural platform extending from the observing floor to well under the first floor, with principal truss planes that box in the telescope yoke. On this platform rest two three-story towers containing rooms and serving as the stowing location for bi-parting shutter leaves. Figures 5 and 6 give cross sections of the building. The entire building, weighing 450 tons, rests on four massive conical steel wheels, 13 cm wide and 91 cm in diameter, which roll on a flat steel track located about 1 m below ground level on a foundation independent of the telescope pier. The wheels are supported by an articulated linkage that allows the full width of the wheel always to be in contact with the track, thus distributing the load evenly, and they are guided by four wheels pushing radially outward against another track. Two of the support wheels are driven by 15-hp dc torque motors in a servo configuration very like that of the telescope itself, with the position information conveyed to the drive by a transducer connected between the yoke and the building.

The telescope is thermally isolated from the outside and from the heated rooms on either side and below by foam sandwich panels. The working spaces are also heavily insulated from the outside. Since all the surroundings conspire to produce temperatures at least somewhat in excess of the nighttime average, however, the telescope and its environs are designed to be actively cooled by means of refrigerated coils in the floor (which is a double slab, with foam insulation between upper and lower parts) and by conventional air-conditioning units that chill and stir the air, maintaining the telescope uniformly at about the expected nighttime temperature. In addition, the telescope floor and telescope yoke are insulated to reduce the temperature differences in the telescope chamber. All warm exhaust air from the building is dumped into the basement, which is carefully sealed by a skirt from the rotating building that dips into a liquid-filled moat at the top of the foundation. The air is then drawn

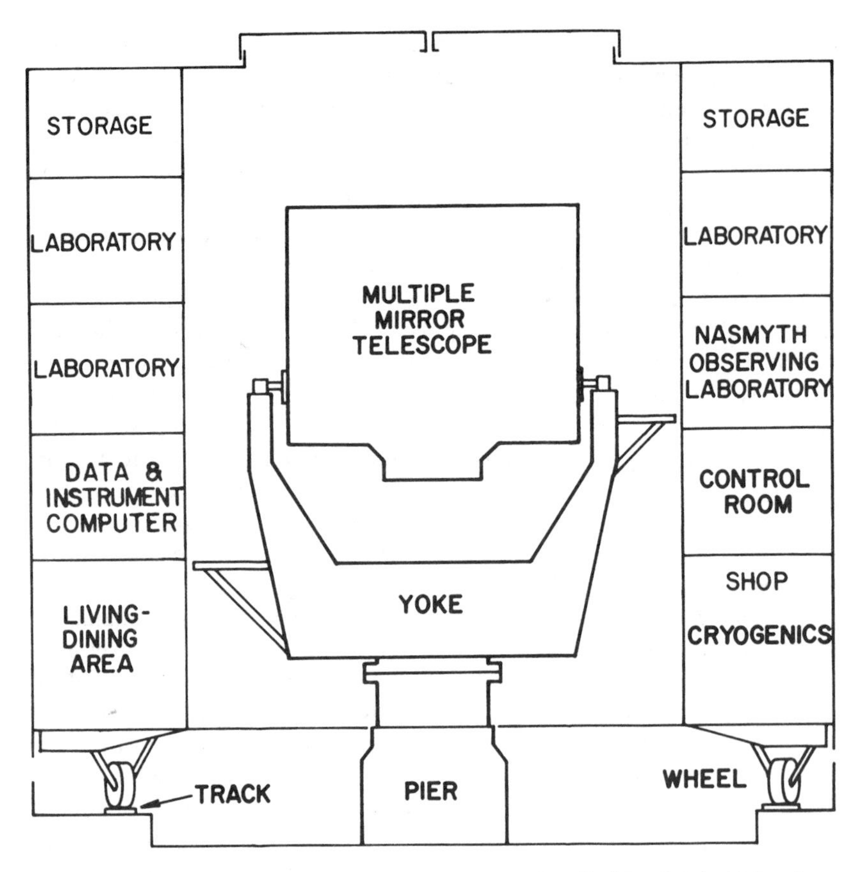

Figure 5 Vertical cross section of the MMT and its building (building height 16.8 meters, width 19.5 meters).

through a tunnel and exhausted about 50 m from the building in the prevailing downwind direction.

Electric power to the installation is furnished by two transformers, one to supply machinery and heavy loads, and one for instrument power. The latter is supplied via a motor-generator set that serves as a low pass filter, which removes rapid fluctuations in the power and which stabilizes the power over the long term. Grounding (always a problem on a dry mountaintop) has been carefully considered both for lightning and for instrument grounds. Copper is buried all around, and particularly in the test-boring holes that were made in the rock. There is a consistent scheme of branching grounds to avoid ground loops.

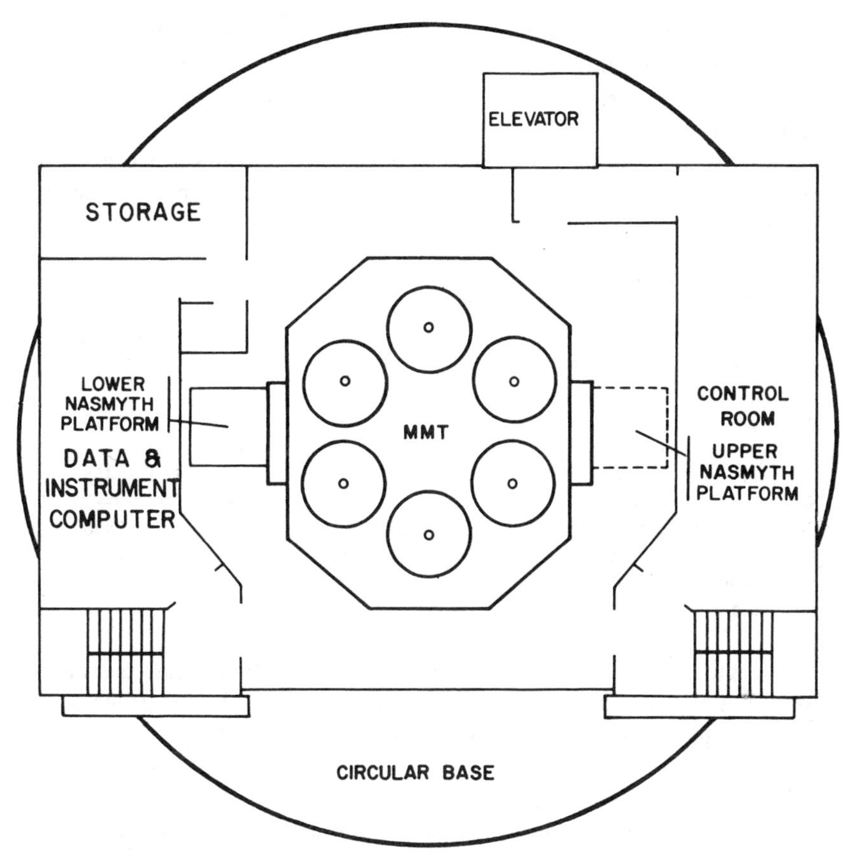

Figure 6 Horizontal cross section of the MMT and its building (building width 19.5 meters, depth 13.4 meters).

5.4 *The Optics*

The optics consists of six main telescope systems (Figure 3), each of which is a classical Cassegrain with a 1.8-m-diameter parabolic primary with focal ratio $f/2.7$, and a hyperbolic secondary producing a final $f/31.6$ for each of the individual telescopes. The focal location is adjusted by longitudinal motion of the secondaries to obtain a focus at various focal locations ranging from the quasi-Cassegrain focus formed by the $f/9$ beamcombiner 787 mm behind the primary vertex ($\equiv$ focus A) to the zero spherical aberration focus 360 mm behind the primary vertex ($\equiv$ focus B). Calculations indicate that the spherical aberration change due over this focusing range is acceptable (< 0.2 arcsec).

In the center of the system there is a 73-cm diameter guide telescope used for offset guiding. This system is a Ritchey-Chrétien type of design with a mild aspheric plate for correction of astigmatism near the focal surface. The focal surface is spherical and the guide star sensors, for examining the wide field of view, track on a spherical surface of radius 56.4 cm. Parameters for both the main and guide telescopes are listed in Table 3.

Table 3 Parameters of MMT optics (for focus A)

Parameter	Main telescopes	Guide telescope
Diameter (mm)	1818	730
Primary focal length (mm)	4935	2110
Primary conic constant	−1.0000	−1.0145
Secondary focal length (mm)	742	455
Secondary conic constant	−1.4358	−2.0677
Effective focal length (mm)	57500	12850
f/ratio (optical secondary)	$f/31.6$	$f/17.6$
f/ratio (IR secondary)	$f/33.8$	
Image scale (microns/arcsec)	279	62

The guide telescope was manufactured using traditional techniques. Since the primary varied only by 1.45% from a parabola, the primary was, in fact, figured using an autocollimation test from a flat to leave a given residual in the interferometer pattern measured from the focus. The secondary received its final figuring in a full autocollimation test of the assembled guide scope. The guide scope primary is a rather thick mirror blank 73 cm in diameter by 15 cm thick and is supported in a strap and multiple mechanical back support mount.

The primaries for the major telescope are of lightweight construction, of egg-crate fused silica form. Since these mirrors were originally intended for a different application, all six mirrors had to be returned to Corning Glass Works to have the front radius slumped to an approximation of the final radius to be placed on the mirrors to obtain the $f/2.7$ parabolas. Optical work on these primaries was carried out on two large polishing machines in the Optical Shop of the Optical Sciences Center. The optical operations involved first generating and polishing the back surface of the mirror convex, edging the front and back plates, and then generating, grinding, and polishing the front surfaces of the mirror to spherical surfaces whose radius matched the final edge radius of the desired parabolas. Interferometric tests at the spherical stage confirmed the absence of astigmatism in the mirrors. The mirrors were tested on top of an air bladder support of somewhat similar form to that to be used in the final telescope. Support of the mirrors during fabrication and testing used an air bladder in the test setup, and two layers of ordinary rug cushioning under the mirror on the fabrication table. The mirrors were moved from the table to the test setup. Matching of the radii of the six mirrors to within 3 mm (± 0.03%) was accomplished using an invar steel tape measurement while doing a knife edge test at the center of curvature on the shop floor. Parabolization was carried out by polishing in the center of the aspheric, leaving the edge of the mirrors essentially untouched. The edge of the mirrors thus served as a continuing radius reference during the fabrication process. In addition, not working the edge portions avoided the usual edge-turn-down problems that occur in a mirror, which is desirable since the edges of the mirrors are to be used in the optical alignment system for the telescope. Initial roughing-in of the parabola required knife edge and wire testing on the shop floor. Final testing involved measurement with a null lens and an interferometric setup in the test tower.

The initial goal for the mirrors was to put 90% of the energy within one-half arcsecond. Most of the mirrors closely approached this (see Table 4). Figure 7 shows a composite of the energy distribution curves for the mirrors. As can be seen, there is some dispersion in quality of the mirrors, with mirror six being the worst of the lot, due partially to some residual astigmatism in the surface. Interferometer photographs of the six mirrors taken through the null arrangement are shown in Figure 8. The basic parabolic figure of all the six mirrors is attested to by the straight fringes. The major distinguishing feature of these mirrors turns out to be a number of fine circular zones that were left in the fabrication process. The geometrical energy distribution curves shown in Figure 7 were obtained after computer reduction of the interferograms on the mirrors. Full diffraction calculations of the primary mirrors show the FWHM (full width at half maximum) of the point spread function to average only 0.14 arcsec. Table 4 contains information on the rms surface error and energy

Table 4 Properties of individual primary mirrors

Mirror number	90% energy diameter (arcsec)	rms surface error (waves at .6328 μm)
1	0.58	0.105
2	0.64	0.096
3	0.56	0.113
4	0.56	0.099
5	0.54	0.102
6	0.86	0.128
Average	0.62	0.107

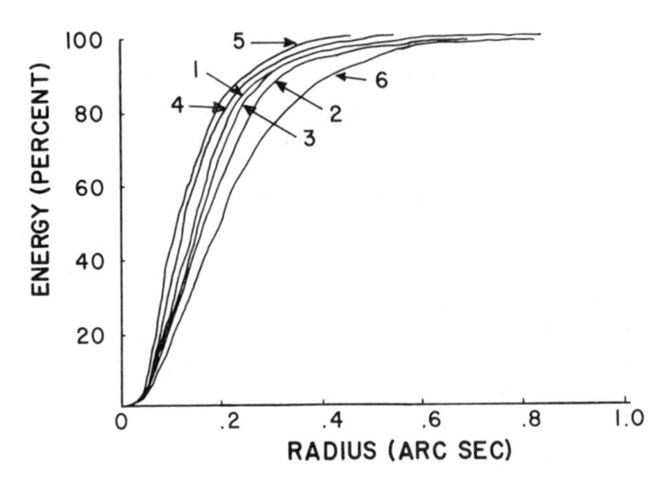

Figure 7 Enclosed energy in a circle of a given radius for the six MMT primary mirrors (based on geometrical optics calculations of the laboratory tests of the optics). 90% of energy falls within 0.6 arcseconds except for mirror number six.

distribution on the six mirrors, both on the test setup and when tested in the final flotation cells at various elevation angles. A full description of the MMT optical design and performance is given by Ruda & Turner (1980).

The MMT construction represents the first time that lightweight structured mirrors of this sort have been used in a variable altitude, gravity environment. The primary mirrors are each located by three axial hardpoints with most of their weight floated on an air bag whose pressure varies with altitude. Seven percent of the total mirror load is carried on the three axial hardpoints. The radial location of the mirror is defined by counterweights through an 11:1 ratio lever system acting on a pair of chains that surround the lower half of the front and rear plates of the mirror. These are adjusted so that the mirror hangs free of astigmatism when one is viewing at the horizontal position. Only a few pounds force is generated against the radial locating points. These radial locating points are referenced from an invar ring cemented to the back of the mirror near its center.

There are actually twelve secondary and twelve tertiary mirrors for the telescope, six of each for visual use, and six for infrared use. The infrared secondary mirrors are smaller (and thinner) than the optical ones to eliminate the radiation coming from the edges of the primary mirrors and to provide a low inertia for the mechanically oscillating infrared secondaries. In fabrication, all twelve secondary mirrors were tested against a concave hyperbolic test plate, which itself had been tested in a null lens interfer-

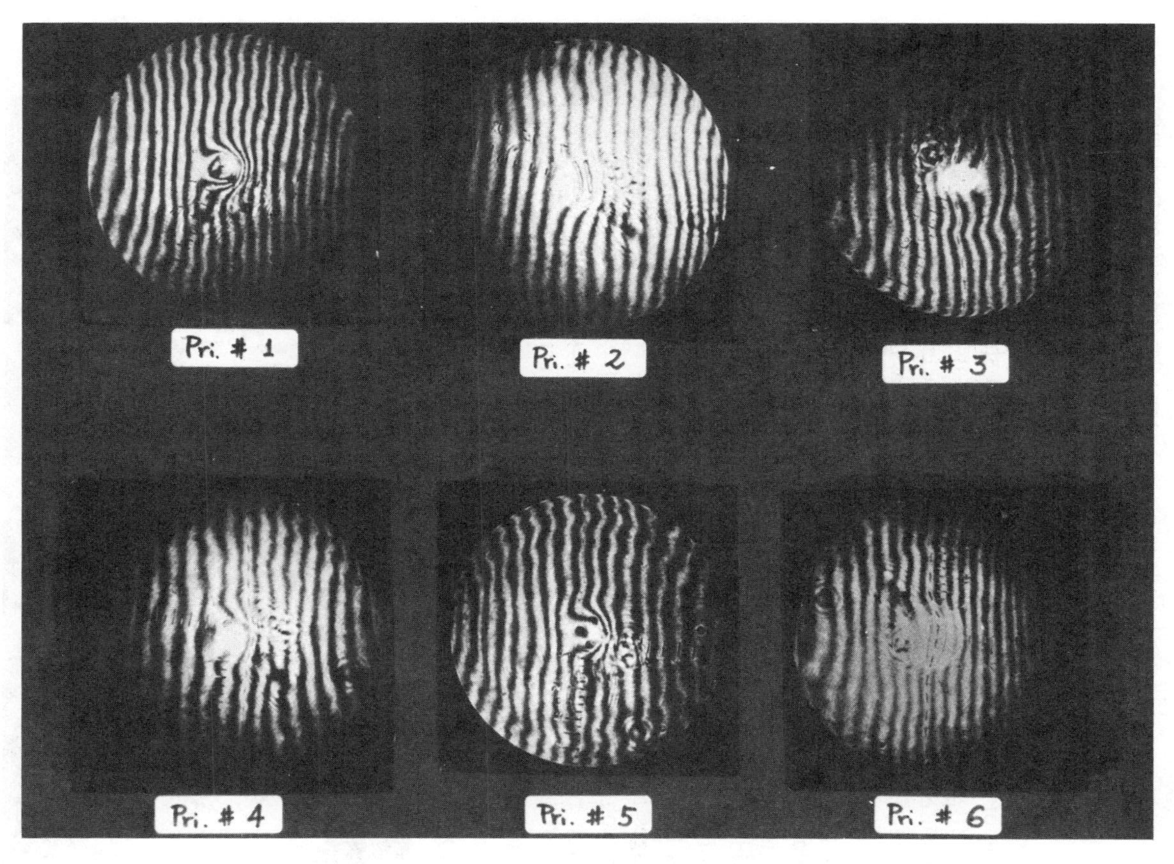

Figure 8 Interferograms of the MMT primary mirrors showing residual figuring errors.

ometer system. Therefore, there is no variation in radius between the secondaries.

The optics were tested while in place on the telescope using star images. Figure 9 shows a Foucault knife edge test of one of the mirrors as well as an interferogram. The circular zones left over after the fabrication process are very evident also in the out-of-focus images for all six telescopes. They are, in fact, the dominant feature in all of these tests, other deformations being of lesser importance. The knife edge test shows, of course, the first derivative (slopes) of the surface irregularities in the direction normal to the knife edge, the out-of-focus image shows the curvature of these irregularities (second derivative). This curvature shows up remarkably well for relatively slow telescope systems like the MMT ($f/31.6$). A closer evaluation of the six telescope knife edge images shows only a very small amount of coma in some telescopes and, except for two telescopes, no astigmatism. Adjustments in the mirror support remain to be made to eliminate this residual astigmatism and coma. The stellar image quality is very good and mostly determined by seeing (Section 5.9). Image sizes with a FWHM of less than 0.5 arcseconds had been reported and sub-arcsecond-diameter images are common. Even so, this image size is mostly determined by atmospheric seeing both inside and outside the telescope chamber and not by the optics. Future improvements in the chamber thermal conditioning will undoubtedly improve the image quality further. Whether then the telescope optics will become the limiting factor remains to be seen.

The average diameter for the non-vignetted field equals 35 arcseconds as compared to 52 arcseconds as the theoretical optimum (Beckers 1980).

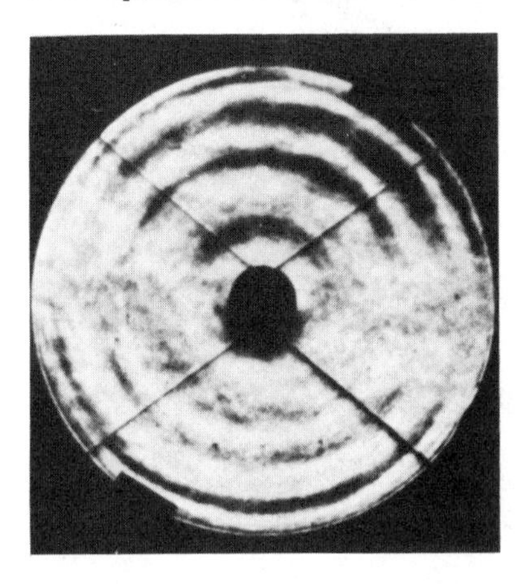

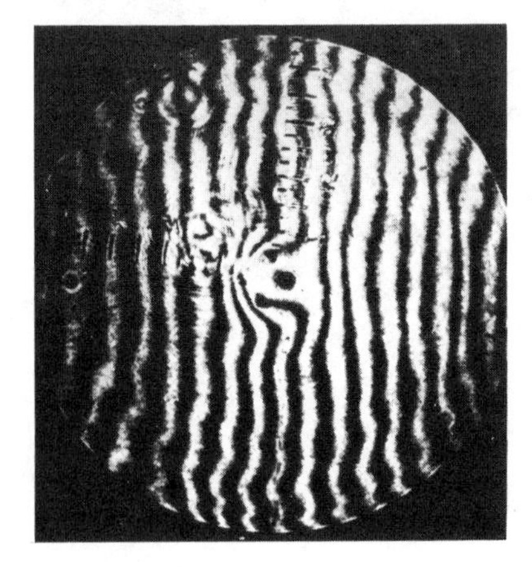

Figure 9 *Left:* knife edge test of the telescope containing primary mirror number 5 using a star image. *Right:* interferogram of this primary mirror. Note the very pronounced circular zones, the worst one having a peak-to-peak amplitude of 1/6 wave.

Future adjustment of the optics will increase this field of view. The vignetting of the telescopes sets in very slowly, however, so that even for a 4-arcminute-diameter field the vignetting is less than 50%.

5.5 *Telescope Coalignment*

REQUIREMENTS In order to produce and maintain an accurately superposed image from the six individual telescopes it is necessary to be able to (*a*) independently adjust each image position, (*b*) determine the initial superposition, and (*c*) continuously detect deviations of individual images from the composite image.

For the MMT, the first is accomplished by remotely controlling the tilt of each of the secondary mirrors about two axes, the second by television monitoring of the focal plane either visually or under computer control with centroid determining routines, and the third by either periodic updates of the second or with an internal artificial star system using laser beams.

The desired accuracy of the coalignment depends on the overall quality of the telescope optics, tracking, and site and the intended uses of the instrument. The individual Cassegrain telescopes as determined from optical shop measurements allowing for primary, secondary, tertiary, and beamcombiner figure imperfections and for spherical aberration and miscollimation provide 90% of the light from a point source in a .7-arcsecond-diameter circle and a FWHM of the point spread function of about 0.2 arcsec. Image quality (FWHM) at the site averages 1 arcsec and has been as good as 0.5 arcsec. Current spectroscopic, imaging, and interferometric uses of the MMT can take advantage of images less than 1 arcsecond, dictating superposition accuracy which does not noticeably degrade the individual telescope performance, e.g. ≈ 0.1 arcsecond superposition error. It should be noted that this provides a more stringent requirement on the coalignment than that specified when the telescope was conceived and designed. At that time, typical seeing was expected to be 2–3 arcseconds, rather than the 1 arcsecond frequently experienced now, and the usefulness and power of the MMT for imaging and interferometry was not fully appreciated.

The range of individual image wander is determined by the rigidity of the supporting structure, which in turn is dictated by the maximum allowable image degradation due to miscollimation. The criteria given in Section 5.1 imply a maximum image motion of 300 arcseconds. The observed maximum relative image motion of the six telescopes is $\pm$ 20 arcseconds over an elevation range 17–88 degrees with typical and maximum drift rates of .05 arcseconds/minute and 0.2 arcseconds/minute during stellar tracking.

ADJUSTMENT OF IMAGE POSITIONS The actuator for the secondary mirror tilt is designed to meet the requirements for image superposition. The secondary actuator characteristics are as follows:

Minimum angular step:	0.059 arcsec motion in focal plane
Range:	600 arcsec
Maximum slew rate:	12 arcsec s^{-1}

The mirror pivot consists of two 12-mm-diameter steel rods to provide both precise angular control and high axial rigidity to allow for a secondary chopper mechanism for infrared use. The tilts are achieved by two spring-loaded micrometer screws coupled to a 100-steps-per-revolution stepper motor by a 110-to-1 harmonic reduction gear. One step of the stepper motor provides a 0.058-μm motion of the micrometer, .23 arcseconds tilt of the secondary, and .067 arcsecond motion of an image at the focal plane.

The actuator assembly also provides for remote focus control using precision ways that assure no change in mirror tilt during focus motion. The torque motor drive provides for a smallest increment in mirror position of 1 μm. Motion of the tilt and focus motors will be under computer control with commands initiated either manually with a control paddle or automatically from a focal plane monitoring and image centroiding device. The current actuators are a modification of the original MMT actuators, which had torque motor drives for all three motions and less precise setting accuracy. Additional coalignment control for precise setting of the direction of the principal rays of the individual telescopes can be achieved by an adjustment of the tilt of the tertiary mirrors. The control system for this has been designed but not yet implemented.

INITIAL SUPERPOSITION The initial superposition of the six images requires accurate determination of the relative position of the center of each of the images and feeding this information to the secondary actuators to provide for the necessary correction. Since the images share some motion due to seeing and tracking errors, it is important that the relative position determination be simultaneous for all six images. A camera consisting of a GE TN2200 series CID camera controlled by a Z-80 microprocessor has been built for this measurement. The scale on the 128 $\times$ 128 array chip equals .16 arcseconds per pixel. In operation, an exposure of the six separated images will be taken, stored in memory, and analyzed for the centroids of each of the images to an accuracy of $<$ .05 arcseconds, the entire operation taking about ten seconds depending on stellar magnitude. The measurements are then entered into the computer controlling the secondary actuators to provide the appropriate corrections. The CID camera, while providing excellent dimensional stability, is currently limited to a small field and relatively bright stars ($m_v = 5$ without intensifier, $m_v = 10$ with intensifier).

A second system for superposition of images utilizing a sensitive intensified focal plane acquisition and guiding TV has been developed (see Section 6.2.2). This provides for viewing of stars of up to $m_v = 18$ continuously in a 4-arcminute field around the slit of a spectrograph or hole into a photometer. The optics provide a focus of the telescope pupils where a set of prisms provides a small deviation of each image and where a pupil selector can be used for examining individual images as desired. With this system, the images can be manually superposed to an accuracy of .2 arcseconds in a few seconds of time. Provisions are being made for automating this with centroid-determining software at a higher accuracy of 0.1 arcsecond. At the same time this system will serve as an autoguider for the MMT.

DETECTING DEVIATION FROM SUPERPOSITION Depending on the telescope orientation, superposition must be updated every 30 to 600 seconds (typically 120 seconds) to maintain .1-arcsecond superposition accuracy with the CID camera on a bright star, or with the focal plane guiding television on a field star either manually or automatically. The optics and computer hardware and software for the guiding television are being designed to make continuous guiding on the individual images straightforward with the capability of updating as frequently as ten times per second.

For instrumentation and conditions (such as daylight observing or no guide star in the field) for which the above approaches are not suitable, the artificial star system using laser beams will be used to provide an internal mechanism for determining telescope misalignment drift. The basic elements of this laser alignment system are shown in Figures 10 and 11. It is described in detail by Reed (1978) and McDonough (1980). In the center of the hexagonal MMT telescope array is a 73-cm $f/17.6$ guide-alignment telescope of Ritchey-Chrétien design, designed for telescope alignment and guiding (Table 3).

For alignment purposes, this telescope serves as a collimator for a laser-generated point source. In order to maintain the desired 0.1-arcsecond collimation accuracy a second laser system provides continuous automatic collimation of this telescope. The outgoing beam is transferred to the periphery of the six large telescopes by three 1.8-meter-long periscopes and thence into the telescope apertures by six roof-prism-90-degree-prism combinations which function as elongated cube corner reflectors. Both the periscopes and the cube corners have the property that the outgoing rays

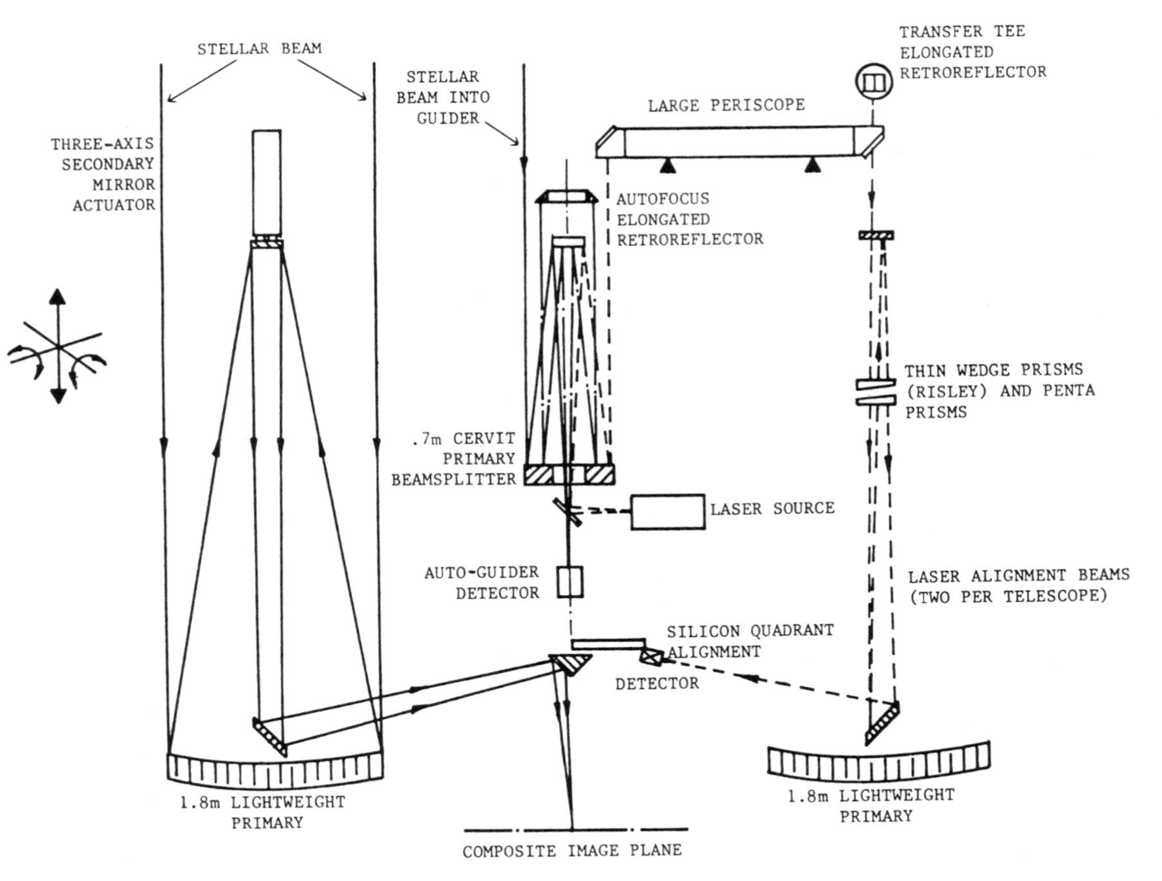

Figure 10 Light paths of the laser coalignment system.

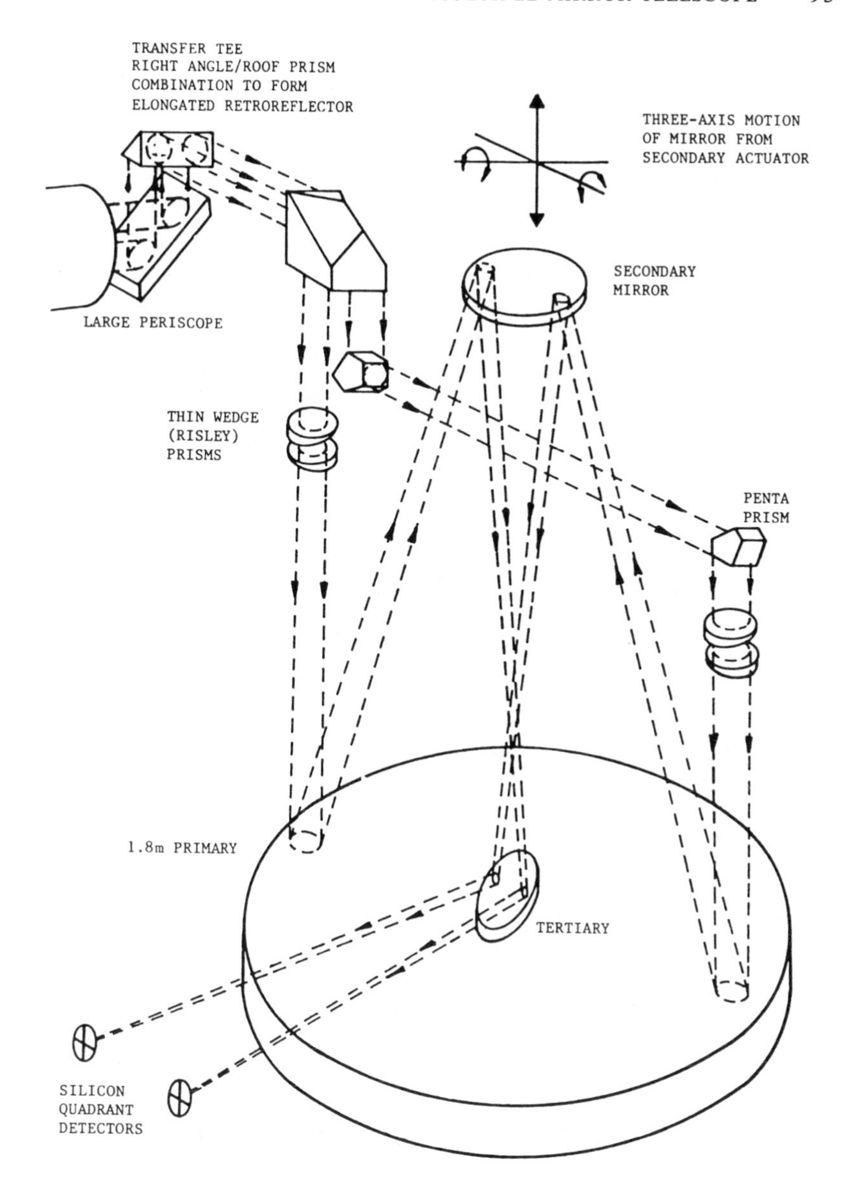

Figure 11 Detail of the laser coalignment system for a single telescope.

remain accurately parallel (or anti-parallel) to the incoming rays if the device is tilted, so long as it behaves as a rigid body. A prototype of the periscope has been tested and found to have a stability of better than 0.05 arcseconds.

Because a single marginal ray cannot distinguish between a change of focus and a tilt, a second beam is introduced into each aperture with a pair of pentaprisms, as shown in Figures 10 and 11. Both beams are then sensed by silicon detectors at the focal plane, the focus beam by a split detector that senses the radial direction only (the pentaprisms giving invariance only in this dimension), and the main beam by a quadrant detector that senses both radial and azimuthal motion. These signals, processed by the computer to remove noise and seeing effects, provide the desired pointing accuracy by averaging over time, and are then used to provide corrections to the tilt and focus of the secondary actuators.

Pairs of thin wedge prisms, shown in Figure 11, provide for remote-controlled two-axis tilt adjustment of each of the beams for initial setting of the telescope focus and image position. To date, the laser system has been operated only with the alignment beams, not the focus beams, directly controlling the torque-motor-driven secondaries without deglitching and averaging. In these experiments the superposition was maintained to 1 arcsecond rms over the short term (1–2 hours), and the long-term drift reduced by an order of magnitude. Direct simultaneous monitoring of the laser beams and stellar images at the focal plane indicate that considerably better performance can be achieved by a fully operating laser system.

The methods of maintaining the image superposition have evolved from the original design since the telescope has been available to use in 1979. Considerable further evolution is likely as experience is gained with this unique instrument.

5.6 *Infrared Capabilities of the MMT*

GENERAL CONSIDERATIONS As the result of several years of development at the University of Arizona, the general principles of design for ground-based IR telescopes were proven by modifying small instruments such as the 0.71-m and 1.55-m telescopes. Low & Rieke (1974) have described these principles in detail and they discuss the limiting performance of instruments using these designs. The MMT represents an effort to extend these concepts to the design of a new type of large telescope that retains all of the performance advantages of proven smaller telescopes without the risk and uncertainty of applying similar solutions to a much larger scaled-up design. In addition, the MMT design adds the fundamental advantage of an unfilled array in those applications where maximum angular resolving power is important. An equally important motivation, though still the subject of considerable debate and future study, was to achieve a design that would reduce significantly the cost of very large telescopes capable of meeting the basic needs of both optical and infrared astronomy. The full performance of the MMT in the infrared has not yet been measured, but certainly no other telescope of competing infrared performance has yet been built. We briefly list here the most important infrared features of the design as they bear on observations in the infrared.

The general distinguishing features of a telescope designed for the IR are: 1. minimum background noise emitted by the telescope, i.e. minimum obscuration, minimum number of uncooled mirrors, and small mirrors with easily maintained coating, 2. stability of the telescope background, i.e. extreme internal rigidity under wind loading and infrequent articulation of uncooled mirrors, 3. a means of background subtraction such as secondary mirror modulation. Smooth, sub-arcsecond tracking and arcsecond pointing are also important.

OPTICAL CONFIGURATION The optical configuration, which is that of six identical 1.8-meter $f/33.8$ telescopes arranged in a circle around a common axis, permits all of the features of the well-proven $f/45$ 1.55-m IR telescope to be preserved. The IR secondaries are undersized slightly to serve as a "cold" stop against the sky. Two of the three uncooled mirrors are easily maintained and can be equipped with special IR coatings; the primaries are comparatively small and are relatively easily maintained with a fresh, clean coating. Thus, it should be possible to achieve and maintain an overall emissivity of the telescope as low as 5%. Clearly, the fourth mirror in the system, the beamcombiner, must be cooled along with the detector, the spectral filters, and the field optics and baffles.

CHOPPING SECONDARIES Because of their importance in determining IR performance, the chopping secondaries merit special attention. It has been shown that no other simple method of background subtraction performs as well as scanning the IR beam on the sky by tilting the secondary mirror. Using a servo-controlled linear motor with sensitive position transducers, it is possible to produce extremely accurate square-wave or linear scans over angles ranging from a few arcseconds to several tens of arcseconds. Because of the small size of the secondary mirrors, chopping speeds above 20 Hz are possible with low power dissipation and without image degradation from mirror bending or from overshoot. In the MMT design the chopping secondary mechanism provides for accurate rotation of the chopping axis around the optical axis so that all "twelve" IR images may be coaligned. The chopping motion is completely independent of the image coalignment system, which also tilts the secondary mirrors.

PERFORMANCE The chopping secondaries and the infrared photometer described in Section 6 were first used at the MMT in October, 1980. The measured performance was exactly as predicted by scaling from the

1.55-m telescope to the full 4.5-m aperture of the MMT. In other words, the MMT performs as well in the infrared as would a fully optimized conventional infrared telescope of 4.5-meters aperture.

The sensitivity of the MMT is best illustrated by the detection of the flat spectrum radio source 0332 + 078 at a magnitude of 18.40 ± 0.25 at 2.2 μm. The measurement required one hour of integration through a 9 arcsec aperture. (The corresponding 1σ level is 20.0 mag.) This source was previously undetected at 2.2 μm in two attempts with the Steward 2.25-m telescope; it is an empty field on the Palomar Atlas plates; and it was not detected in the visible on a deep plate obtained with the Kitt Peak 4-meter telescope. It is the first example of an extragalactic infrared source that is undetectable in the visible. Its measured flux level is four times fainter than any object detected previously at 2.2 μm.

5.7 *Interferometry*

Because the MMT is the largest optical/IR telescope now in operation, when measured from edge to edge (6.9 meters), it potentially has the highest resolving power at these wavelengths. This fact, coupled with the additional consideration that the MMT is the first of a possible series of large optical/IR arrays to be built in the future, suggests that the MMT will be very important as a tool for spatial interferometry both at optical wavelengths and in the IR. An important incentive to workers in this field is the newly discovered result that the "seeing" at the MMT is remarkably good compared to that at other large telescopes in the Tucson area. To achieve the full possible angular resolution of the MMT, it is necessary both to "phase" the individual telescopes with respect to each other and to eliminate differential changes of the polarization state of the incident lightbeams. So far this has been achieved both at optical and infrared wavelengths with pairs of opposite telescopes. Because of the off-axis reflections on the telescope tertiaries and on the beamcombiner, the polarization state of the light will be modified, but this modification is identical for these pairs. The pathlengths of the two telescopes can be adjusted until an interference pattern appears.

D. W. McCarthy and F. J. Low have already operated the MMT as a two-element Michelson interferometer at 5 μm and have measured both the pathlength equality of the three pairs and the stability of a single pair. Work is proceeding toward the goal of operating the MMT as a phased array at the longer IR wavelengths. The pathlength stability, as shown by McCarthy's data in Figure 12, is inherently adequate for phasing at 20 μm and may be adequate for 10 μm. The coherence length ($\lambda^2/\Delta\lambda$) is a function of both wavelength and bandwidth and can therefore be adjusted within the detecting system to extend operation to shorter wavelengths

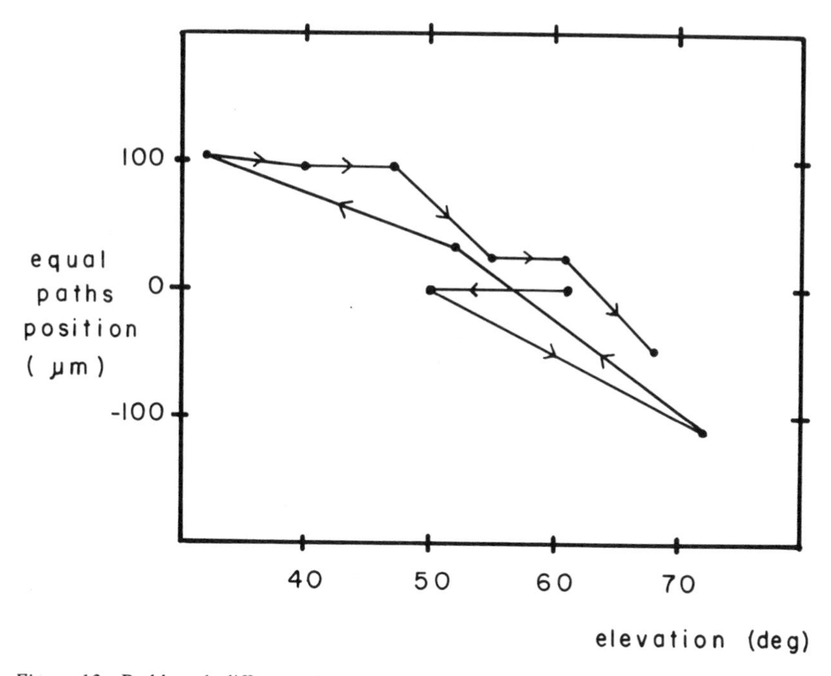

Figure 12 Pathlength difference between an opposite telescope pair as a function of telescope elevation angle.

with reduced bandwidth. K. Hege, indeed, obtained optical fringes for $\Delta\lambda$ as large as 0.01 μm. Ultimately, however, it will be necessary to devise a method for continuously measuring and controlling the pathlength for each of the telescopes.

5.8 *Digital Telescope Control and Data Collection Systems*

Two computer systems are used at the MMT for routine observational programs. A third will shortly be added for coalignment control. The first system, called the mount computer, is primarily dedicated to driving the telescope control servos. As shown in Figure 13, it is composed of a Data General Nova 800 minicomputer with several input and output devices. Mass storage is accomplished with two floppy diskettes. External signals are fed directly into the computer where they are processed and used to update the command signals to the telescope drives. The second computer system is called the instrument computer, and it is shown in Figure 14. The same type of computer is used here (Nova 800) along with terminals similar to those used with the mount computer. The instrument computer is used to remotely control instrument functions and to record, calibrate, and display data. A hard disk is available with 50 M byte storage,

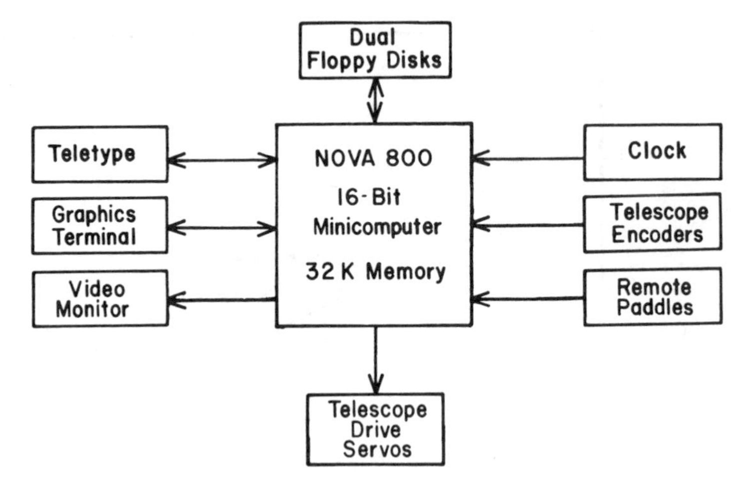

Figure 13 Block diagram of the MMT mount control computer system.

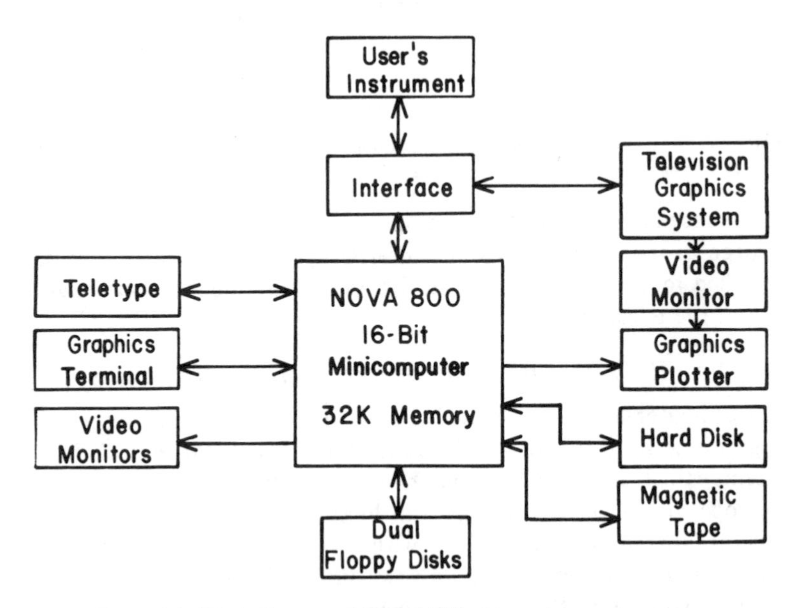

Figure 14 Block diagram of the MMT instrument computer system.

and a 9-track tape drive is used for permanent storage of data. Three-dimensional graphics (gray-scale pictures) may be produced, displayed, and reproduced as hard copy. In addition, two bidirectional, optically isolated communication channels link the two computer systems together.

The MMT is pointed by positioning the altitude-over-azimuth mount under control of the mount computer system. Every 100 milliseconds the absolute shaft-angle encoders are read to 0.077 arcsecond precision, and these readings are compared with the calculated source position. The position errors are used to generate correcting torque commands in a Type II position/velocity servo system for each axis. These servo systems are composed of four electronic loops which control torque, velocity, and position by means of negative feedback and which also provide improved acceleration performance by means of a positive source velocity feedforward loop. The torque and velocity loops are closed in hardware, but the position loop is closed in the mount computer software, which integrates the position error while tracking to produce a velocity command signal. The desired rate is converted to an analog voltage in a digital/analog converter and applied to the hardware velocity loop. When changing sources, another algorithm is automatically applied to produce the optimum motion. In this case, the commanded velocity is made proportional to the (signed) square root of the magnitude of the position error. Thus, the telescope will tend to move in a parabolic path. Software limits on maximum velocity and acceleration result in smooth ramping up of telescope velocity to its maximum slewing speed as well as ramping down to just meet the source. The algorithm used for repositioning results in optimum telescope motion to track even a rapidly moving source.

A dynamic model of the telescope servo system has been developed in the Laplace complex-frequency domain. In addition, mechanical and electrical measurements were made to determine the relevant physical characteristics of the drive system, such as moment of inertia, spring constant, viscous damping coefficient, and motor torque constant. Computer simulations were then used to derive hardware and software parameters to optimize the performance of the position control system. In particular, this model has proved to be very useful in reducing the tracking errors caused by wind torque disturbances. Frequency spectra of errors due to such disturbances have been measured by injecting an electronic signal that disturbs the servo in the same way as varying wind or friction torques, and the agreement with the model calculations is good over several decades in frequency. A detailed description of the mount control system is given by Ulich & Riley (1980).

While tracking at sidereal rates, the encoder position readout follows the commanded position to 0.07 arcseconds rms, which is the size of the least significant bit in the encoder readouts. After correction for systematic encoder errors we find that these encoder positions track the stellar positions to accuracies of better than 0.25 arcsec for periods of 10 minutes and longer. Of course, due to foreshortening, the error on the azimuth

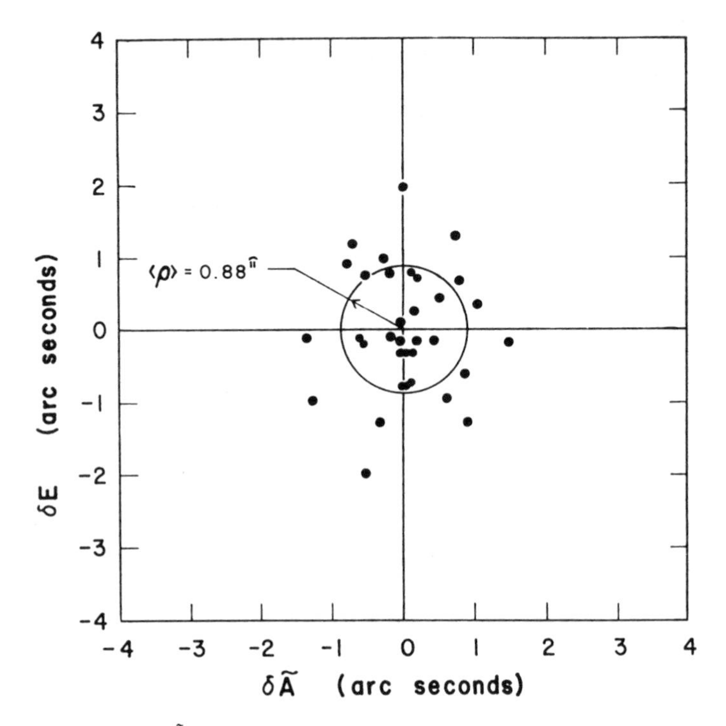

Figure 15 Azimuth ($\delta\tilde{A}$ = pointing error in sky) vs elevation (δE) pointing error for a number of stars at many sky positions.

encoder is reduced by the cosine of elevation to yield a smaller true angle on the sky. The tracking actually improves near the zenith, where a zone of avoidance with a radius of about 0.16 degrees exists because of the limited maximum velocity in azimuth. The tracking errors are larger when a significant wind exists in the telescope chamber, with an average error of about 0.5 arcsecond in a 50 km/hr gusty wind.

The true source position must be corrected for atmospheric refraction, instrumental defects, and zero-point offsets in order to determine the desired encoder readings. Optical refraction is sufficiently well understood that its correction is straightforward and can be calculated with high accuracy. The mount control software uses ambient temperature and barometric pressure inputs to calculate the refraction coefficient and to display the value currently being used. Other corrections such as gravitational flexure, tilt of the vertical (azimuth) axis from the astronomical zenith, nonorthogonality of the telescope axes, collimation errors, and offsets are automatically applied to the calculated apparent source position. The coefficients that represent these errors are determined by performing a least-squares fit to a large number of measurements of pointing errors. Figure 15 shows the residuals left after such a fit was carried out for one night's observations with a CID camera (Radau & Ulich 1980) at the focus of the guide telescope in the MMT. The mean radial pointing error, which would have been observed if the optimum coefficients had been used, was 0.88 arcseconds, and the standard deviation of the radial pointing error about zero was 1.12 arcseconds. Over a period of weeks the pointing characteristics change slowly, and relatively frequent calibrations are necessary to maintain accuracy similar to that shown in Figure 15. In addition, several periodic encoder errors have been detected by special experiments, and when these are corrected the pointing accuracy should be improved. Thus, even though with conventional techniques the MMT has more accurate "dead-reckoning" pointing than other telescopes, it may be improved significantly in the future.

5.9 *The Site, Image Quality, and Seeing*

The general selection of Mt. Hopkins as an observing site was described in Sections 3 and 4. The decision to put the MMT on the absolute summit, though having a certain inevitability, was not made without scrutiny. Seeing tests consisting of the measurement of star-trail widths from a 10-cm telescope indicated that the summit and two slightly lower knolls to the east and north had comparable (and good) seeing conditions. The summit was chosen because the prevailing fair-weather winds are from the west, and because the summit itself, though small in area, was still larger than the other knolls.

In an effort to preserve the natural mountainside cover of scrub vegetation on the southern half of the summit, we chose to tolerate a very steep (22% grade) pitch at the end of our road, so that the road could follow the ridge line off the northeast side of the peak, rather than follow an easier grade, which would have exposed a lot of road surface and cut bank to the direct heat of the sun. The parking and walk-around area in which the building sits is covered with light-colored crushed stone. This essentially places the building on a pinnacle with steep (45°–90° inclination), undisturbed slopes on all sides except the northeast. Our expectation was that these surroundings would heat fairly uniformly and have short thermal response time, giving us the best possible conditions consistent with a low placement of the telescope (about 9 m above ground level), which was forced upon us by budget limitations.

As mentioned above, a short thermal response time was in our minds at all stages of design of the telescope and building, and we have succeeded pretty well in the open optical support structure and in the lightweight building skin backed up everywhere by good insulation. The observing floor and the yoke are necessarily massive, and, therefore, are intended to

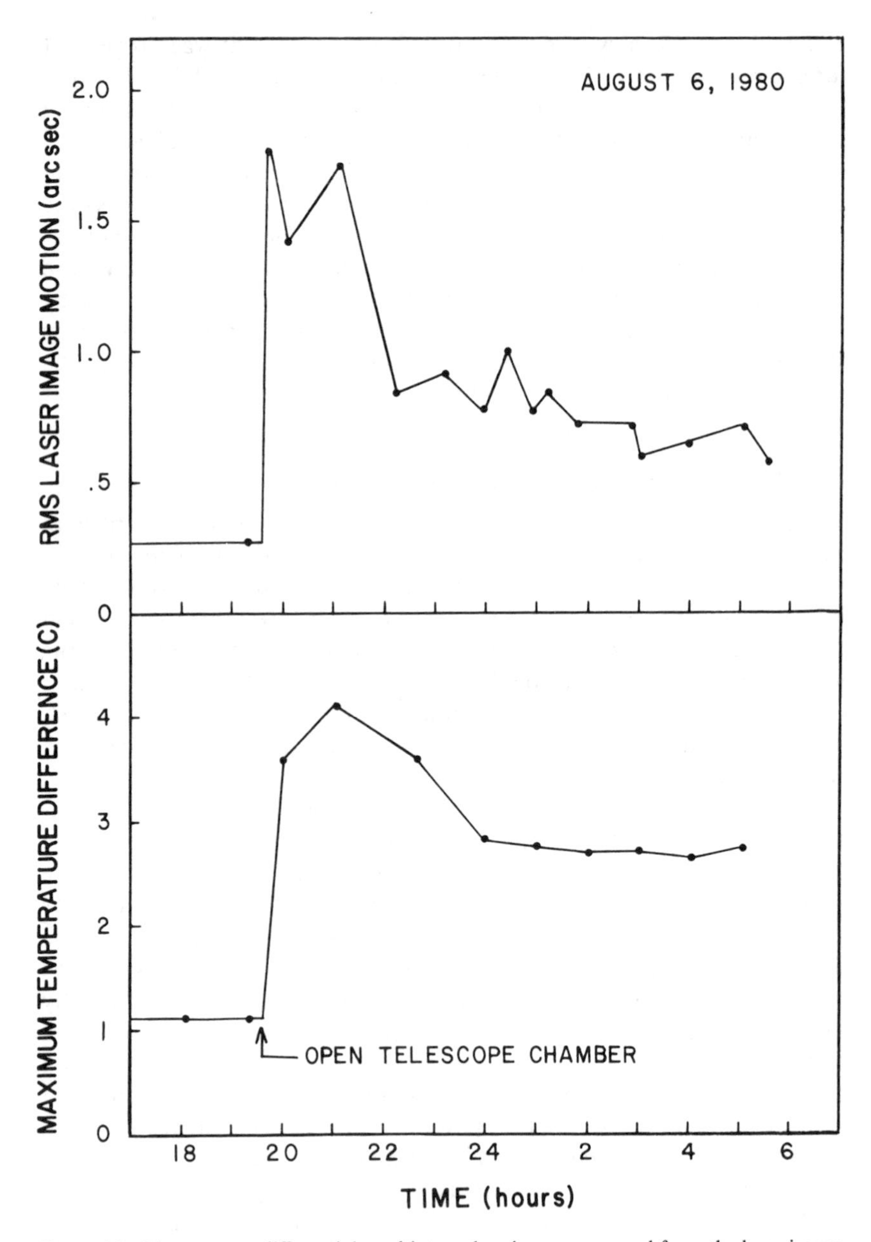

Figure 16 Temperature differentials and internal seeing as measured from the laser image as a function of time. The maximum temperature differential usually occurs between the telescope chamber floor/yoke and the air.

be actively cooled. The yoke arms and the floor are insulated with a layer of styrofoam and plywood to decrease the thermal time constant of their surfaces. As yet, our main cooling system is not active, because the refrigeration machinery has not been purchased, yet the seeing at the MMT appears to be at least equal to, if not better than, that at Kitt Peak and Mt. Lemmon. One additional thermal feature that is doubtless helping us is the extreme openness of the observing chamber, which gives a very free circulation of air even without fans, and a large solid angle open to the sky for radiative cooling.

The stellar image quality at the MMT is routinely measured by means of a CID camera placed at the quasi-Cassegrain focus. One arcsecond corresponds to six pixels on this camera. Based on the limited observing of about one year, the MMT experiences typically one arcsecond seeing (FWHM images). Occasionally, the image quality exceeds 0.5 arcsecond and seeing worse than 2 arcsecond is uncommon. Much of this seeing appears to be internal to the telescope and telescope chamber. Because of the existence of the laser coalignment system (see Section 5.5), we have a way of measuring the internal seeing in the MMT by means of the recording of the varying error signals from the laser detectors. The position fluctuations of the laser images is substantial. It typically amounts to 1–2 arcseconds peak to peak, and the rate of fluctuation falls in the 0.2–4 Hz range with the slower fluctuations dominant when the chamber is closed and with the fast fluctuation dominant when the chamber is open and when it is windy. Figure 16 shows the laser beam fluctuations for a full observing night. Also shown is the maximum temperature differential in the chamber (between the yoke and the air). Both quantities are large. We suspect a relation between the two. It is possible to estimate the importance of the internal seeing with respect to the total stellar image size by making the approximation that its effect can be evaluated purely by geometrical optics effects. The schlieren causing the image deterioration are larger than the 25 mm of the telescope aperture used to image the laser beams. Visual examinations show most of them to be smaller than the telescope aperture. If so, the displacement of the laser images can directly be related to the FWHM of a point spread function. Assuming a Gaussian distribution this relation equals $\mathrm{FWHM}_{\mathrm{laser}} = 2.35 \times$ rms laser image motion. Most of the laser image motion originates in the main telescope rather than in the guide telescope. Assuming the different sources of image deterioration to add quadratically one has

$$\mathrm{FWHM}^2_{\mathrm{star}} = \mathrm{FWHM}^2_{\mathrm{optics}} + \mathrm{FWHM}^2_{\mathrm{laser}} + \mathrm{FWHM}^2_{\mathrm{other}}.$$

Figure 17 shows a scatter diagram of measurements of $FWHM^2_{star}$ against (rms laser motion)^2. It is obvious that often much or most of the image deterioration occurs in the telescope so that further efforts to re-

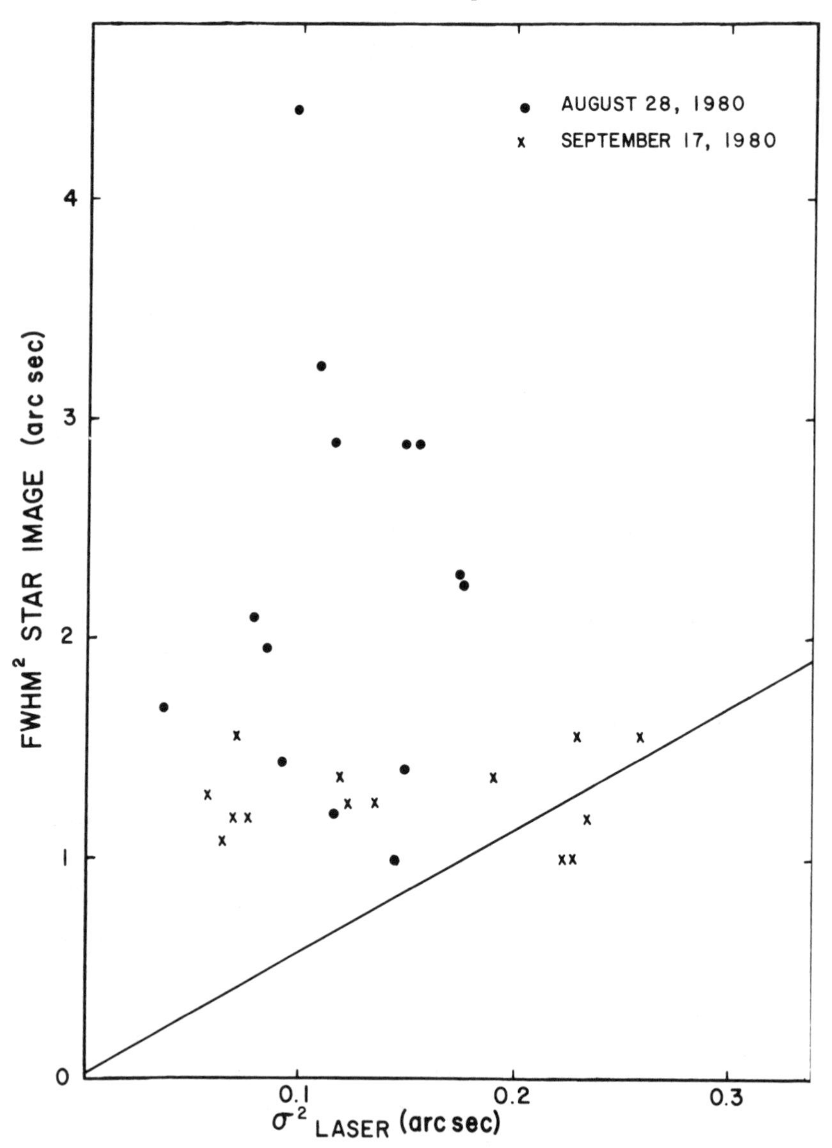

Figure 17 Full width at half maximum of stellar image diameter vs rms laser image motion (σ). The solid line represents the condition when all of the stellar image deterioration is caused by the optics and internal seeing.

duce the telescope and telescope chamber thermal effects are very much needed.

6. MMT INSTRUMENTATION

6.1 *General*

Auxiliary instrumentation for use with the MMT can be located both at the quasi-Cassegrain focus and at one of the two Nasmyth foci. The optics to feed the Nasmyth foci do, however, not yet exist so that all MMT instruments used now, or planned for the near future, use the quasi-Cassegrain focus. Because of the unconventional illumination of the images in the focal plane, an $f/9$ beam with a filling factor of $\sim$ 50% by the six $f/31.6$ beams (see Figure 3), special techniques are often needed to optimize the coupling of the instruments to the telescopes. The technique used for the MMT spectrograph has been described by Fastie (1961, 1967). Table 5 lists the seventeen instruments that are or shortly will be used with the MMT. Some of these are instruments that are constructed specially for the MMT and are, therefore, optimized for this telescope (e.g. Far-IR Photometer, MMT Spectrograph, IR Spatial Interferometer). These will, when finished, become MMT facility instruments. Others are instruments built for other telescopes but adapted to the MMT to conform to the MMT pupil configuration (e.g. Fast Photometer, MHO Echelle Spectrograph) whereas still others are coupled to the MMT without further optimization.

In the rest of this section we describe in more detail the instruments that are specifically being built as facility instruments for the MMT.

6.2 *The MMT Spectrograph*

GENERAL CONSIDERATIONS The most basic function of a modern faint object spectrograph is to record quantitative spectra of stellar or small objects at moderate dispersion, with subtraction of sky background. In the past decade, a variety of electronic detection schemes have been used for this purpose. For example, the Wampler-Robinson scanner was an early and very successful device to read electronically the output of image intensifiers. While the basic star-minus-sky mode of operation will continue to be a key function, in designing a spectrograph for the coming decade we have aimed to take full advantage of the two-dimensional formats of the new generation of electronic detectors. Observations that need area format detectors are cross-dispersed spectra of good resolution and broad cover; long-slit spectra of extended objects; and simultaneous spectra of multiple objects in the field of view of the telescope, obtained through the

Table 5 Instrumentation for use with the MMT

Instrument	Scientist(s)	Spectral range	Spectral resolution ($\lambda/\Delta\lambda$)	References
SAO CCD camera[a]	J. Geary H. Gursky R. Schild T. Weekes	400–1000 nm	~10	Leach et al. (1980)
UAO CCD camera	P. Stockman	400–1000 nm	~10	—
Fast photometer[a]	J. McGraw	300–900 nm	2	—
8-color photometer	B. Zellner	320–1050 nm	~10	—
Near-IR photometer[a]	W. Hoffmann F. Low G. Rieke	1–5 μm	~10	Low (1979)
Middle-IR photometer	W. Hoffmann F. Low G. Rieke	5–30 μm	~10	Low (1979)
IR photometer	J. Huchra	1.2–3.4 μm	10	—
Optical polarimeter	R. Angel P. Stockman	300–860 nm	10	—
MMT spectrograph[a]	R. Angel D. Latham R. Weymann	300–1000 nm	1000–6000	Angel et al. (1979)
SAO Z-Machine[a]	M. Davis D. Latham	400–800 nm	1000	Davis & Latham (1978)
MHO echelle spectrograph[a]	F. Chaffee D. Latham	400–800 nm	10^4–10^5	Chaffee (1974)
Agassiz echelle spectrograph[a]	F. Chaffee R. McCrosky	400–800 nm	10^4–10^5	—
SAO FTS (FIRS)[a]	N. Carleton W. Traub	8–14 μm	8000	—
UAO Moderate-resolution FTS[a]	H. Larson U. Fink	0.8–5.6 μm	4.10^3–2.10^4	Larson & Fink (1975)
UAO High-resolution FTS	H. Larson U. Fink	0.8–5.6 μm	2.10^5–10^6	—
Optical speckle camera[a]	K. Hege	300–800 nm	5–250	—
IR Spatial Interferometer (prototype)[a]	F. Low D. McCarthy	2–20 μm	~10	McCarthy et al. (1977), Low (1979)

[a]Instrument has been used successfully on the MMT.

use of an aperture plate with several holes corresponding to the positions of objects in the focal plane. In designing the spectrograph we, therefore, want to provide a versatile instrument incorporating both the basic linear and area format capabilities, while maintaining mechanical stability and simplicity of operation.

Designing for the MMT rather than a conventional telescope presents unique problems and advantages. The plate scale is large, 3.6 arcsec/mm, which means that a stigmatic spectrograph will give excellent resolution along the slit. The field of view is 2 arcminutes for no more than 10% vignetting by the beamcombiner (Beckers 1980). We have taken as a design goal that the spectrograph should be unvignetted for a slit length of 5 cm or 3 arcminutes. When stigmatic imaging is not needed, the large plate scale and unique geometry of the pupil make it practical and attractive to construct a simple image slicer for the *f*/9 focus. A scheme originally proposed by Fastie (1961) has been adopted and is described in the next section. In a conventional spectrograph the slit jaws or entrance apertures are reflective, and a sensitive TV system is used to identify the star field, to focus, and to center the program object on the aperture. Offset guiding may also be accomplished by field stars on the TV screen. At the MMT as it now stands, the TV will also be used to maintain the accurate coalignment of the images from the six primaries, by correction of the secondary tilts (see Section 5.5).

No single detector yet covers the full optical range efficiently. One must still reckon to use magnetic or proximity-focused image tubes in the ultraviolet and blue, where cathode quantum efficiencies of 25% are achieved. Because of the large scale of the telescope, we set as a goal that the spectra should match the largest commonly produced size of image tubes of 40 mm. We also decided that the camera should be constructed to allow the use of magnetic intensifiers, which have the best efficiency, resolution, and distortion characteristics. Readout of the image tube will ultimately be by an electronic detector with two-dimensional format. However, as a practical matter we made it our immediate goal to put the spectrograph into operation with the Smithsonian photon-counting Reticon system.

OPTICAL DESIGN Because the two best detectors we want to use, magnetic image tubes and CCDs, are complementary in their wavelength coverage and rather different in their formats, we decided early on to divide the beam below the slit with a dichroic mirror, and operate red and blue channels simultaneously. Ultraviolet and blue light is reflected to a large collimator and optics optimized for a 40-mm image tube, while the remaining light passes through to a smaller refractive collimator and smaller optics tailored for a CCD, as shown in Figure 18. In this way the complete spectrum can be observed at once.

The scale of the instrument is set by our goal of matching detector formats. We have adopted a folded Schmidt as the best camera for use

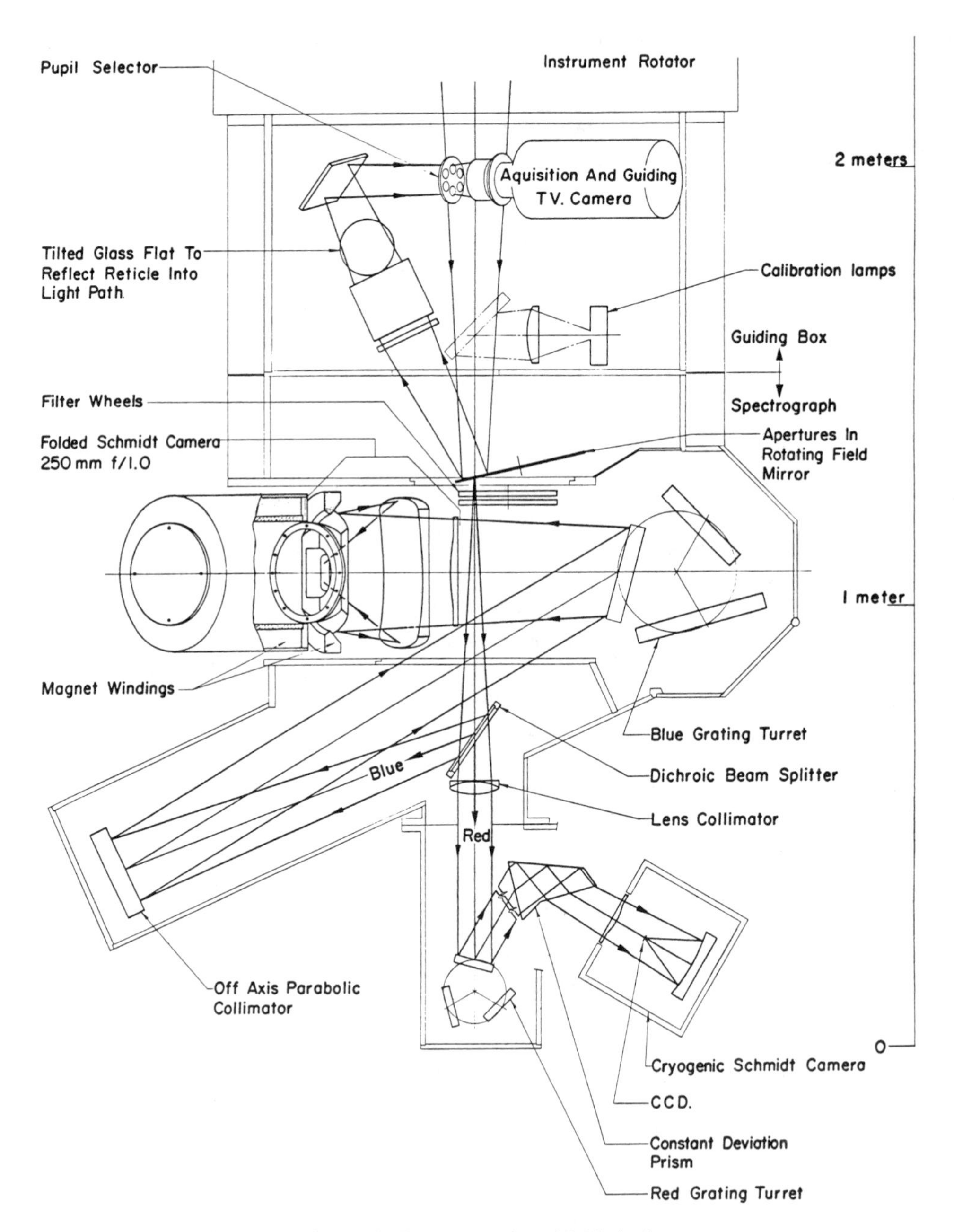

Figure 18 Schematic of the MMT spectrograph and field viewing system.

with a bulky image tube. The field of view for good images in this type of camera is no more than 10°. Thus, the focal length of the camera with a 40-mm format should be no less than 230 mm. Based on considerations of using the fastest possible camera ($\sim f/1$) to get the maximum scale reduction, we adopted for the blue camera a focal length of 250 mm, and a beam size of 150 mm, requiring a collimator focal length of 1.35 m. The CCD detector for the red channel is small, $<$ 18-mm diagonal, and the above dimensions are scaled down by a factor of two in the design for the cryogenic CCD camera. This camera has no folding flat, and the CCD is supported by a spider directly at the camera focus. For both cameras the image scale at the detector is 53 μm/arcsecond, large enough that spatial structure in sub-arcsecond seeing will be resolved by good area detectors.

Another key element that led to the geometry of Figure 18 was our incorporation of cross-dispersing prisms that can be optionally introduced into either the red or blue channels. To satisfy our goal of being able to switch quickly from direct imaging to spectroscopy, and to allow quick interchange of gratings, the design calls for gratings to be mounted on turrets in both channels. The red channel can thus be set up with silvered flats at both the grating and prism positions for direct imaging at 20 arcseconds/mm, alternatively with either the grating or prism giving dispersion, or with both together for a cross-dispersed format on the CCD.

Four gratings are currently available or on order for the blue channel. For low dispersion, a 300 gr/mm grating in first order gives dispersion of 130 Å/mm and covers the spectrum from 0.3 to 0.6 μ at a resolution of $<$ 10 Å. The highest dispersion grating that can be accommodated by the 30° included angle of the spectrograph without substantial vignetting has about 45° blaze angle. Such a grating is being ruled at Hyperfine Inc., with 250 gr/mm for use with the cross-dispersion prism to give full spectrum cover. The dispersion ranges from 9 Å/mm in the UV to 13 Å/mm in the blue, and resolution considerably less than 1 Å will be achieved, depending on the detector being used. The red channel will also be equipped with gratings for a range of dispersion. For the cross-dispersed mode we propose to use a grating blazed at 21° with 125 gr/mm which will give full cover from 0.43–1.1 μm on the CCD, with dispersion ranging from 50 Å/mm in the blue to 100 Å/mm at 1 μm. There is adequate separation between orders in both channels so that a slit of 10 arcseconds can be imaged without having the orders overlap. There is also room to accommodate the elongated images produced by separate image stackers for star and sky spectra, as now described.

A unique feature of the design is the arrangement for reformatting a normally round stellar image into a long thin image to improve spectral resolution. The method for image slicing takes advantage of the MMT's optical configuration. The secondaries are slightly displaced so as to arrange the six images stacked in a line. At the focal plane the images are formed on small segments of a field lens. These are laterally displaced from their original position in the parent lens, so as to form a common overlapping image of the telescope pupils. A negative lens then serves to give a virtual star image that is three times smaller than the focal plane stack, brings the focal ratio of the common beam to $f/9$, and restores the pupil to its original position at infinity. This method was originally described by Fastie, and is explained in detail by Angel et al. (1979). We call this arrangement an image stacker, and the spectrograph will be provided with a pair (for star and sky) that can be introduced at the focal plane when desired.

A good focal plane monitoring TV camera can resolve the entire useful guiding field of the MMT, about 4 arcminutes, with arcsecond resolution. We, therefore, designed the TV to be boresighted on the telescope axis, but with interchangeable lenses to allow for high magnification of the central field when desired. The full field (70 mm) is reflected by a plane field mirror inclined at $12\frac{1}{2}°$ to the telescope axis. A large Cooke triplet lens (125-mm diameter) is used as a collimator, and reimaging onto the TV is accomplished by one of several camera lenses located close to the exit pupil of the collimator. The shortest focal length camera gives a reduction factor of 5.4, bringing the full field within the TV camera format of 11×15 mm. In order to distinguish the overlapping images formed by different primaries, it is only necessary to insert stops at the exit pupil, where the six primaries are imaged as a cluster. Alternatively, a circle of six wedge prisms will be available to separate the six images on the TV, even though they are together at the spectrograph entrance aperture. With the aid of a projected reticle image, which will be in the form of dot-like artificial star, it will be possible to maintain accurate co-alignment and guiding without disturbing an observation with the spectrograph.

DETECTOR The first observations with the MMT spectrograph utilized a photon-counting Reticon detector system. The basic approach of this detector, to centroid individual photon events with a Reticon diode array at the output of a high-gain image intensifier package, was pioneered by Shectman (1976). The MMT detector includes several major improvements over earlier SAO versions called the Z-machine (Davis & Latham 1978).

The heart of the MMT detector is a custom dual 1024 Reticon and associated electronics for high-speed readout and signal processing. The

Reticon head is modular so that it can be interchanged between image intensifier packages. The first MMT applications used a package of three Varo military tubes, all with fiber-optic windows. The first is a special Varo 8605 40-mm electrostatic diode inverter with custom thin S-20 cathode for enhanced blue response. It is fiber-optically coupled to a standard Varo 8605 selected for high green sensitivity, which in turn is coupled by a 40- to 25-mm fiber-optic reducer to a Varo 3603. This 25-mm MCP inverter with special fast phosphor is fiber-optically coupled to the Reticon with a special split faceplate made by Galileo Inc. The split faceplate effectively eliminates the 1.0-mm dead space between the two diode arrays and thus minimizes the effects of distortion by mapping the sensitive areas of the Reticon arrays nearly onto a single diameter of the intensifier output. It also makes it much easier to align the detector along the spectrum, an important practical feature on the echelle spectrograph. On this package not only was the first tube carefully selected for maximum sensitivity, but also we tried to maximize the number of photoelectrons per event at the second cathode. This is what sets the shape of the pulse-height distribution, which in turn influences the counting efficiency attainable. With too little effective gain from the first stage it is possible to pick up extra dark events from the second cathode and to get persistence from areas that have just been brightly illuminated, such as comparison lines, despite the frame subtraction in the signal processing. Both these effects were a problem with the earlier Z-machine detectors but are essentially eliminated in the MMT detector. The spectral sensitivity of this system is basically a classical S-20 cathode response with an ultraviolet cutoff just short of the H and K lines because of the fiber-optic input window. The peak absolute counting efficiency was measured to be nearly 15% at 5000 Å.

One of the most encouraging aspects of the MMT detector is the improvement in the fixed patterns. Raw data from the Z-machine detector have approximately 50% modulation with most of the power in every 2-, 4-, and 8-pixel pattern. The new system has little power in these frequencies and an overall rms modulation of about 3%. The residual patterns appear to be mostly cathode nonuniformities. Under ideal conditions flat-field corrections can attain a signal to noise of at least 200. The operational limit is not quite so good, apparently because of the way slightly different areas of the cathode get illuminated due to mechanical, magnetic, and electrostatic flexure, and also because of differences in the way stellar and lamp exposures illuminate the slit.

PERFORMANCE In May and June of 1980 a large number of exposures were taken with the MMT spectrograph and its detector. A resolution of

35 μm FWHM at the center of the diode arrays was routinely achieved, with some degradation at the edges of the intensifiers. Because of the fiber-optic demagnifier, this corresponds to about 60 μm at the first cathode. Observations of spectrophotometric standards through a 1.5 arcsec slit generally gave overall system efficiencies of about 1%, including the atmosphere, slit losses, spectrograph, and detector. The observed dark rates were generally less than 10 counts/cm^2 s^{-1}. This is several times higher than was attained with the same system during wintertime tests at Agassiz station where the cathode temperature was probably some 10°C less. In Section 7.1 we describe some of the scientific results obtained with the MMT spectrograph.

6.3 *Infrared Photometer*

The first MMT infrared photometer was designed and built by George Rieke, Bill Hoffmann, and Frank Low. The instrument is shown schematically in Figure 19. The InSb detector is cooled to 4.2 K (Rieke 1980) and will be used primarily at 2.2 and 3.4 μm. Note the long $f/32$ cooled collimators, the cooled beamcombiner, and the dichroic beamsplitter (all at 77 K). When used with the servo-operated, secondary mirror modulation system, it is expected that background-limited performance will be achieved with very small ($<$ 3 arcsec) apertures. In addition to selectable cooled apertures, the system includes a large cooled filter wheel and provision for a cooled circular variable filter. Thus both broad-band and narrow-band absolute photometry can be performed efficiently.

Don McCarthy and Frank Low are developing a 10- and 20-μm photometer which incorporates features that allow the six optical pathlengths to be made equal, thus allowing operation as a phased array. When "seeing" is exceptionally good, this system should operate fully phased with continuous integration and extremely high signal to noise. When less than ideal conditions prevail, the data may be reduced in the less efficient "speckle" mode. In either case, the angular resolving power is expected to be diffraction-limited.

6.4 *The SAO CCD Camera*

The concept of a true two-dimensional detector having close to unity quantum efficiency over a broad region of the visible and near-IR spectrum and with both low noise and sharp images is, of course, attractive, both for direct imaging applications and for spectroscopy. Although no commercially available imager can claim to satisfy all these desired characteristics at present, recent developments in large solid-state arrays such as charge-coupled devices (CCDs) promise an approach to the ideal in the

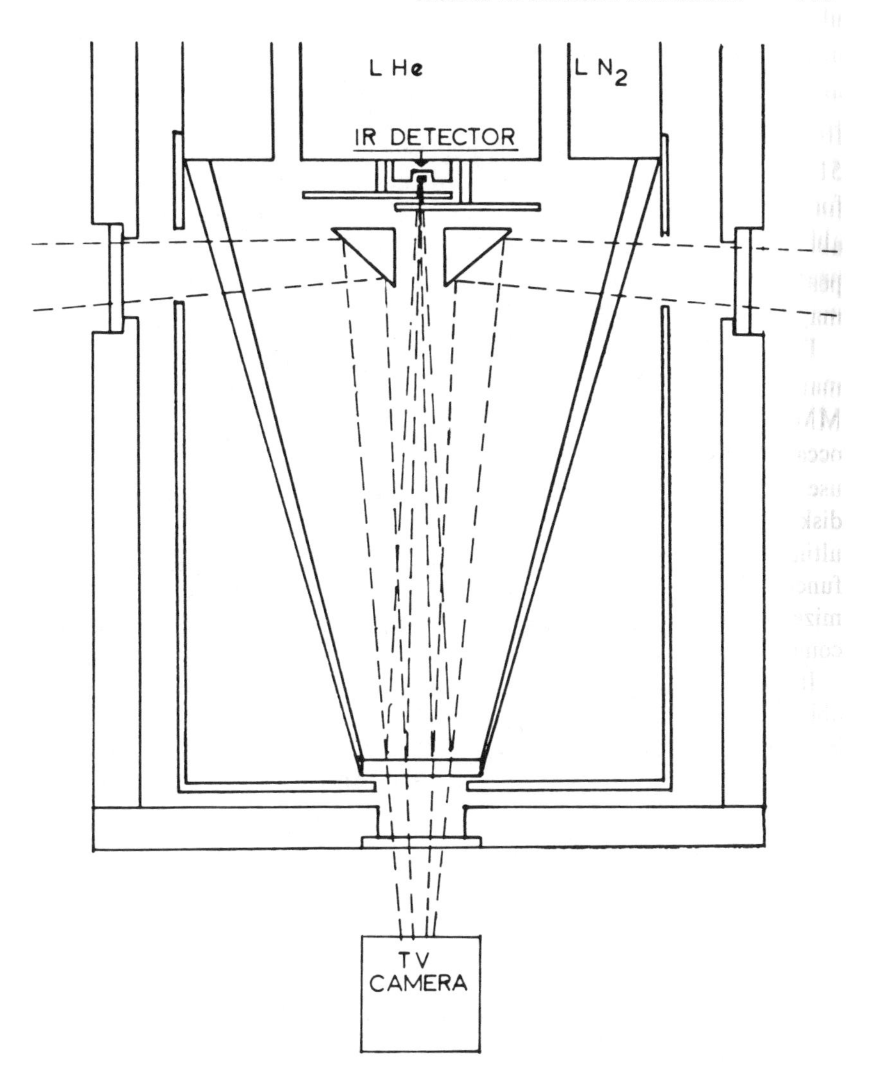

Figure 19 The MMT infrared photometer.

near future. Development of a CCD camera at SAO during the past year has been motivated by the desire to use devices that are presently available for actual astronomical observations, both to see what the characteristics of this new technology are in this setting and, we hope, also to do some scientifically interesting projects in the process.

The system that has resulted consists of an upward-looking dewar container to maintain the CCD at a lower temperature (typically below

150 K) for dark current suppression and a self-contained electronic module to scan the imager on command, process the resulting video signal, and transmit the digitized data to the instrument computer. At present we are using the large CCD manufactured by RCA and available in both frontside- and backside-illuminated versions. This device is organized into 512 rows of 320 elements, each pixel being 30-μm square. The resulting format measures 9.6 mm $\times$ 15.4 mm, one of the largest presently available in a solid-state imager. Quantum efficiencies have been measured to peak as high as 75% and, in the case of the backside-illuminated version, may be quite flat from blue to near-IR wavelengths.

Despite the large pixel size of this particular CCD, they are still poorly matched for direct photography to the 278 μm per arcsec scale of the MMT without some sort of optical demagnification, except perhaps on occasions of truly exceptional seeing. We are presently investigating the use of simple lens systems to provide a better match of pixel size to seeing disk to prevent signal degradation due to oversampling. It may, indeed, ultimately prove desirable to design special-purpose optics to perform this function, both to produce better images over wider passbands and to minimize scattered light, as well as perhaps to modify the present beam-combiner to produce a wider field.

It is not yet clear whether the RCA CCD or any other presently available device will prove useful as a detector for spectrographic applications. Such a detector must have exceptionally low readout noise in addition to high charge-transfer efficiency at low signal levels to prevent loss of spatial resolution. Part of the development effort using the present camera system will focus on defining the nature of the problems in these respects, in hopes that the information can be used by the developers of future advanced solid-state imagers to improve performance in these critical areas.

7. ASTRONOMICAL RESEARCH WITH THE MMT

It is a difficult task to describe possible future astronomical research with an instrument before the instrument in question is fully operational, and even before the full array of instrumentation has been defined, much less constructed. Nevertheless, some of the rather unique properties of the MMT and the characteristics of the site suggest the direction that astronomical research with the MMT is likely to take. We describe, in turn, several of the instruments now operating or soon to be operating on the MMT and the astronomical research program to be carried out with them, as well as some results already achieved in the brief period the MMT has been operating.

7.1 *Spectroscopy*

As described in Section 6.2, spectrographs on the MMT can operate in either a conventional mode, with all six images superposed, or aligned along the slit in the "in-line" mode. To date, only the conventional mode has been used except for some very preliminary tests with the Mount Hopkins Observatory echelle spectrograph, but some interesting results have already been obtained in this mode.

Following the discovery of the twin quasars (0957 + 561 A, B) by Walsh et al. (1979), Weymann et al. (1979) used the Z-machine spectrograph to set a limit of $\sim$ 15 km s^{-1} between any differences in the absorption line redshifts in A and B, which substantially strengthened the case for the gravitational-lens hypothesis suggested by Walsh et al. In May 1980, during the first trial of the MMT spectrograph in the conventional mode, Weymann et al. (1980) discovered the triple quasar 1115 + 08, the second example of a gravitational lens (Figure 20). During this first test, Liebert, Latham & Steiner (1980) also observed the magnetic white dwarf in AM Herculis during a rare faint "off" phase and were able to measure a field strength of 20 MG from the Zeeman patterns in hydrogen lines (Figure 21). The excellent seeing at the site, together with the small amount of vignetting over a large spectral range and very good image characteristics of the spectrograph make it well suited for moderate-resolution spectroscopy of faint objects where breadth of coverage is important.

A unique feature of the MMT is its use in conjunction with spectrographs operating in the in-line mode. This feature, combined with the frequent occurrence of 1-arcsec and better seeing and the utilization of echelle gratings, allows unprecedented slit transmission even for quite high spectral resolution: In the 15-cm echelle spectrograph for the MMT under preliminary design, an R-2 echelle grating feeds a 75-cm camera which matches a 1-arcsec entrance slit to a 25-μm-resolution detector yielding $\sim$ 5 km s^{-1} FWHM and frequently allowing most of the light to pass through the slit. Typically, a conventional coudé spectrograph under average seeing yielded comparable resolution with a slit transmitting only $\sim$ 5% of the light (Dunham 1956). Again, considering as an example QSO absorption lines, objects at $m_v \sim 16.5$ can be studied at a resolution that will, judging from the 21-cm data to date, fully resolve the line profiles in these objects and allow definitive abundance analyses.

It is startling to realize that we can now contemplate spectra at 5 km s^{-1} resolution of objects at $m_v \approx 16.5$; only two decades ago, before the introduction of image tubes, such resolution was obtainable only for objects of about 9 magnitudes brighter with existing coudé spectrographs (Dunham 1956).

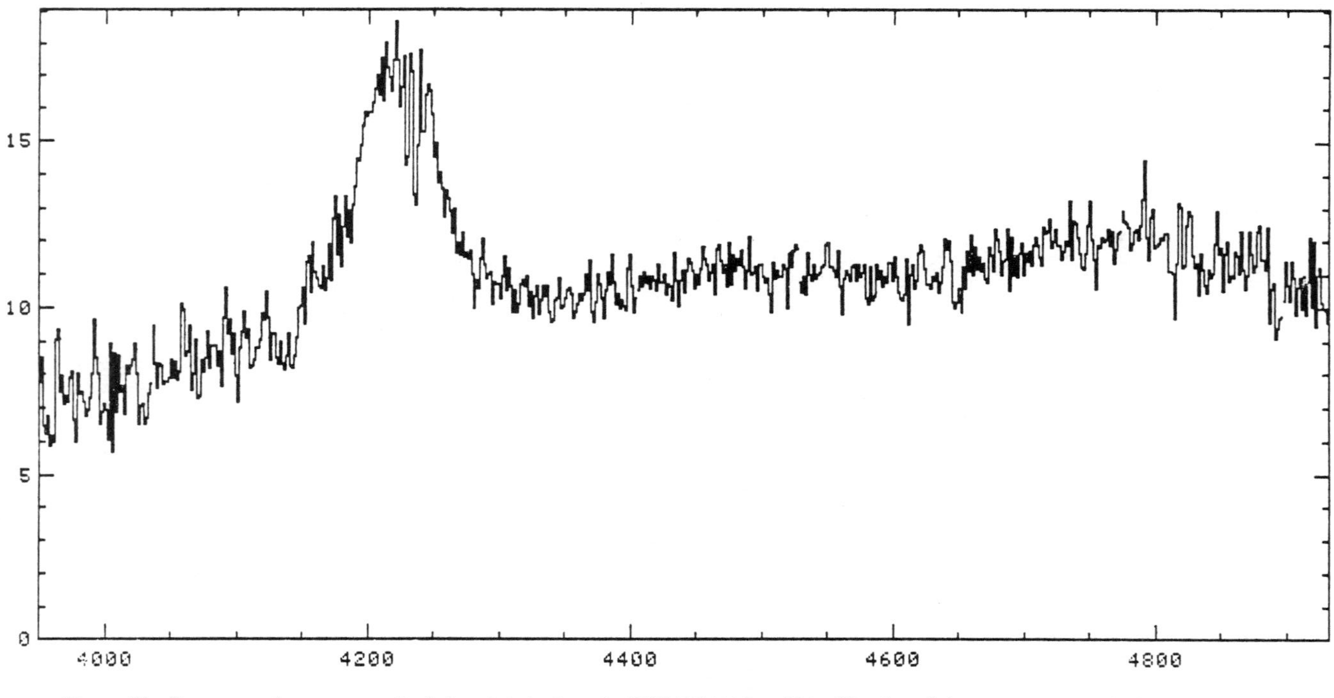

Figure 20 Spectrum of component C of the triple/quintuple QSO PG1115 + 08A. Wavelength in angstroms, vertical scale in counts.

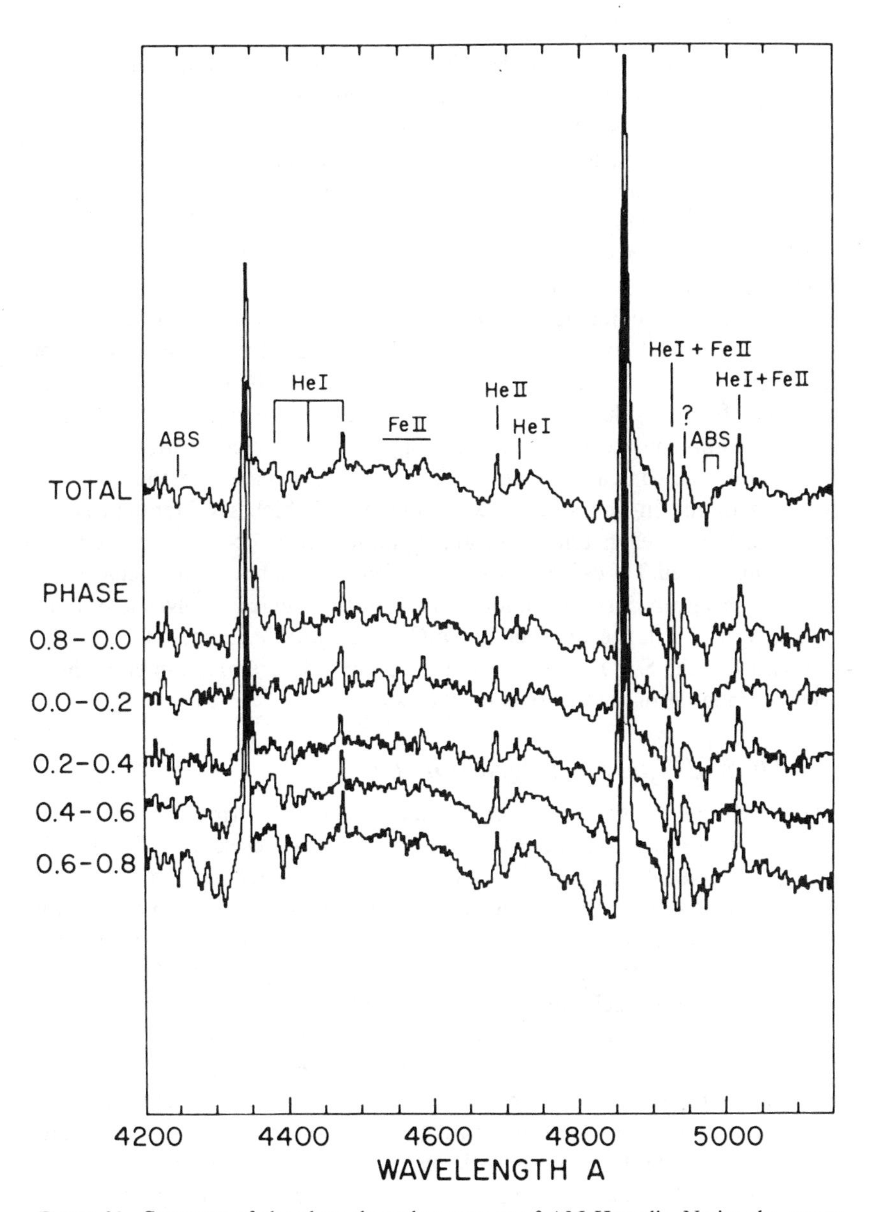

Figure 21 Spectrum of the phase-dependent spectra of AM Herculis. Notice the strong broad Zeeman components of the H_β line in absorption at 465 nm and 480 nm resulting from 20 megagauss magnetic field. Courtesy J. W. Liebert, D. W. Latham, and K. G. Steiner.

7.2 *Direct Imaging*

Astronomical research at the MMT involving direct imaging was *not* one of the prime research applications envisioned for the MMT. It is, therefore, of some interest that one of the most noteworthy recent results involved direct imaging with an intensified vidicon of the triple quasar, which showed distinct elongation (incipient splitting?) of the brighter of the three images (Figure 22). This image was obtained using only one of the six telescopes to obtain the best possible resolution and one-second exposures were centroided and shifted and then superposed to increase signal to noise, achieving FWHM of $\sim$ 0.6 arcseconds. For maximum signal to noise, all six telescopes should be utilized, but of course it is not necessary for the six images to be coincident. To distinguish real image structure from residual telescope aberrations and seeing effects, it is, in fact, advantageous to separate the images on the detector and to coadd them later electronically. A device has been built, based on the concepts involved in the in-line spectroscopic mode, in which six identical small fields, one from each telescope, are arranged in a 3×2 format on the single intensified TV camera without adding of the sky background radiation. A program currently under way with this device involves a systematic search of a large number of QSOs for multiple images. Other programs using the SAO CCD camera (Section 6.4) are also aimed at high quality imaging of small areas on the sky.

7.3 *Optical Speckle Interferometry*

The technique of speckle interferometry allows information on image characteristics down to the diffraction limit of the telescope. (For a general review, see Labeyrie 1978.) It is essential to record the data over times such that the speckle patterns are frozen in position (i.e. a small fraction of a second). In practice, this is achieved by a high-gain-intensified TV camera readout onto video tape (Hege et al. 1980).

The use of this technique with the MMT is of particular interest for two reasons. First, it appears that the seeing characteristics of the site often result in a substantial fraction of the energy residing in a small number of speckles, so that the limiting magnitude of the technique may be somewhat fainter (perhaps down to $m_v \sim 18$) than initially thought. Second, experiments to date with opposite mirror pairs show that the optical support structure is sufficiently stable that speckle information from the six mirrors acting coherently is obtainable. The diffraction pattern is, thus, very similar to that from a single 6.9-m telescope and information down to scales of $\sim$ .02 arcsec available. This opens up a large number of astronomical research possibilities in solar system astronomy

Figure 22 Image of the triple/quintuple QSO PG1115 + 08A. The angular resolution of the image is 0.6 arcseconds. The components are separated by approximately 2 arcseconds. Courtesy K. Hege.

(e.g. properties of Pluto's satellite and the diameters of several asteroids), stellar astronomy (studies of preplanetary discs, inner regions of circumstellar shells), and extragalactic astronomy (structure of Seyfert nuclei and searches for sub-arcsec image splitting in the brightest QSOs).

7.4 *Research in the Infrared*

From its inception, the MMT was designed with infrared research in mind. To this end, the secondary support structures offer a minimum amount of emissive surface area and a set of undersize secondaries for use during infrared work is provided. A variety of IR research programs are well suited to the instrument.

Considerable experience has already been obtained using opposite mirror pairs as a Michelson interferometer (see Section 5.7). The fringe amplitude is measured by rapidly scanning the pattern over a grid placed over the detector. In this way, the visibility function for a given separation has been determined for a number of late-type giants and supergiants in the 2–10-μm region. The Michelson technique has the disadvantage that only one Fourier component of the image is measured at any one time. In principle, direct imaging and speckle interferometry can be carried out in the IR just as it can in the optical. In practice, until a two-dimensional imaging IR device is available, the technique is somewhat different: the image is scanned very rapidly across a single slit placed over the detector and a (one-dimensional) power-spectrum analysis is performed; the intrinsic telescope plus atmosphere modulation transfer function is obtained from nearly simultaneous observations of a nearby point source. The applications of this technique are, as above, to resolve late-type supergiants and possible preplanetary discs. Details on the IR structure of the nucleus of NGC 1068 should be possible. An interesting proposal has been made to operate the MMT as a phased array at 20 μm, not simply with the objective of obtaining increased resolution but to obtain a 6.9-m diffraction-limited image so that very small apertures can be used to reduce the background to an absolute minimum for 20-μm photometry. A photometer with the six IR secondaries chopping in phase will shortly be placed in operation and will be equipped with an indium antimonide detector for work in the 1–3-μm region and a bolometer for work at 5–20 μm. At 2 μm most QSOs are readily observable, and the brighter ones can be studied with low spectral resolution. Giant ellipticals with $z > 1$ ought to be measurable. A circular variable filter spectrometer will be able to observe the broad emission lines of active nuclei and the brightest QSOs. Fourier transform spectrometers operating in the 50–25-km s^{-1} range will be used to examine spectral features in very red objects of special interest (e.g. MWC 349) and to study processes of star formation (e.g. 2-μm spectra of the stellar objects embedded in the ρ Ophiuchus complex).

8. FUTURE MULTIPLE OBJECTIVE TELESCOPES

The supergiant telescopes of the future, either in space or on the ground, will almost certainly be multiple objective telescopes. In space, the need to send up the telescope in moderate-size pieces will preclude a monolithic primary. Even then, the cost of such a telescope, at present, is prohibitive.

The supergiant ground-based telescope is within reach and is extremely worthwhile. The cost per second of dark time observing with a 15-m ground telescope and a 2.5-m space telescope is similar, though the overall cost, annual cost, and amount of dark observing time are an order of magnitude less on the ground. The ground telescope will, for example, be advantageous for spectroscopic observations of any object brighter than the sky, studies of extended objects of low surface brightness, in multiple object spectroscopy (e.g. galaxies in a cluster), in spectroscopic observations between 1 and 3 μm where detectors are limited by internal noise, in high angular resolution infrared and optical (speckle) observations, and in the rapid development of new equipment. It would be useful for submillimeter astronomy at 0.3 mm. For sub-arcsecond seeing conditions it would compete successfully with the Space Telescope in optical sky background-limited research. Two studies, by Faber (1980) and Strittmatter (1980), have explored many of these options in detail. Some graphs that demonstrate the multiple object advantage are in Woolf & Angel (1980).

A number of proposals for building giant multiple objective telescopes are discussed in Hewitt (1980). There currently seems to be some hope for construction of multiple objective telescopes between 10 and 25 meters for the USSR, European Southern Observatory, USA, and the state of California. Kitt Peak Observatory and the Universities of Arizona, California, and Texas are attempting a joint program of technology development to lead to a $\approx$ 15-m telescope. It has been accepted that it is inappropriate to attempt to build a monolith of this size, and arrays and bowls are not suitable for the infrared. Thus, the technology development program has among its main tasks to explore the reasons for choosing between a multiple objective telescope of Type A or B (Type A is segmented primary, Type B is multiple objective). The Kitt Peak NGT (next generation telescope) program explored both of these options at the 25-meter size as the steerable dish and the large MMT. Further developments of the Type A system have been conducted at the University of California by J. Nelson (1980) and colleagues, while a study of a large Type B system has been carried out at the University of Arizona.

The University of California study has been concerned with developing a primary mirror that retains its shape regardless of applied forces. The relative positions of 1.4-meter hexagons are sensed, and position actuators correct any errors. The group has developed sensors and actuators as well as a method of bending and polishing precise off-axis paraboloids, which has been successful for a 0.35-meter circular prototype. The University of Arizona study, in contrast, has explored the possibilities of adaptive optics, in so far as current knowledge of seeing parameters permits. Also, they have explored the use of light pipes, and the use of the separate subtelescopes for highest visual efficiency and smallest visual images by instantaneous wavefront tilt corrections. Both studies have explored some techniques that are applicable to both Type A and Type B systems. The experience with the MMT has pointed to two major changes that should be incorporated in future instruments. First, the experience with phasing the telescope for infrared interferometry has encouraged us to expect that we can make an MMT where all the elements are held in phase, that reaches the diffraction limit of its entire aperture rather than of that aperture's component parts. Second, we have seen the separate star images being moved around by differential seeing so that they looked like a swarm of bees. We believe that we can often use fast coalignment on stellar images to make these coalesce and so obtain higher angular resolution than we would on a rigid telescope of the same area.

Type A and Type B systems have different faults and merits so that at this time the choice between them is not obvious. A Type A system starts with a lightweight primary because the segments are thin, and this saving in weight propagates through the entire telescope. To reduce cost still further, the telescope and dome can be shrunk by reducing the focal ratio. This makes alignment tolerances very tight, though this is not a problem if alignment is carried out on a star image. More important, the provision of a large, high quality, achromatic field with low light losses becomes difficult. Currently, one can attain acceptable performance at a primary ratio of $f/2.0$, and ingenious optical design might push as far as perhaps $f/1.6$. A Type B system has a much shorter tube for the same mirror area, so the minimum dome can be smaller around a Type B system. If dome cost is proportional to $(\text{size})^{2.5}$, then the minimal dome around a Type A system of $f/1.7$ will cost twice as much as one around an 8-mirror Type B system of $f/2.7$. In existing observatories, domes are typically one third the project cost, so that such factors are important. The smaller size could in principle be used in a Type B system for telescope weight and cost reduction, but it is not yet clear whether the tubes of Type A and Type B systems have the same demands for rigidity and high frequency response.

The provision of the high quality wide field is more easily met in an $f/2.7$ primary in Type B. For visual imaging, one should probably use multiple CCDs and electronic image combining so that use of the individual foci is no disadvantage. For multiple object spectroscopy one can as easily lead light pipes from separate foci as from a single one. Thus, in wide field visual use Type B is at an advantage. However, when one comes to developing a phased field for coherent imaging, Type B suffers from a small amount of vignetting at quite modest field angles and for some purposes this may cause problems. Thus, for phased work with wide field, Type A is advantageous. Phasing a Type A system is initially very time consuming if there are many facets. However, a Type B system has the telescope structure enter into the phasing, so that although the phasing operation is fast, it must be carried out more frequently. It is possible that

Table 6 Diffraction structure for a Type A and various Type B configurations[a]

Mirror system	Fraction of energy inside first zero[b]	Central resolution	Peak IR signal/noise
Type A (segmented mirrors)			
Single circular disk	.836	1.0	1.0
Type B (multiple mirrors)			
6 at points of regular hexagon (MMT)	.620	1.25	0.86
8 around a square (MT2)	.580	1.28	0.85
7 at points of regular heptagon	.554	1.28	0.83

[a]Mirror spacing equals 0.2 times mirror diameter.
[b]For Type B systems, integration was stopped at 3rd zero of individual mirror patterns.

laser interferometer distance sensing might eliminate these problems with both types of systems.

The biggest questions concern the primary optics and the extent to which the telescope figure must be operable by "dead reckoning" rather than by setting up on a star or continuously using an offset reference star. Dead reckoning probably needs very stable mirror materials like ULE or Cervit. The blanks are relatively expensive, and there may be both large costs and manufacturing delays for a Type B system. On the other hand, for a system with partially adaptive optics, thin pyrex mirrors or pyrex honeycomb mirrors may be perfectly acceptable, and their use might notably reduce the cost of the Type B system. Honeycomb is not appropriate for bend and polish, and unless one keeps solid pieces very thin, they might not work out well for segments in a Type A system. Further, the focal tolerances are so tight in a Type A system that Cervit may be needed for this reason alone. So far, pyrex honeycomb has only been built in small sizes, so it is not yet an established technique.

The diffraction patterns of phased Type A and B systems are quite different, with a Type A more closely resembling that of a filled circular disk. Assuming perfect optics, all telescope systems of the same mirror area should produce the same central intensity in their diffraction patterns. Thus, as the width of a dilutable mirror system increases, and the angular resolution goes up, less of the energy finds its way into the central maximum, and one expects to find that (energy inside 1st zero) $\times$ (resolving power)^2 is constant from system to system. In Table 6, we show the result of calculating these quantities, and also the comparative signal-to-noise ratio for an infrared telescope with an optimum small diaphragm. It can be seen that a Type B system will differ from a Type A system in that the Type A will have a somewhat higher signal-to-noise ratio for IR observations and a somewhat lower resolution. It is not easy to put these relatively minor differences into perspective. At wavelengths shorter than $\sim$ 3 μm, the light concentration will be set by seeing and there will be no appreciable difference between the systems. The highest sensitivity IR observations will be made from space, and ground observations will mainly refine these with measurements of image position, structure, and time variability. Here, the high resolution should be advantageous. On the other hand, for IR spectroscopy the energy concentration is likely to be more important.

If it should turn out that costs distinctly favor one kind of system, or, rather unlikely, that a major problem should be found in one kind of system, then the choice may be easy. At the time of this writing, it does not seem likely that the choice will be so painless. Thus, while it is clear that the future lies with either the Type A Segmented Primary Telescope, or Type B Multiple Mirror Telescope, the crystal ball is cloudy when it is asked which should be preferred.

9. CONCLUSION

With the construction of the Multiple Mirror Telescope a major new and novel facility has been added to the capabilities of the world of optical nighttime astronomy. The MMT is being used not only to gather high quality astronomical data but also to test new techniques for astronomical telescope construction. It is quite remarkable that a telescope like the MMT with the many deviations from conventional optical telescope technologies has, in fact, become a very high quality, very usable facility in a very short time after it was nominally completed. The credit for that goes to the many competent and dedicated scientists, engineers, and others who participated, and who participate now, in the design and implementation of the telescope. The MMT Dedication Symposium (*SAO Special Report 385*) listed over 340 people who are associated with this venture. We do not attempt to give such a full list. We do, however, want to specially acknowledge the contribution to the MMT by some who have contributed in a major way, who have dedicated a major part of their professional lives toward the realization of this unique facility, and who are not co-authors of this chapter: T. E. Hoffmann, A. Meinel, M. Reed, G. Sanger, F. Whipple, and J. T. Williams. The help of such capable administrators as J. Gregory and P. Sozanski from SAO and R. Kassander and A.

Weaver from the University of Arizona was essential to accomplish the MMT. The love and dedication towards the project of these and of many other participants are the reason for the success of the Multiple Mirror Telescope. We gratefully acknowledge the help of M. Green, V. Tersey, and G. McLoughlin in the preparation of this manuscript.

Part of the material presented in this publication is based upon work supported by the National Science Foundation under Grants AST 76-20822, AST 76-21732, and AST 79-25421.

References

Angel, J. R. P., Hilliard, R. L., Weymann, R. J. 1979. *The MMT and the Future of Ground Based Astronomy, SAO Special Report,* ed. T. C. Weekes. 385:87–118

Barr, L. 1980. *Optical and Infrared Telescopes for the 1990s,* ed. A. Hewitt, pp. 23–53

Basov, N. G., Dimov, N. A., Gvozdev, M. I., Kokurin, Yu. L., Lomakin, V. N., Steshenko, N. V., Tarasov, G. P., Ustinov, N. D., Vasiljev, A. S., Zarubin, P. V. 1979. *The MMT and the Future of Ground Based Astronomy, SAO Special Report,* ed. T. C. Weekes. 385:185–89

Beckers, J. M. 1980. *MMTO Technical Report No. 1*

Burke, J. J., Kirchoff, W. 1968. *Sky and Telescope* Nov:284–86

Chaffee, F. H. 1974. *Astrophys. J.* 189: 427–40

Carleton, N. P., Hoffmann, W. F. 1978. *Physics Today* Sept 1978:30–37

Davis, M., Latham, D. W. 1978. Instrumentation in Astronomy III. *Proc. Soc. Photo-Optical Instrumentation Engineers* 172:71–76

Dunham, T. Jr. 1956. *Vistas in Astronomy* 2:1223

Faber, S. 1980. *A Scientific Case for a 10 Meter Telescope.* Preprint

Faller, J. E. 1978. *Optical Telescopes of the Future, Proc. ESO Conf.,* ed. F. Pacini, W. Richter, R. N. Wilson, pp. 301–11

Fastie, W. G. 1961. *J. Opt. Soc. Am.* 51: 1472

Fastie, W. G. 1967. *Applied Optics* 6:397–402

Fotescu, G., Heilbrunn, G. 1909. *UK Patent No. 21456*

Grainger, J. F. 1979. *The MMT and the Future of Ground Based Astronomy, SAO Special Report,* ed. T. C. Weekes. 385:199–207

Green, J. R. 1979. *Literature Search on Multiple Objective Telescopes.* Steward Observatory in-house publication

Grundmann, W. A., Richardson, E. H. 1980. *Optical and Infrared Telescopes for the 1990s,* ed. A. Hewitt, pp. 11–22

Hall, D. N. B. 1978. *Optical Telescopes of the Future, Proc. ESO Conf.,* ed. F. Pacini, W. Richter, R. N. Wilson, pp. 239–49

Hanbury Brown, R. 1964. *Sky and Telescope,* Aug: 64–69. See also *Space Physics and Radio Astronomy,* ed. H. Messel, S. T. Butler, Chap. 5

Hege, K., Hubbard, E. N., Strittmatter, P. A. 1980. *Applications of Digital Image Processing in Astronomy, Proc. Soc. Photo-Optical Instrumentation Engineers* 264:29–37

Hewitt, A. 1980. *Optical and Infrared Telescopes for the 1990s, Proc. KPNO Conf.*

Hoffman, T. E. 1978. *Optical Telescopes for the Future, Proc. ESO Conf.,* ed. F. Pacini, W. Richter, R. N. Wilson, pp. 185–208

Hoffman, T. E. 1980. *The MMT and the Future of Ground Based Astronomy, SAO Special Report,* ed. T. C. Weekes. 385:23–36

Horn-d'Arturo, G. 1935. *Publ. Oss. Astron. Univ. Bologna* 3(3):19–35

Horn-d'Arturo, G. 1950. *Publ. Oss. Astron. Univ. Bologna* 5(11):3–15

Horn-d'Arturo, G. 1953. *J. Brit. Astron. Assoc.* Jan: 71–74

Horn-d'Arturo, G. 1955. *Publ. Oss. Astron. Univ. Bologna* 6(6):1–18

Horn-d'Arturo, G. 1956a. *Publ. Oss. Astron. Univ. Bologna* 6(18):1–6

Horn-d'Arturo, G. 1956b. *Optik 13* Vol 6: 254–58

Ingalls, A. G. 1951. *Scientific American* Jan: 60–63

Jacchia, L. 1978. *Sky and Telescope* 55: 100–102

Labeyrie, A. 1978. *Ann. Rev. Astron. Astrophys.* 16:77–102

Larson, H., Fink, U. 1975. *Applied Optics* 14:2085–95

Leach, R. et al. 1980. Submitted to *Publ. Astron. Soc. Pac.*

Learner, R. C. M. 1978. *Optical Telescopes of the Future, Proc. ESO Conf.,* ed. F.

Pacini, W. Richter, R. N. Wilson, pp. 275–88
Liebert, J., Latham, D. W., Steiner, K. G. 1980. Preprint
Low, F. J. 1979. *The MMT and the Future of Ground Based Astronomy, SAO Special Report,* ed. T. C. Weekes. 385:79–86
Low, F. J., Rieke, G. H. 1974. *The Instrumentation and Techniques of Infrared Photometry, Methods of Experimental Physics, Vol. 12, Astrophysics, Part A: Optical and Infrared.* New York: Academic
McCarthy, D. W., Low, F. J., Howell, R. 1977. *Optical Engineering* 16:569–74
McDonough, D. F. 1980. *MMTO Technical Report No. 5*
Meinel, A. B. 1970. *Applied Optics* 9:2501–04
Meinel, A. B. 1978. *Optical Telescopes of the Future, Proc. ESO Conf.,* ed. F. Pacini, W. Richter, R. N. Wilson, pp. 289–99
Meinel, A. B. 1979. *The MMT and the Future of Ground Based Astronomy, SAO Special Report,* ed. T. C. Weekes. 385: 9–22
Meinel, A. B., Meinel, M. P. 1980. *Optical and Infrared Telescopes for the 1990s,* ed. A. Hewitt, pp. 95–107
Mertz, L. 1970. *Optical Instruments and Telescopes Conf., Reading, England,* ed. J. H. Dickson, pp. 507–13. Reading, England: Oriel
Mertz, L. 1980. *Optical and Infrared Telescopes for the 1990s,* ed. A. Hewitt, pp. 957–69
Nelson, J. 1980. *Optical and Infrared Telescopes for the 1990s,* ed. A. Hewitt, pp. 11–22
Oort, J. H. 1980. *Astron. Astrophysics.* Submitted
Oxmantown, Lord. 1828. *Edinburgh Journal of Science* IX(17):25
Radau, R. E., Ulich, B. L. 1980. *MMTO Technical Report No. 6*
Reed, M. A. 1978. *Optical Telescopes for the Future, Proc. ESO Conf.,* ed. F. Pacini, W. Richter, R. N. Wilson, pp. 209–26. See also *The MMT and the Future of Ground Based Astronomy, SAO Special Report,* ed. T. C. Weekes. 385: 65–78 (1979)
Richardson, E. H., Grundmann, W. A. 1978. *Optical Telescopes of the Future, Proc. ESO Conf.,* ed. F. Pacini, W. Richter, R. N. Wilson, pp. 251–68
Rieke, G. H. 1980. *Applied Optics.* Submitted
Ruda, M., Turner, T. J. 1980. *MMTO Technical Report No. 2*
Shannon, R. R. 1979. *The MMT and the Future of Ground Based Astronomy, SAO Special Report,* ed. T. C. Weekes. 385:57–64
Shectman, S. A. 1976. *Publ. Astron. Soc. Pac.* 88:960–65
Steshenko, N. V. 1979. *The MMT and the Future of Ground Based Astronomy, SAO Special Report,* ed. T. C. Weekes. 385:191–97
Strittmatter, P. A. 1980. *Optical and Infrared Telescopes for the 1990s,* ed. A. Hewitt, pp. A1-A66
Synge, E. H. 1930. *Phil. Mag.* 10:353–60. See also *J. Brit. Astron. Assoc., Annual Report 1930* 40:406
Thomas, W. 1965. *US Patent No. 3507547*
Ulich, B. L., Riley, J. T. 1980. *MMTO Technical Report No. 4*
Vaisala, Y. 1949. *Urania Barcelona* 220: 89–93
Walsh, D., Carswell, R. F., Weymann, F. J. 1979. *Nature* 279:381–84
Weymann, R. J., Chaffee, F. H., Davis, M., Carleton, N. P., Walsh, D., Carswell, R. F. 1979. *Astrophys. J. Lett.* 233:L43–L46
Weymann, R. J., Latham, D. W., Angel, J. R. P., Green, R. F., Liebert, J. W., Turnshek, D. A., Turnshek, D. E., Tyson, J. A. 1980. *Nature* 285:641
Woolf, N., Angel, R. J. P. 1980. *Optical and Infrared Telescopes for the 1990s,* ed. A. Hewitt, pp. 1062–1150

Telescopes for the 1980s, pp. 129–93

THE SPACE TELESCOPE

C. R. O'Dell

NASA/Marshall Space Flight Center, Alabama 35812

1. INTRODUCTION

This article, in contrast to others in this monograph, deals with an observatory still under construction. It belongs here, however, because the Space Telescope will assume a role along with the Very Large Array, The Multiple Mirror Telescope, and The Einstein Observatory, and should be the premier optical telescope of the 1980s. No nice pictures can be shown (Figure 1 is an artist's conception); we'll have to wait until 1985 for that. But I can point out what we can expect in its performance and the reader can draw his own conclusions as to the complementary nature of the Space Telescope. Since this project is more expensive and complex than any other astronomical enterprise, it is also worth understanding how it is being brought to completion.

The Space Telescope will provide a tremendous leap forward in our ability to do optical astronomy. It is impossible to reduce the performance of a multiple capability instrument to only a few parameters, but when this is done, we see just what is in the offing. The leap in imaging capability is something that we do not foresee being done for faint objects observed from the ground. The ability of the Space Telescope to reach fainter objects will be so much greater than that of even the most ambitious successors to the multi-mirror telescope. Finally, we know of no other way to open the wavelength range other than going above the Earth's atmosphere.

Section 2 of this chapter on the Space Telescope describes the basic constraints imposed on ground-based telescopes. Section 3 traces the history of the Space Telescope from its conceptual origin through start of actual construction, a period of over 50 years. Section 4 presents the design considerations that enter into determining what the Space Telescope should look like, while Section 5 describes the actual spacecraft being developed. The scientific instruments turn the images formed by the telescope into scientifically useful information and they are discussed in

8243-2902/81/0820-0129$02.00

Figure 1 A depiction of Space Telescope as it will appear in orbit early in 1985. In actuality, the aperture door will not be opened while the Space Shuttle is in the vicinity.

Section 6. Space Telescope is a remote observatory with special requirements and solutions for operations; these are described in Section 7. Finally the object of all this activity is to be science; I describe today's views in Section 8.

2. CONSTRAINTS ON GROUND-BASED OPTICAL TELESCOPES

There are three ways of measuring the performance of an optical telescope: its ability to image, its ability to detect faint objects, and its wavelength coverage. This section describes what natural phenomena limit the performance of ground-based telescopes as measured by these characteristics.

Imaging means the ability to form direct images onto usable detectors and it is characterized by best image quality and the usable field of view. Usually the best images are found at the center of the field and there will be some radius off axis within which the images will be still indistinguishable from the best images. This radius is a practical definition of the radius of the field of view. The theoretical limit to the image quality of a telescope is set by the diffraction of light passing through the principal aperture. Rayleigh showed that the limiting quality image is characterized by a bright central image core of diameter 2.44 λ/D radius (where λ is the wavelength and D the aperture) surrounded by successive fainter and larger rings of light. The diameter (full width at half maximum or FWHM) at one half of the central brightness will be FWHM = 1.03 λ/D radians. The perfect limit to the resolution of two stars is arbitrarily set at the condition (1.22 λ/D radians) where the first dark ring of one star falls under the central image of the other. A clear aperture telescope of 5.1 meters (200 in.) used at 633-nm wavelength would then give images of 0.03-arcsecond diameter (FWHM) and resolution. As we shall see, the atmosphere imposes limitations long before this point is reached. This smearing widens the effective field of view; thus a larger quantity of lower quality information is traded for less abundant high quality information.

Detection of faint objects is primarily a question of signal to noise. Since I am trying to identify fundamental limits, let us describe the case of a faint image seen against a bright sky. The signal (S) is the total number of photons recorded from the point source of flux f_λ by a telescope system of known aperture (D), quantum efficiency (Q_λ), and effective bandpass W_λ over the total observing time (t), $S = f_\lambda \cdot (\pi D^2/4) Q_\lambda W_\lambda t$. In the limiting case the noise (N) will be the square root of the signal from a section of sky of the same angular area ($\pi\, \phi^2/4$) as the star image. If the surface brightness of the sky is given the general value s_λ, then $N = \sqrt{s_\lambda \cdot \pi\phi^2/4 \cdot \pi D^2/4 \cdot Q_\lambda \cdot W_\lambda \cdot t}$. The limiting detectable star flux is then

$$f_\lambda = \frac{1}{\sqrt{Q_\lambda W_\lambda}} \cdot \left(\frac{S}{N}\right) \cdot \frac{s^{1/2}\phi}{Dt^{1/2}}.$$

This equation shows that a fixed efficiency system with a constant sky brightness will improve linearly with decreasing image size and only linearly with increasing aperture.

Astronomical seeing or "seeing" is an extremely important limit on ground-based telescopes. It essentially brings the giant down to the level of the dwarf. It is the great equalizer. It has driven site evaluation groups to the corners of the globe and the astronomer to the mountain top. It is always present and seems to be at its best at a given site just before the first large telescope is completed.

What determines the quality of seeing is easy to understand. As we look out through the Earth's atmosphere, a line of sight to a point source (star) passes through many inhomogeneities of temperature, pressure, and, most important, index of refraction. This means that a ray coming from a star to an observer undergoes a random-walk process and when it reaches the aperture of the telescope it will appear to have come from a slightly different direction than is really the case. When the aperture is much smaller than the inhomogeneities (seeing cells), then the image of the star will be as small as the telescope allows through its size and quality, but will move about as the seeing cells change position. The human eye is an example of such a small telescope and this is why stars twinkle, since sometimes the images are moved completely off the eye's pupil. The fact that seeing depends upon the zenith distance is also obvious to the naked eye, as twinkling increases rapidly towards the horizon, since the number of seeing cells along the line of sight is increased.

In large telescopes, seeing manifests itself in a different, more serious fashion (Figure 2). When the telescope aperture is significantly larger than one meter, which is characteristic of the seeing cell size, then the star's image is composed of many small images or speckles. Each speckle is a sharp image of a quality determined by the telescope. The problem is that there are many such speckle images, each faint and moving, so that

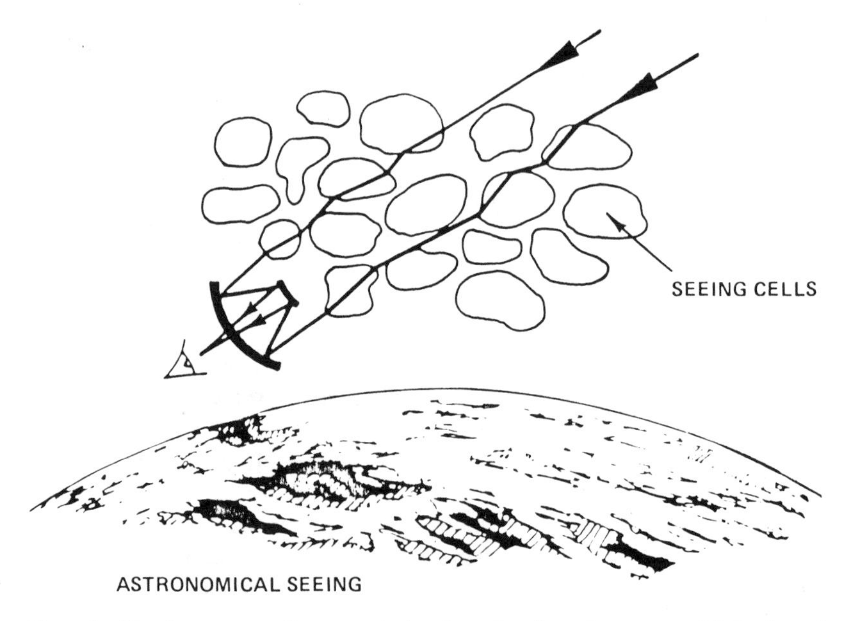

Figure 2 The phenomenon of astronomical seeing arises from the passage of light through temperature variations in the terrestrial atmosphere. Each temperature element has its own index of refraction.

time exposures average the speckles out into a nonmoving but fixed direction image called the seeing disk.

The size of the seeing disk varies with time at any given site and the best and average conditions vary between sites. Since Equation 1 shows that the seeing disk size (ϕ) is just as important as the aperture (D) in the detection of faint objects, it is not surprising that observatories are located at the best seeing sites, even if they are remote. Figure 3 gives the relative occurrence of images of different size with the Hale Telescope at Palomar Mountain, California. This is not one of the best sites for seeing, but it is characteristic. What we see is that a variety of conditions is found and a most common value is about 2 arcseconds. Sometimes it is appreciably better, but other times it is so much worse that observations stop. Using our equation for the FWHM, and inserting values of 633 nm for the wavelength and 2 arcseconds for the seeing disk size, we find that the Hale telescope characteristically produces images no better than those from a seven-centimeter-aperture telescope. They are just brighter! This fundamental limit is placed on all large telescopes and means that the gains achieved by increasing the aperture are only linear, while the construction costs vary at a much larger power.

Construction of even larger ground-based telescopes is certainly worthwhile, and the thrust of our efforts is to keep the cost from increasing dramatically with aperture size. However, the performance advantage lies with observatories in space, for a system like Space Telescope will produce images some ten times better than the Earth allows. When new technology is developed for ground-based telescopes, it can also be applied to space telescopes.

The new field of speckle imaging promises very high angular resolution from the ground through high time resolution recording of images followed by data processing of each speckle and superimposing them onto a single, composite image. There is much room for growth in this technique, but practical limits will keep this approach from being used for faint sources.

Geography affects the usefulness of a given site in ways beyond the seeing. Sites away from the equator usually have weather patterns of rainy and dry seasons. Sometimes these can be out of phase with the seasonal change of length of the nights, e.g. southern California has its clear season in the summer, when the nights are shortest.

All sites are affected by the lunar cycle. The sky brightness is increased about 25 times at full moon, which means that observing programs are often tailored around this reduction of observing capability. Traditionally, the bright sky period near full moon has been the time for high resolution spectroscopy where the sky brightness is not an important factor.

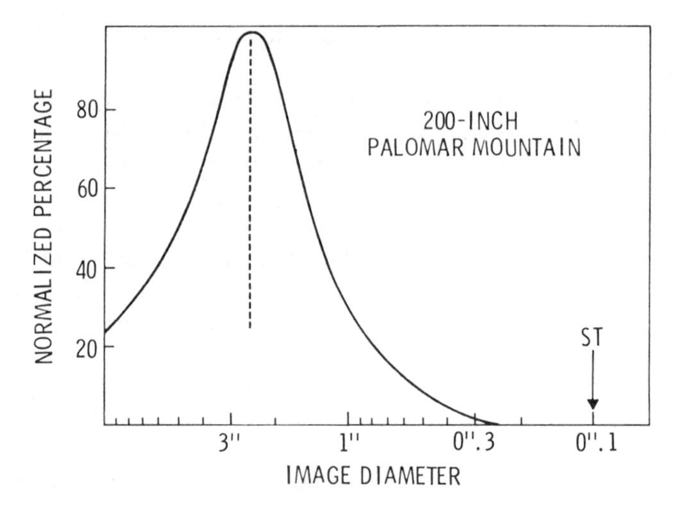

Figure 3 Seeing disk size for the Palomar 200-inch telescope. The seeing disk size varies considerably at any one site. This distribution is perhaps characteristic of most good sites. Less complete data indicate better average conditions apply at a few other sites such as Cerro Tololo in Chile.

The refraction of light through even a perfectly quiescent atmosphere limits observations through two effects, chromatic and differential aberration. The amount of refraction is greater for shorter wavelengths. The difference in refraction is one-half arcsecond between the blue and yellow at 30° zenith distance and increases rapidly from there. This distortion is superimposed on the seeing disk. Differential refraction is the second-order effect due to differences across the field of view. One-half arcsecond degradation of images occurs across a 1/3° field of view over a range of distances from the zenith to 50°. This effectively limits such observations to about four hours.

3. HISTORY OF THE PROJECT

It is surprising to many of the enthusiastic, young people working on the Space Telescope program to realize that the idea goes back at least fifty-seven years, when Herman Oberth (1923), the rocket pioneer, pointed out the advantages of such an instrument. His perception was visionary, for almost none of the technology existed then. It is not surprising that the prophet should be a rocket expert, for the transportation system was the principal obstacle. The lead in rocketry clearly belonged to Germany during the late 1930s and during World War II. When the United States found itself in 1945 to be the proprietor of a large supply of captured V-2

Figure 4 Prof. Lyman Spitzer of Princeton University is shown in 1946, about the time that he laid the groundwork that led to the Space Telescope. For a period of 35 years he has been the most eloquent scientific advocate of space astronomy.

rockets and a team of experienced rocket engineers, there followed a flurry of American scientist involvement and thinking about the future of astronomy in space.

Lyman Spitzer of Princeton University (Figure 4) began his work on the Space Telescope at that time. He published a report (originally classified) that not only gave the advantages of the space environment, but even listed scientific problems that should be addressed (Spitzer 1946). After only mixed success with the V-2 rockets, most of the stellar astronomers dropped their interest in space astronomy, and the attitude of astronomers in general was that there were more important things to be done. After all, the 200-inch telescope on Palomar Mountain had just become operational and modern radio astronomy was getting started with the detection of 21-cm radiation from interstellar hydrogen. Spitzer remained the senior credible advocate of space astronomy during the succeeding decade. As bigger rockets became available in response to pressures of the "cold war," the potential for space astronomy became greater, and by the time the National Aeronautics and Space Administration (NASA) was formed, it was clear to at least a visionary few, including Arthur D. Code of the University of Wisconsin and Lyman Spitzer, that astronomy should be part of the science program of this fledgling agency. As we shall see later, both of these men took on responsibilities for major astronomical satellites which made the Space Telescope possible.

The National Aeronautics and Space Administration called for a senior group of scientists sponsored by the National Academy of Sciences (NAS) to lay out a program for space science. This group met at Iowa City, Iowa, in 1962, and amid its other recommendations identified the Space Telescope as a natural long range goal of the space astronomy program (National Academy of Sciences 1962). This recommendation was repeated (National Academy of Sciences 1966) at a similar study at Woods Hole, Massachusetts, in 1965. Following this recommendation, the NAS established a special committee, chaired by Lyman Spitzer, to establish the scientific goals and utility of a 3-m class space telescope. The report of this group was issued in 1969 as a slim volume (National Academy of Sciences 1969, Spitzer 1968) which laid out the definitive guidelines eventually used by NASA.

Later the Space Telescope was supported (National Academy of Sciences 1972) by an NAS committee chaired by Jesse L. Greenstein of the California Institute of Technology. This group surveyed the needs of astronomy in that decade. At the time, the nature and cost of the Space Telescope was only partially defined and it was viewed as a more distant prospect than has proven to be the case.

Once the Space Telescope began being studied on the recommendation of these advisory bodies, there was a specific need for scientists to be involved. To accomplish this a series of groups was formed, each appropriate to the particular stage of development of the project.

In the autumn of 1971, NASA Headquarters established a Space Telescope Steering Committee chaired by Nancy G. Roman, then head of the Astronomy and Relativity Office. Dr. Roman invited six other astronomers to join her in providing scientific guidance to the feasibility studies being conducted by two NASA field centers, the Goddard Space Flight Center near Washington, DC, and the Marshall Space Flight Center near Huntsville, Alabama. This group met regularly for one year, identifying in more detail than the NAS committee had the specific applications of the Space Telescope and responding to design features proposed by the field centers and their supporting contractors.

This committee was succeeded by the Space Telescope Mission and Operations Working Group, which was chaired by the newly appointed Project Scientist (the author). This group was based at the Marshall Space Flight Center, which had recently been given lead center responsibility. The members included a few NASA employees, appointed according to their key assignments on the Space Telescope project, but the majority were selected through open competition. In the winter of 1973, NASA issued a call for participation, and by June of that year the selections had been made and the scientific structure was established. There was a tiered structure, with the Working Group at the top, composed of NASA project (field center level) and program (Headquarters) people, leaders of Instrument Definition Teams (leading activities studying scientific instrument preliminary design), and members at large, who were to be concerned with the overall characteristics of the observatory. This was a large structure, including 40 scientists from 26 different institutions.

The formation of these two committees was typical of the way in which NASA developed new programs. There are always many more ideas than there is time and money; however, a few ideas are subjected to feasibility studies (Phase A in NASA parlance), which are intended to establish in approximate form the practicality of undertaking a given project. Many approaches are usually examined, with the best identified. Only a small fraction of the Phase A studies advance to become Phase B studies, where preliminary designs are established. During Phase B, different approaches are often examined, but the end product is a basic design that NASA can use in establishing a Phase C/D contract for Design and Development. Phase B also establishes a much more accurate estimate of the necessary development schedule and costs. Again, only a few of the projects submitted to Phase B study advance to Phase C/D. The low rate of survival is due to lack of funding. Phase A and Phase B studies are usually funded out of general use funds; i.e., funds which NASA always has available to study projects.

Major projects entering Phase C/D must be specifically included by name within the NASA budget and are subject to many reviews beyond the agency. A project must be accepted by the executive branch of government through its arm, the Office of Management and Budget. Otherwise, it will not appear as an item in the NASA budget forwarded by the President to the Congress. Within the Congress, the House of Representatives and the Senate must both pass separate authorization and appropriation bills that include the project. Thus, it is hardly surprising that so few projects survive the jump from study to hardware.

As Space Telescope became ready for the design and development phase, an Announcement of Opportunity was issued by NASA Headquarters (March 1977) calling for proposals to participate during the hardware phase of the program. The process initiated by the Announcement of Opportunity resulted in the selection of 60 scientists from 38 institutions to participate in several activities. There was appointed a senior scientific body, now known as the Space Telescope Science Working Group, composed of a few key NASA employees, the Investigation Definition Team

Leaders (sometimes called the Principal Investigators), Interdisciplinary Scientists, Telescope Scientists (who have particular involvement with the telescope portion of the observatory), an Astrometry Science Team leader, and a Data and Operations Team leader. The Investigation Definition Team leaders and the Astrometry Science Team leader all have groups of six to ten scientists working with them, while the Data and Operations Team has a separately chosen leader, but its membership is made up of representatives from each of the teams.

3.1 *Establishing the Technology*

Space Telescope builds upon two decades of astronomical space technology. When NASA was created, it was clear that space astronomy was one of the disciplines that would benefit. The question became one not of "if" but "how." One extreme approach would have been to build a series of rocket payloads of progressively greater sophistication, and then take on the challenge of satellites; after all, this was the very successful approach adopted by the scientists studying the upper atmosphere and the sun. The constraints were much greater on the astronomers. For the much fainter stellar sources, longer observing times are necessary to gain useful information. High levels of pointing and control stability are also required of the telescope. These considerations caused the fledgling agency to make a big first step, to construct a series of spacecraft known as the Orbiting Astronomy Observatories (OAO). Three spacecraft were planned, each providing a basic power, pointing, and communications capability to an enclosed scientific instrument package. This meant that problems of control, data handling, and guidance had to be solved in taking the first step. It is hardly surprising that the initial launch occurred later than was expected. The first spacecraft failed before obtaining any scientific data. It was replaced in the OAO program by another functionally identical spacecraft which flew a set of photometric experiments prepared by the University of Wisconsin, and a rudimentary survey camera system prepared by the Smithsonian Astrophysical Observatory. The third OAO launch was unsuccessful while the fourth placed into orbit a 32-inch-aperture telescope-spectrometer system developed by Princeton University. This OAO operated successfully for over seven years. The scientific results obtained by these two spacecraft (Figure 5) were significant (NASA 1972, Spitzer & Jenkins 1975), and the technology base established was equally important. The critical missing element was a highly precise optical system, which was not required by any OAO experiment. That technology had to be established elsewhere.

The most significant optical telescope for stellar imaging was flown in the Stratoscope II program. This program was to fly a balloon-borne 36-

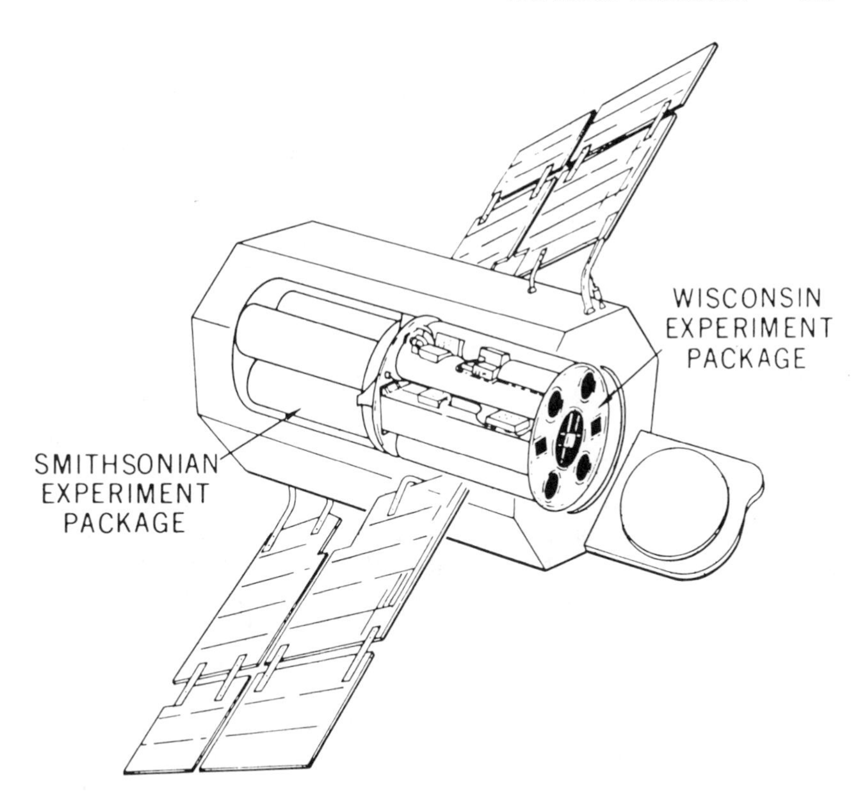

Figure 5 Phantom view of OAO-A2 experiments. The second OAO spacecraft was characteristic of the four that were developed. A standard spacecraft supplied power, pointing control, and data capability to an experiment(s) packaged inside. The last OAO spacecraft (Copernicus) ceased operation in early 1981 after $8\frac{1}{2}$ years of operation.

inch-aperture telescope into the stratosphere to obtain spectra and fine resolution images. After several only partially successful flights, Stratoscope II finally produced a series of images of stars, planets, and galactic nuclei of about 0.2 arcsecond FWHM value, much better than had ever been achieved on the ground for faint objects and approaching the diffraction limit for a telescope of this size (Danielson, Tomasko & Savage 1972). In some ways, Stratoscope II was more difficult to build than Space Telescope, for the designers had to contend with a small residual atmosphere inside the telescope tube. Since it took only short exposures, the overall challenge was less, but it was an important step in preparing for Space Telescope.

A natural successor to the spectrometer experiment on OAO-3 is the International Ultraviolet Explorer (IUE, see Figure 6). This is a specially designed aircraft containing a 0.45-m telescope feeding another spectrometer. An important difference in performance is realized by using a two-dimensional television detector so that many wavelengths can be studied simultaneously, instead of the one-at-a-time look provided by OAO-3's photomultiplier detectors. A very important design difference came from the fact that IUE was placed into a geosynchronous orbit. This presented the very real operational advantage of continuous contact with the spacecraft, from either the Goddard Space Flight Center or the European Space Agency, the two partners in this enterprise. The Orbiting Astronomical Observatory spacecraft, on the other hand, were placed in low Earth orbit, with only brief intervals of contact whenever they passed close to a ground station.

During the same time as these actual spacecraft were being built, there were studies being conducted of the possibilities for large, long-lived, imaging telescopes. The range of ideas was enormous. Orbits considered ranged from low Earth orbit (like the OAO's) through geosynchronous (like IUE), to the lunar surface. A telescope on the lunar surface appeared to be attractive, but it was rejected because of the high transportation costs and the bright conditions and wide thermal excursions on the moon's surface. A telescope in geosynchronous orbit would have been ideal from a designer's point of view, but the initial transportation cost would have been high (it would have required a Saturn class rocket) and the cost to revisit it for refurbishment would have been prohibitive. The degree of involvement of man considered also covered a wide range. At one end, it was argued that the presence of man (even for occasional servicing visits) placed prohibitively expensive safety requirements on the design, and that it was cheaper to build a throw-away satellite of high reliability components. The other extreme position studied was one where astronaut crews actually operated the observatory from orbit and lived within or nearby the facility. As we shall see later, this issue was resolved by the capability of the available transportation system.

The Space Telescope design had settled down to a 3-m-aperture instrument in low Earth orbit by the time Phase A (feasibility determination) was initiated (1971). At that time, NASA had not committed itself to the development of the Space Shuttle, so that the design had to be compatible with the Titan launch vehicle, and manned involvement was a very open subject. Several alternative design approaches were studied for many parts of the system, including the mirror (thick and rigid, or thin and flexible), the telescope structure (metal with thermal control, or plastic-graphite material with a low thermal coefficient of expansion), and scientific instruments (a single multipurpose instrument, or several separate ones). Solutions, trade-offs, and pitfalls to be avoided were the main prod-

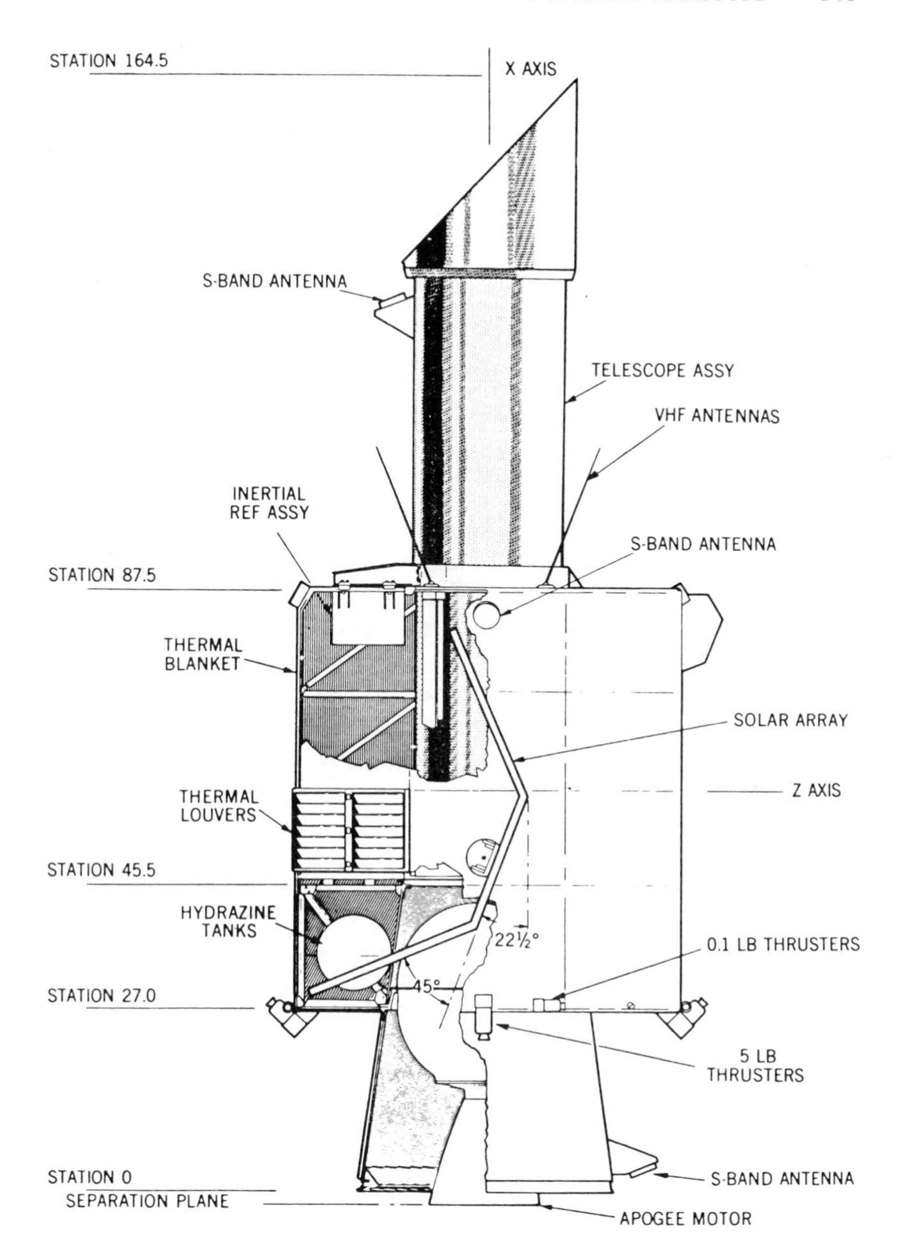

Figure 6 The International Ultraviolet Explorer spacecraft is a geosynchronous orbit observatory housing an echelle grating spectrometer-telescope system. The station numbers give the distance in inches from a reference datum.

uct of the Phase A studies. The work also allowed a better set of guidelines to be developed.

Phase B studies establish preliminary designs, but they do this through the study of many options. The concept of Space Telescope was now defined much more clearly, and NASA in-house activity began to be heavily supplemented by contractor work. In Phase A most of the work was done by NASA personnel at the Marshall Space Flight Center and the Goddard Space Flight Center. In Phase B we continued the role of NASA people, but also added optical and aerospace contractors. Not only did this bring to the project new people with fresh ideas and varied relevant experiences, it also expanded the level of activity enormously. Parallel telescope preliminary designs were developed by the Perkin-Elmer Corporation of Danbury, Connecticut, and the ITEK Corporation of Lexington, Massachusetts. Parallel preliminary designs of spacecraft systems were carried out by the Boeing Aircraft Corporation of Seattle, Washington, the Martin-Marietta Corporation of Denver, Colorado, and the Lockheed Missiles and Space Corporation of Sunnyvale, California. Contracts were let with all of these groups, but the products far exceeded what we paid for. It is a standard and legal practice for such corporations to use up to a certain fraction of earnings from one program to fund activity that will lead to competitive positions when bidding for future work. Since Space Telescope was clearly desired as a design and development contract, the corporations each put into their studies several times the value of the government study contract.

During this time, the concept for Space Telescope became well defined. A delicate balance was established so that as ideas came forward, they were incorporated by everyone, but in a way that did not compromise the proprietary nature of the data developed by corporations with their own funds. Early in Phase B NASA was given approval to develop the Space Shuttle, which became the vehicle around which we were to design the observatory. Obviously, this constrained our design, but it also eliminated many resource-diverting possibilities.

Up to this point, NASA had always used as the size the 3-m aperture recommended by the National Academy of Sciences Committee in 1969. The size of the aperture was not arbitrarily derived, but it was also not necessarily optimum. The Space Shuttle could accommodate an observatory as large as 3.2 m. A design calling for the maximum aperture demanded a spacecraft configuration where the optical telescope was at one end of a cylinder and the supporting spacecraft systems were at the other, behind the scientific instruments (Figure 7*a*). This configuration had a high moment of inertia which required the development of a powerful and unavoidably expensive stabilization system. If the aperture was reduced to

2.4 m, then there would be space around the primary mirror (the center of gravity) to package the spacecraft systems in a configuration having a much lower moment of inertia and allowing well-proven concepts to be used for pointing control (Figure 7*b*). Of course, anything smaller would be able to use this same technology and the overall costs would be less. As

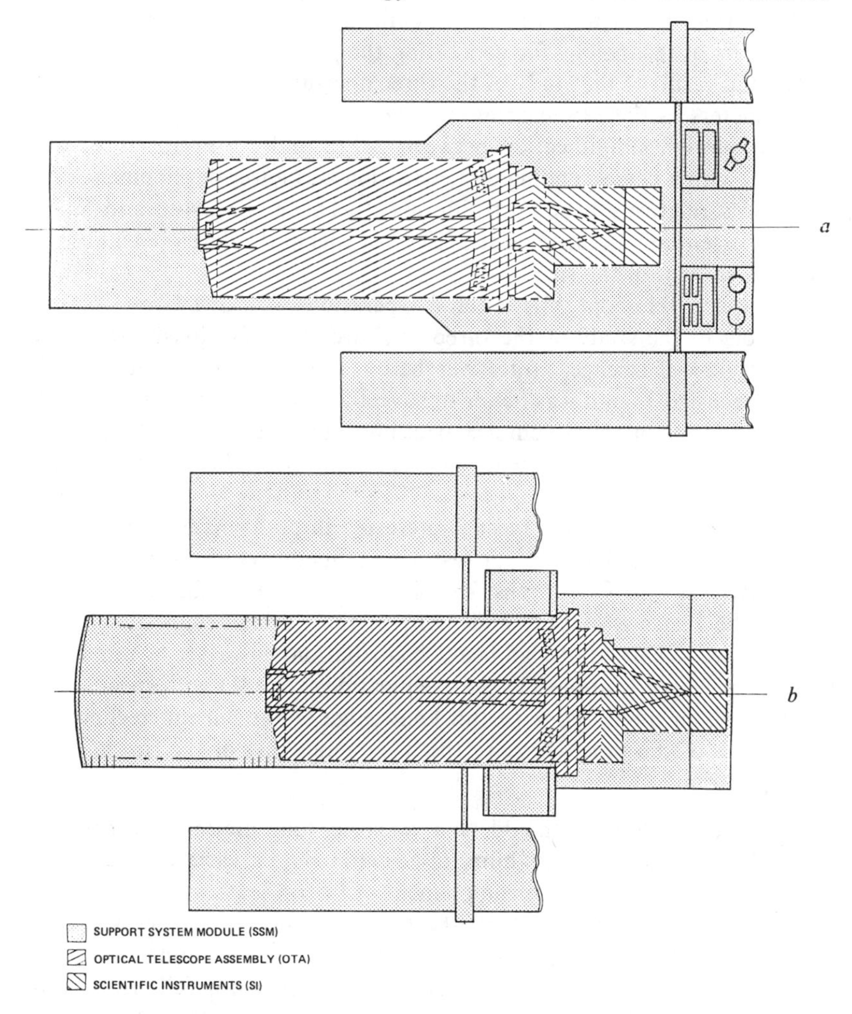

Figure 7 (*a*) The high moment of inertia configuration of Space Telescope that resulted from using the same configuration demanded by a 3-m aperture. The aft location of the heavy command and control components necessitated a powerful control system. (*b*) The low moment of inertia configuration for Space Telescope that resulted from packaging the heavy command and control components in a toroidal ring around the primary mirror. This confined Space Telescope to an aperture of 2.4 meters.

will be discussed below, ITEK had developed a test mirror of 1.8-m aperture similar to Space Telescope's in design. These considerations caused NASA to request the study of three possible aperture sizes for Space Telescope: 3 meters, which was the historical baseline for the program; 2.4 meters, which would allow a different and inherently less expensive configuration; 1.8 meters, which was both simpler and at a clearly established level of technology for generating the primary mirror.

The end result of the study of the three aperture sizes was the selection of 2.4 meters as the new baseline for the Space Telescope program. The expected design simplification between 3.0-m and 2.4-m apertures was realized, with a corresponding cost reduction and lower development risk. There was relatively little cost difference between the 2.4-m and 1.8-m designs, essentially because most of the cost was in the development of the components, not in the size.

The Space Telescope Mission and Operations Working Group played a key role in the study of the three aperture sizes. Obviously, from the point of view of science, larger was better, but if larger meant prohibitively expensive, then it was better to have a 2.4-m design than none at all. This was a difficult and complex decision, but in the end the Working Group endorsed the proposed 2.4-m design, and this size became the official baseline.

The scientific users were mostly interested in performance rather than engineering design. This did not restrain the members of the Space Telescope Mission and Operations Working Group from being an integral part of the Phase B activity and many of their design suggestions were incorporated. Their most significant contribution was a set of minimum performance specifications. These specifications were set with specific scientific goals in mind and with full knowledge of the current technology capabilities of the time; they are the basic guidelines of the Space Telescope program:

1. The Space Telescope should be a versatile, long lifetime observatory; i.e., it must have the capability to accommodate a variety of scientific instruments and vary the complement of instruments with time.
2. The optical image should satisfy the following requirements in the visual and near-vacuum ultraviolet wavelengths: resolution using the Rayleigh criterion for contrast of 0.10 arcseconds; a full width half intensity diameter of 0.10 arcseconds; 60 percent of the total energy of a stellar image must be contained within a diameter of 0.15 arcseconds.
3. The overall Space Telescope system must work efficiently down to wavelengths permitting the study of the Lyα line at 1216 Å, thus

requiring a short wavelength limit of about 1150 Å. Likewise, it must allow efficient observations at infrared wavelengths longer than those readily accessible from the ground.
4. The system should accommodate at least four scientific instruments.
5. It must be capable of measuring objects appreciably fainter than those accessible from the ground. At present this means going to about magnitude 27 with a signal-to-noise ratio of 10 in 4 hours of observing time.
6. It must be capable of measuring extended sources of surface brightness 23.0^{m} arcsec^{-2} with a signal-to-noise ratio of 10 in 15 hours.
7. The Space Telescope must have the capability of using scientific instrument entrance apertures comparable in size to the image.

The work done during Phase B was not limited to concepts and paper activities. A series of hardware tests were also performed. Three principal tests addressed the most critical design areas. 1. The ITEK Corporation asked the Corning Corporation to fabricate an egg-crate construction mirror blank upon which they then generated a $\lambda/64$ quality mirror surface. This work determined that ultra-precise mirrors of lightweight construction could be made without serious difficulty. 2. The Martin-Marietta Company tested the Reaction Wheels proposed to control the pointing of Space Telescope on a seismically isolated table having the inertia characteristics of Space Telescope with a stability of 0.007 arcseconds (root mean square average). 3. The General Dynamics Corporation constructed a 1.5-m-diameter graphite epoxy shell that modeled one of the Phase B designs for the structure that holds the primary and secondary mirrors in precise alignment. This work established that a precise structure of this new material could be fabricated that had the thermal coefficient of expansion and mechanical properties necessary for Space Telescope. There were many other feasibility tests that could have been done, had funds permitted, but these three resolved the major issues.

3.2 *Struggles on the Hill*

NASA is a federal agency, receiving its direction and funding through authorizations and appropriations by the Congress and the President through laws. Between the word and the deed can come a world of adventure. As we are taught in grade school, laws are bills that have been passed by both sides of the Congress (the House of Representatives and the Senate) and have been signed by the President. If the President refuses to sign a bill, then his veto can be overridden by a two-thirds majority vote of both sides of the Congress. Plans and budgets are controlled on the basis of fiscal years (FY), which run from October through September, e.g. FY 82 runs from October 1, 1981, through September 30, 1982.

NASA prepares its budget well before the start of the fiscal year. Since NASA is run by presidential appointees, it is intended to be responsive to the executive branch of the government. As such, its budget is first submitted to the Office of Management and Budget (OMB) in September of the preceding year. It will usually have taken about four months of intensive activity to get this draft budget prepared. This means that the initial budget activity for FY 82 started in May 1980 and went to the OMB in September 1980. The OMB determines if this draft NASA budget is compatible with administration policies and consistent with the overall government budget that they are preparing. It is very common for several iterations to occur, with NASA trying to justify and defend its proposals and OMB trying to reduce costs. (It was not always thus. When NASA was perceived in the eyes of the world as the agency for winning the cold war, things were much easier and NASA was encouraged to bring in bold, innovative ideas. Now it is very closely scrutinized and only supported in a very conservative way.) The final product of the OMB, the Federal Budget, is submitted by the President to the Congress in January.

The total budget is very complex and the Congress has evolved a system of committees that allows examination of each part by specialists. There are parallel structures for both authorization and appropriation, with a large central committee and many smaller subcommittees. A bill normally must be passed by the subcommittee before reaching the full committee, who in turn must pass it before it goes to the floor of the House or Senate. This means that the House and Senate subcommittees have the largest say about what is authorized and what is appropriated and the chairmen of these committees are very influential. To simplify, it can be said that the budget for NASA comes down to a decision by four subcommittees, each having about a dozen members. The advantage of this is that each subcommittee member can take the time to study and be well informed about the agencies that he decides upon. The disadvantage is that one vocal and negative person can be very disruptive to an otherwise well-structured system. It is fair to say that the authorization subcommittees are "bullish" about their agencies while the appropriation subcommittees examine their budgets with skeptical eyes and sharp scissors.

During Phase B, the Space Telescope project received unusual attention from the Congress. The institutional caution of NASA and the lack of support from Congress for the space program was such that NASA chose to include the Space Telescope as a specific item in the budget submitted for FY 75, even though such studies were usually done under the anonymous umbrella of future programs. The Congress was accustomed to reviewing specific items only as they were ready to go into the hardware phase. Since it is wasteful to start a program and then stop it, Congress looks very carefully at each new program that specifically appears in the proposed budget. When the subcommittee saw a new item called Space Telescope that could potentially cost a half billion dollars they looked very hard at it. In fact, the House appropriation subcommittee eliminated it! If it had been left this way, the program would have been dead even before it got started. This subcommittee also was responsible for the National Science Foundation budget and its members were well aware of the expenditures for astronomy there. Funds had recently been allotted for the Very Large Array and the chairman thought that this was enough big facility funding for a while. Fortunately, the Senate appropriations bill included Space Telescope study money, which meant that when selected members of the House and Senate appropriations committees met jointly to reconcile the differences in the bills, the Space Telescope was reconsidered. It was reinstated at a level less than NASA had requested but was included in the final budget and thus still remained alive.

This sequence of events set a pattern that was to mark the Space Telescope program. It created an impression that the Space Telescope could be made to cost less if you just applied more pressure and offered less funds. It created an early opposition to the program by a few Representatives and Senators whose positions were overridden. It alerted working scientists that they could be influential in decisions made in the Congress. The spontaneous and sometimes confusing series of letters, telephone calls, and visits from scholars surprised and impressed the subcommittee members and their staffs. This did not replace the detailed work done by NASA, but it did bring home the fact that scientists wanted this observatory. Not only did such activity prove useful on this occasion, but it also provided the familiarity with people and the system to allow similar responses in future times of need (Spitzer 1979).

The cost of the Space Telescope was even more ill defined than the program itself, which is not surprising, since one needs a good idea of what is to be done before determining what it will cost. As the engineering approach for Space Telescope became identified, the costs could be better determined. These costs appeared to exceed the choke limit of Congress and efforts were made to reduce the identified costs to an acceptable level. At this point costs for the three size options were studied and the reduction to a 2.4-m aperture was made. This decision reduced the cost and the technological uncertainties of the project. Other cost reductions were less clear, for the budget was changed to cover only the original development costs through orbital verification; the operations and refurbishment costs were put into a less well-defined budget to be requested later. Charges of obfuscation and unreasonableness were traded, each with exaggeration and a grain of truth.

The Phase B activity on Space Telescope was planned to allow a start of construction in FY 77; this did not happen. The original new construction date was even earlier, but the Phase B program was stretched out to adjust to the insufficient funding level which was all that was being approved by the Congress. When the FY 77 budget was being prepared, NASA did not include a new Space Telescope construction, much to the concern of the people working on the project, the scientists who had already contributed several years support with no guaranteed reward, and the Phase B contractors, who were spending their own money at a 3–5 to 1 ratio. The hue and cry grew as the unveiling of the FY 77 budget revealed that Space Telescope was only in for a stretched-out completion of Phase B. NASA people could do nothing since the NASA administration had accepted this budget, but the scientists and contractors were not so constrained and they started lobbying the Congress to put authorization to go ahead with planning and construction of the Space Telescope into the NASA bills. Adding and funding such a new start that was not in the administration's budget would be an unusual step, but it was not without precedent, and the telephone calls, letters, and visits began anew. It almost worked, but in the end the reluctance of a few Representatives and Senators (who had objected to funding of Space Telescope each previous year) and the NASA administration made the difference and no formal attempt to get the new start was made. One compromise was reached, which allowed the selection process of the scientists and contractors to begin during FY 77. This would allow a rapid start to occur in FY 78.

All of the work done during Phase B produced results in the FY 78 budget. NASA supported the Space Telescope, the Office of Management and Budget let it survive, and the Congress supported it. In fact, it all seemed rather anticlimactic after the life and death struggles that had been encountered in each of the previous years. The opponents still had their say, and an agreement was made that the total cost of the project should not exceed $575 million in FY 82 dollars. The bargain was struck, but the last shots were still to be fired. It is the right of a senior Congressman to request audits of federal programs by the General Accounting Office, and a special audit was immediately begun. There have now been two audits, the second looking into detail at the total runout costs for the Space Telescope program. Since there are still many uncertainties in how future agency issues will be handled, Space Telescope is very long lived, and inflation may continue for many years, the final costs of operation turn out to be much larger than the development costs! Many of the scientists involved felt the General Accounting Office report read more like a supermarket newspaper cover article than a cold-blooded appraisal, but it gave the Congressional nay-sayers what they wanted.

The European Space Agency (ESA) established itself as an integral part of the Space Telescope program during Phase B. This started out when ESA placed an observer on the Space Telescope Mission and Operations Working Group and NASA sent a delegation to several locations in Europe to generally educate scientists and administrators about the Space Telescope program. The ESA observer was already involved when the budget crisis for FY 75 developed. An outspoken critic of big money for science asked if any effort at sharing costs with ESA was being made. The NASA answer did not satisfy him, and NASA was asked to give international involvement a thorough study. This was done, but it took several years. International agreements always take time and this was no exception, for both organizations had to thoroughly study what options were realistic for them. The final agreement was that ESA would make three major contributions: one of the scientific instruments, namely, the Faint Object Camera; the solar array, which provides spacecraft power; and continuing manpower support to the scientific facility operating Space Telescope. The reward for this collaboration was to be 15 percent use of Space Telescope observing time, a figure based on ESA's share of the total project cost. This has proven to be a very positive part of the Space Telescope program for the European engineering and scientific contributions are clearly on a level with those of NASA. In addition to the steps taken to contribute directly to the NASA-led part of the program, ESA is also taking steps to make sure that Space Telescope data are easily available to all scientists in ESA member nations.

4. DESIGN CONSIDERATIONS

The Space Telescope has been designed with certain performance goals in mind but within the constraints of many engineering and cost considerations. In this section I describe the design considerations that were identified during Phase A and Phase B studies and the early observatory design period. It is these considerations that established the extreme limits to the possible engineering designs. Cost was always a major additional element, but the one most difficult to quantify, and it will not be presented in detail here. The lack of accurate cost information is largely a result of insufficient time, experience, and resources of government people running the Space Telescope project, but it is also due, in part, to having to rely on competing potential contractors for cost figures. At a time of competitive parallel studies, it is difficult to get complete (and candid) cost estimates from even the most experienced, altruistic, but profit-oriented, company.

The major design constraints begin and end with the transportation system that defines the launch and refurbishment possibilities. They include the communications system available to control and receive data from the observatory. They, of course, also include the more obvious features of optics, pointing control, and general configuration.

NASA decided that the Space Shuttle was to be the only available United States low-Earth-orbit transportation system for the 1980s. This is basically a policy-economics decision. The Shuttle is an excellent system for many tasks, but it is not ideal for all payloads. Nevertheless, it was decided that it would be the only system available since it would not be economically feasible if it only flew a part of the NASA payloads, and economics was the major consideration of the post-Apollo period. This means that Space Telescope and all other major contemporaneous satellites have to be designed around the characteristics of the Space Shuttle. Fortunately, these characteristics are relatively benign.

Many elements of the Space Shuttle Transportation System are well described elsewhere. It is a rocket engine–propelled aircraft capable of flight into low Earth orbit and ballistic reentry into the atmosphere, followed by a landing on an airport with a large runway equipped with specific electronic approach aids. Each aircraft in the fleet is nearly identical and capable of multiple spaceflights. The three principal characteristics of the Shuttle that affect Space Telescope design are the orbit capabilities, the volume available, and the configuration and loads imposed during launch and landing. All of these characteristics are related.

The Space Shuttle was designed as a low-Earth-orbit transportation system. The boost given by its solid rockets and main fuel tank reservoir will give an orbit useful for many, but not all missions, and is of relatively short lifetime before atmospheric drag brings it back into the lower atmosphere. Since the propulsive energy is fixed, the exact orbital altitude will vary according to the inclination of the orbit. Auxiliary fuel tanks can be added in the aft portion of the payload bay. The flexibility added by this augmentation is shown in Figure 8.

The optimum orbit for Space Telescope would be geosynchronous, about 42×10^3 km radius; unfortunately, this is not possible for something so massive. If we cannot have this orbit, then the next best orbit is much lower. The new altitude must reflect a balance of two factors, orbital lifetime and radiation shielding. Low altitudes are easy to reach and many missions fly there (an advantage for servicing the Space Telescope), but the ballistic drag of the residual atmosphere causes a continuous de-

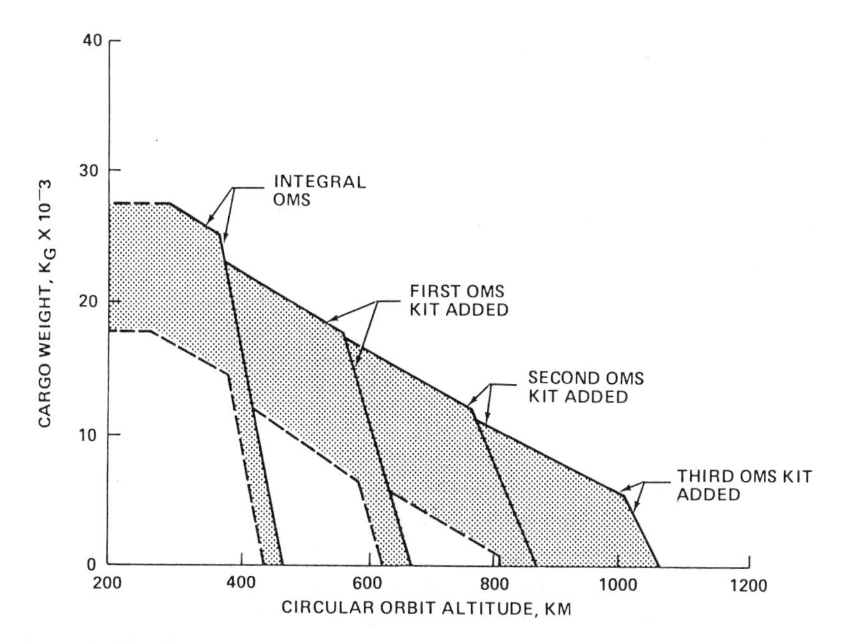

Figure 8 The Space Shuttle delivery capability is shown. Higher orbital altitudes can be reached by augmenting the Orbital Maneuvering System (OMS) fuel in discrete steps, at a penalty in payload weight.

cay in the orbit. The orbit needs to be high enough so that the orbital decay never leads to the Space Telescope's actually heating up and being destroyed before a periodic revisit and orbit boost can be made. However, higher is not always better, because the radiation environment becomes more hostile as one approaches the trapped particle belts that surround the Earth.

The orbit has constrained the design of the Space Telescope. The residual volume available is a cylinder of 4.6-m diameter and 14-m length, capable of carrying a total weight of about 17,000 kg. As Figure 8 shows, the auxiliary fuel needed for the best orbit (about 500 km) reduces the weight and volume available for Space Telescope.

The Space Shuttle will have a maximum loading due to propulsive acceleration of about three times the force of gravity. This is the most obvious of the launch/landing load constraints; others are more insidious. The launch acceleration loads are longitudinal and along an axis of natural strength for a telescope. Landing loads are vertical, across the weak axis, and can be as high as 4.2 gravity forces. Moreover, the structural strength must be such that no component will break loose during a crash landing and imperil the crew. In addition to all of this, there is a high vibration condition at launch, especially while both the liquid-fuel main engines and the solid-fuel rocket boosters are just beginning to burn. All of these constraints must be observed and they demand a much more rigid design than that which is needed for the actual construction or operation of the Space Telescope itself.

4.1 *The Optical System*

The optical system for the Space Telescope must meet a variety of constraints of length, diameter, focal ratio, manufacturability, and field of view. These constraints are more limiting than those on a typical large ground-based telescope. Like the design of any other complex system, the final result will be the best compromise.

The volume available in the Shuttle payload bay determines the overall length and diameter available. In the discussion of the history of the project I explained how considerations of moment of inertia advised that as much mass as possible be placed close to the center of gravity. Since the primary mirror is the heaviest single component, and there must be room behind the mirror for scientific instruments, this meant packaging the support equipment in a torous surrounding the primary mirror. This method of packaging limits the aperture to about 2.4 m. A folded optical design (Cassegrainian in general configuration) is demanded by many considerations of compactness, but the full 14 m cannot be used for the optical system. It is necessary to extend a light shade beyond the secondary mirror and for there to be a usable distance between the back of the primary mirror and the focal surface. It is also necessary that there be a usable distance behind the focal surface for the scientific instruments.

The overall focal ratio was primarily confined by the scientific instrument design. Most scientific instruments require some internal operations to be performed, such as collimation and dispersion in spectrographs or magnification in cameras. Too large a focal ratio requires too large a space needed for the scientific instruments, while too small a focal ratio makes it very difficult to share the limited field of view. With a sectioned field of view, there is little space for relay lenses or collimators of reasonable size.

The other length consideration is directly related to the focal ratio of the primary mirror and the focal ratio of the system. Meinel (1960) demonstrated clearly how a delicate balance must be struck. A low-magnification secondary mirror means that the immediate stability requirements are less, but the larger separation distance (between primary and secondary mirror) means that greater constraints are placed on the alignment structure. This consideration drives the design to smaller primary mirror focal ratios. The limit set on the focal ratio is essentially set by the manufacturing techniques of generation and testing, especially the latter. Very fast mirrors can now be made but they are very difficult to test. A very precise mirror like Space Telescope's requires use of a shearing interferometer for testing to better than $\lambda/50$ and the test system becomes difficult to design and operate at very small focal ratios.

The available field of view is a primary consideration. Obviously, bigger is better, as long as the best images are not compromised in quality. The two-mirror system that provides the best field of view is the Ritchey-Chrétien system, which gives coma-free images over a wide field, but at the price of astigmatism at the edge of the field, curvature of the focal surface, and slightly more complex primary and secondary mirror figures. Because the astigmatism in stellar images at the edge was acceptable for use as fine guidance signals, the Ritchey-Chrétien system was a clear choice for the optical design.

A final consideration is wavelength range. An all-reflective system was needed to give the ultraviolet through infrared efficiency that was needed. Unfortunately, optimizing the performance at both ends of the range is expensive. Since the Space Telescope operates at temperatures far above cryogenic temperatures, the instrumental background signal is very high. The best way to minimize the effect of this is to rock the secondary mirror so that the background can be "chopped out." This is ideal, but it is a significant complication in a secondary mirror mount that needs to be ultra-stable for superb optical images. The wavelength coverage at the other end is determined by the mirror reflectivity. The ideal situation here is high reflectivity right down to the Lyman limit at 912 Å, where the interstellar medium becomes very strongly absorbing. A surface of bare aluminum comes close to this in its reflectivity properties (Figure 9), but is impractical since oxidation occurs almost immediately when it is exposed to air and problems of depositing a reflecting coating in space are formidable. There are two protective coatings available, MgF and LiF. MgF is a practical substance, easily deposited, but it becomes absorbing just below Lyman α (1216 Å), thus cutting out many interesting interstellar absorption lines found by OAO-3 (Copernicus) (Spitzer & Jenkins 1975). The alternative (LiF) extends the wavelength range but is highly hygroscopic, and special handling of the optics is necessary after the coating is deposited. The LiF coating was seriously considered for Space Telescope, but the additional costs due to the long assembly and test procedures were prohibitive.

4.2 *Pointing and Stabilizing the Observatory*

Because the Space Telescope will have much sharper images than any ground-based optical telescope, it demands greater pointing and stabiliz-

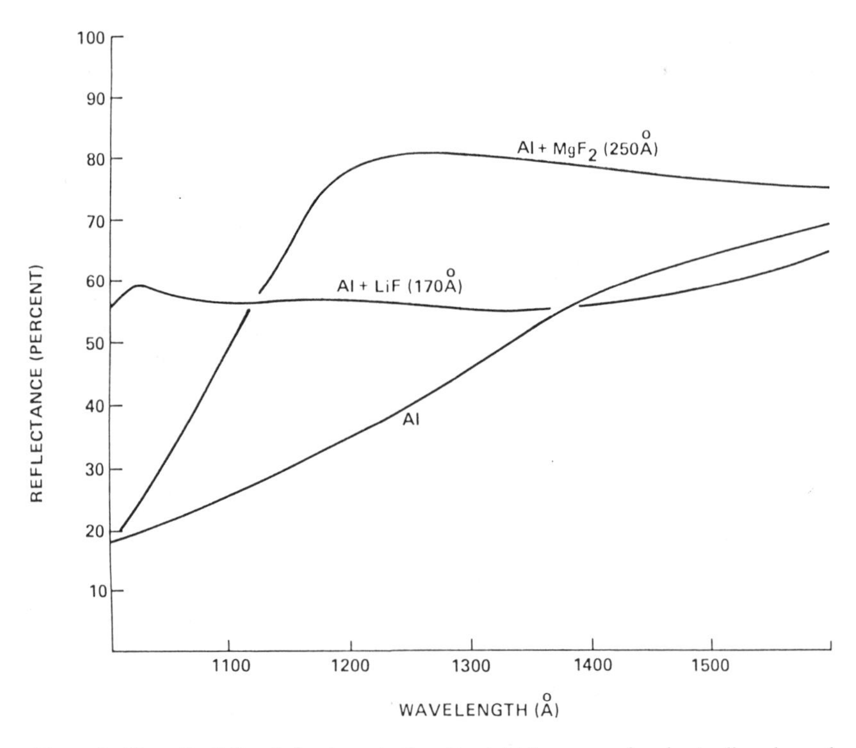

Figure 9 The reflectivity of aluminum in the ultraviolet is preserved and actually enhanced by overcoating with transparent materials. Unfortunately, LiF is very hygroscopic, and expensive to use as compared with MgF_2.

ing accuracy than astronomers and designers usually require. Pointing and stabilizing requirements are usually satisfied on the ground by the telescope mounting and tracking drive. The situation is very different in a floating spacecraft that has nothing to hold on to and must continuously determine its orientation.

The Space Telescope images of about 0.1-arcsecond size will often be used with entrance apertures of comparable size, in order to obtain the best contrast. A low resolution spectrograph being used to study the stellar nucleus of a galaxy is an example of this. In order to reliably use information from such small apertures, the Space Telescope must point with an accuracy only a small fraction of 0.1 arcseconds. The design requirement is to point to ± 0.01 arcseconds accuracy. Obtaining this latter specification necessitates a series of steps.

The two basic sources of attitude information are gyroscopes and the stars. Since gyroscopes continuously drift they require periodic updating by reference to the stars, so that both systems are actually required. Several levels of star reference are possible. The primary level is through wide field of view telescopes providing nearly all-sky coverage. These systems are called fixed-head star trackers and can give positions of stars down to about fifth magnitude with an accuracy of about an arcminute. The next level of star information must come from the telescope itself. Very high precision is possible there and the 0.01-arcsecond information is provided by the same device that provides the stability information.

The tracking system (fine guidance) is the most demanding part of the Space Telescope system. The telescope must be able to track on an object with an accuracy of 0.007 arcseconds with a time constant of much less than one second. The only way to reach this level of accuracy is to use images formed by the Space Telescope itself. Rather than risk interference with the portion of the field of view being used for observation, images at the edge must be used. The Ritchey-Chrétien optical design gives astigmatic images at the edge of the field of view. The tracking system must accommodate this.

Two approaches to the fine guidance system have been advanced. The first is a system of orthogonal prisms, with apex separations of about one arcsecond, adjusted so that the apex of one set of prisms falls on the tangential best focus image and the other set's apex falls on the sagittal ray's best focus; thus each prism bisects a narrow image along its apex. By means of tilting plates or other devices a guidestar's image is brought up to a nearby prism intersection and the four separate beam intensities are used to tell how the image is moving off the pyramids. The second approach is interferometric. By mechanical and optical lever arms a potential guidestar is sighted through an entrance aperture. The beam is collimated, then split into orthogonal components. Each component then illuminates a Koester's prism (Figure 10). When the illuminating beam is exactly perpendicular to the prism base, the interference patterns seen by the photomultipliers are the same and the signal is balanced. As the guide star angle changes, the fringe patterns and the subsequent intensities change, with extreme sensitivity. Both of these complex systems can work, and each has its own advantages.

Not only must the fine guidance system be able to track stationary targets, it must also follow moving objects. In fact, even fixed stars require continuous movement because of velocity aberration. We are familiar with aberration of starlight caused by the tangential velocity of the observer vis à vis the line of sight to a star. As the Space Telescope orbits the Earth the stars move back and forth with respect to their true directions. This is not a problem for directions near the poles of the Space Telescope's orbit because the guidestar and object to be observed are

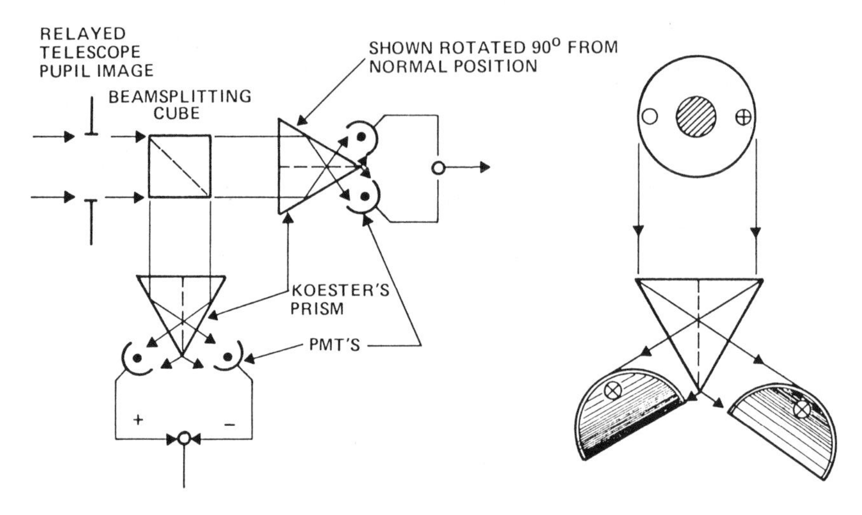

Figure 10 The Koester's prism interferometer that provides the guiding signal is shown in schematic form. As a star moves in angle, the intensity pattern seen by the photomultipliers changes rapidly.

moved by the same amount. Near the plane of Space Telescope's orbit the line-of-sight dependence of the aberration is such that the guidestar moves with respect to the object. This means that small adjustments of relative position must continuously be fed in. Fortunately, the magnitude of this correction can be calculated very accurately, but the system must be able to react with enough sensitivity. These considerations are then compounded when one looks at solar system objects, where the source is moving because of the orbital motion of it and the Earth and also because of the parallactic motion. The requirement has been set that Space Telescope should be able to track Halley's Comet at its most rapid angular rate during the 1986 apparition (0.21 arcseconds/second of time).

All of these position signals must be converted into telescope pointing changes. Two basic approaches are available: reaction wheels (the wheel turns faster in one direction, the Space Telescope turns in the other) and small rockets. Because of the sensitivity to possible contaminants, only reaction wheels or their functional equivalents could be considered.

4.3 *Controlling the Observatory*

Left to its own, the Space Telescope is not a very intelligent machine. It knows how to protect itself if a serious malfunction occurs, and it can decide when to drop into a hibernating condition while waiting for the

Shuttle to arrive. Aside from these few protective functions, all other commands must come from an operations control center. Similarly, all scientific and engineering information about the observatory must be transmitted to a ground station and nothing is retrieved by the Shuttle. These constraints demand a highly reliable communications system.

Satellites in low Earth orbits have usually been controlled through transmissions to and from Ground Station Tracking and Data Network stations as they passed within view. NASA has decided that this system of ground stations will be eliminated as the Tracking and Data Relay Satellite System (TDRSS) becomes operational, prior to Space Telescope launch. The TDRSS is composed of two highly versatile relay satellites located above the Atlantic and Pacific Oceans in geosynchronous orbits, together with their communicating ground stations. Each satellite can handle simultaneously 20 channels of Space Telescope–type data at 50 kps and numerous other channels. This capacity would seem adequate, but since there will be far more potential users of the high data rate channels than there are channels available, the Space Telescope must be able to operate without a continuous high rate of communications. This requirement can only be met if there are on-board tape recorders to store scientific data as it is acquired and if there is the on-board storage capability for spacecraft commands. This requirement is even more important because the Space Telescope can only see at least one of the relay satellites about 85 percent of the time. The other constraint is placed on the Space Telescope antenna system by the high data rate channel. To use this channel requires either a very large, fixed omnidirectional antenna or a high gain steerable antenna.

4.4 *Scientific Instruments*

The approach to the scientific instruments for the Space Telescope has been one of the most difficult issues to resolve. Serious compromises have had to be made, since no acceptable simple approach was ever found.

The first major issue was the nature of the scientific instruments. One view was that there should be one large scientific instrument at the focus. This instrument would share collimators between different dispersion spectrographs. It would have multiple relay systems for providing different plate scales to detectors for imaging, and do photometry of the sources as a useful side function. The alternative was a set of separate scientific instruments, each being independent. The super scientific instrument approach made reasonably efficient use of the restricted field of view of good quality images and could be managed by one organization. The individual scientific instruments could each be more efficient for specific

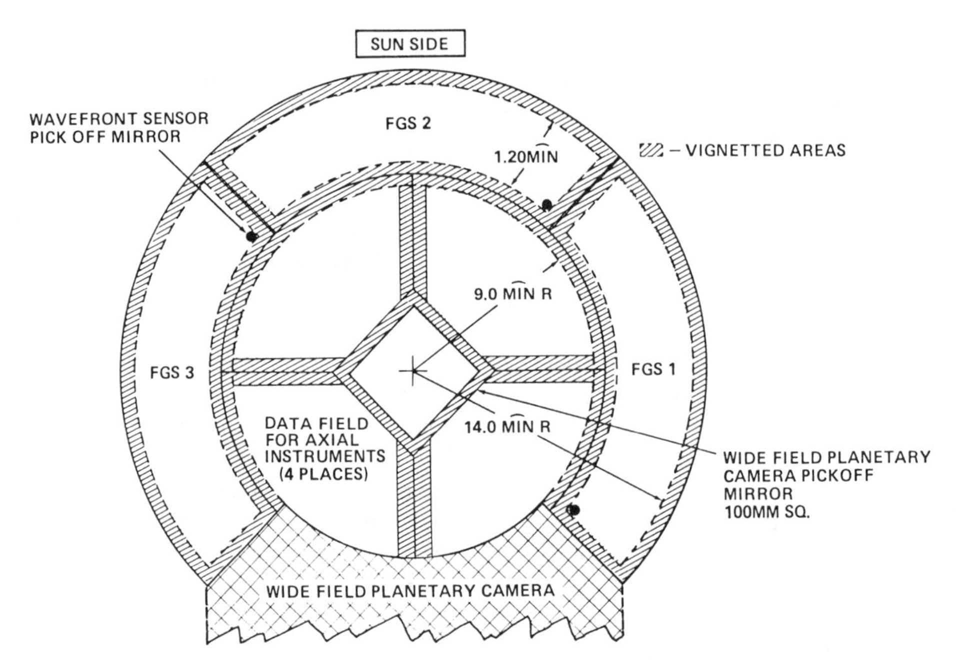

Figure 11 Focal plane map. The focal plane (actually a curved surface) is divided into dedicated segments for use by the scientific instruments and the Fine Guidance Sensors (FGS).

applications, but packaging could make it difficult to obtain access to the focus. This limitation could be eliminated if a moveable turntable could be used, but this idea was dropped as too costly because of the very high precision that was demanded. The super scientific instrument approach was dropped because of its inefficiencies and its lack of flexibility for modification as improved detectors become available and research emphasis changes. The final decision was to divide the focal surface into eight areas (Figure 11). The central region is diverted by a flat mirror to be perpendicular to the optical axis (the radial scientific instrument location). This region is surrounded by four on-axis sections feeding separate locations (the axial scientific instrument locations). The outer, astigmatic images are picked up by diverting mirrors and fed to the three Fine Guidance System locations. All locations are well defined, both in the focal surface and the volume available for the scientific instruments. This approach allows the scientific instruments to be developed, operated, and replaced separately, an important feature in a long-life observatory.

4.5 *Straylight Control*

Maximum use of a telescope in low Earth orbit demands the ability to make observations even when the instrument is illuminated by earthlight, moonlight, and sunlight. This goal can be reached, but demands more attention to straylight control than does a ground-based telescope. The ideal level of suppression is such that the effects of straylight are small compared with the light from the natural environment.

The naturally occurring background radiation is primarily from the zodiacal light, which is sunlight scattered from interplanetary particles. Since the particles are concentrated to the ecliptic plane, the brightest band is there, increasing monotonically towards the Sun. The minimum brightness occurs slightly away from the ecliptic zenith and is the equivalent of the light from a 23rd magnitude star over every square arcsecond (23^m arcsec^{-2}). The design requirement for the Space Telescope is that the effects of straylight never exceed this value.

Sunlight and moonlight are relatively easy to calculate and control, since they are compact sources producing nearly parallel light bundles. Earthlight is more difficult to control since it is both high intensity and comes from over a wide range of angles. All of these must be brought down to the level of the darkest part of the zodiacal light. Since this can never be done when looking close to a very bright source, we established the requirement that 23^m arcsec^{-2} or better must be reached at 50° from the Sun, 15° from the moon, and 70° from the bright limb of the Earth.

There are three basic approaches to straylight control: masking, diverting, and absorbing. Masking means the blocking of lines of sight by opaque surfaces. This is the technique used when installing an outer tube around a telescope and when using a baffle around the primary and secondary mirrors. Directing means controlling where the light goes. This technique recognizes that it is never possible to keep all light out, so it creates a design where the reflections are well defined and do not reach critical parts of the telescope. The key element here is the degree of specular reflection. In general, the surfaces with the lowest total reflectances have very diffuse reflection patterns, so that important amounts of light go in all directions. There are some surfaces already developed that have a low total reflectance and almost all of the light is reflected specularly. Absorbing means the use of very low reflectance surfaces which essentially "eat" the photons on the spot. Reflectances of less than one percent are now possible.

4.6 *Maintenance and Refurbishment*

Maintenance and refurbishment are what make the Space Telescope a true observatory and not just a project. The distinction between maintenance and refurbishment is that the former refers to alterations and repairs made while in orbit and the latter refers to activities carried out on the ground following a return to Earth.

The principal determinant of maintenance and refurbishment policy has been the availability of the Space Shuttle. If the Shuttle were available upon demand and at no cost, the Space Telescope instruments would be designed using low cost, high risk components, because they could be easily serviced. If the Shuttle use must be paid for on a case-by-case basis, then a balance must be struck between development costs and visiting costs. The original use plan (mission model) for the Shuttle indicated that there would be so much demand that a Shuttle would not be available frequently, regardless of the cost. As NASA has failed to develop the payloads for utilization of the Shuttle, the availability has gone up. Unfortunately, official recognition of this availability came too late for us to prepare a design using the Shuttle to the maximum extent.

The availability of the Shuttle was a major concern in the long range planning of the focal surface sharing and the detectors. The issue came down to one of whether or not film would be used on the Space Telescope. Film has the great advantage of existing with very large numbers of picture elements and could have used essentially the full field of view of diffraction-limited images. The low quantum efficiency could be circumvented by image intensifiers. The obvious operational problem is the delay between exposure and ground-based development of the images. The critical issue was the fogging of film caused by energetic particle bombardment of the spacecraft. Even modestly sensitive film would be fogged in about two months. If the Shuttle were available at low cost on a monthly basis, use of film could be considered; if not, we were forced to go to remote detectors (which may be better anyway). It was decided that the Shuttle would be too busy to visit Space Telescope monthly, and, moreover, it would be prohibitively expensive to do this. Thus it was decided not to use film.

The Space Telescope is designed so that it can be deployed and retrieved by the Shuttle. This means that the system must be closed to prevent contamination by the small maneuvering rockets on the Shuttle. Moreover, it means that all appendages must be stowable during launch and must be collapsible or jettisonable prior to retrieval.

Maintenance planning is carried out on two levels. The first involves balancing cost and risk in component development. It may be cheaper to pay for quality than to plan for maintenance. The second is the question of how to make an item either serviceable or replaceable in orbit. The use of astronauts imposes safety constraints and they can only handle large components in the environment of a pressurized space suit.

Refurbishment is something we must plan on (nominally every five years) but we can hope that it is never really necessary. There are certain tasks, such as cleaning and recoating the primary mirror, which we do not foresee as being practical in orbit. This means that return to Earth is necessary at times. However, the contamination and misalignment during an Earth return must be minimized so that the refurbishment time is not more than one year.

5. THE PLANNED SPACECRAFT SYSTEM

The previous section described the factors that constrained the Space Telescope design. In this section we discuss the design actually being developed. In some cases the hardware already exists, in others we are still finalizing the design. Like any satellite, the Space Telescope is designed as a whole, so no part of what is described here should change very much between now and launch.

Understanding the Space Telescope system demands an understanding of how the government develops new systems and procures major hardware. Over the years there has been acrimonious discussion of what is the correct balance of in-house (government) work and that done out-of-house by contractors. The proponents of heavy in-house involvement argue that it is less costly because there is no profit to be paid, while the out-of-house proponents argue that the greater efficiency of specialists more than compensates for profit and that a smaller civil service is a better civil service. The balance struck is where there is a small body of study people working for the government who are supplemented during the development phase with contracting specialists as necessary. There is a body of engineers, scientists, and managers who direct and supervise the contracted activity. Ideally, the best of these are as good as the best of the contractors. The contractors outnumber the civil servants about ten to one. The challenge is to orchestrate this activity so that the various parts of the work are done at the right time.

Within the government there is a hierarchy of control. At the top is the administration, which deals with budget and policy. Then there is the Program Office, which represents the project to the administration of NASA and to outside groups. The field centers have specific responsibility for the project and a Project Office is set up for this purpose. In the case of the Space Telescope there are four major components, the Optical

Telescope Assembly, the System Support Module (the spacecraft proper), the scientific instruments, and operations. The Marshall Space Flight Center has overall responsibility for developing the Space Telescope, and the Optical Telescope Assembly and the System Support Module, in particular. During the development phase, the Goddard Space Flight Center supports Marshall Space Flight Center by assuming responsibility for the scientific instruments and operations.

There are literally hundreds of major contracts on Space Telescope, but two dominate. These are with Lockheed Missiles and Space Corporation in Sunnyvale, California, for the Support System Module, and with the Perkin-Elmer Corporation of Danbury, Connecticut, for the Optical Telescope Assembly. It is worth describing how these contractors were selected, as this process defines how the project is managed and is characteristic of all large procurements.

The government first issues a Request for Proposal, a document that defines what is to be built in terms of performance requirements and interfaces. The Request for Proposal does not tell the bidders how to do their job. It just tells them what the product should do and what conditions it must meet in terms of interacting with other hardware or operations systems. The competitive bids are evaluated by a highly controlled system known as a Source Evaluation Board. This board is supported by committees of engineers and managers who provide factual interpretation and evaluation. The recommendations of the Source Evaluation Board then go to the selection official, usually a Center Director or the Administrator, depending on the magnitude and scope of the procurement, who actually makes the choice of the contractor. More accurately, choice is made of a contractor with whom to start holding negotiations for a specific contract. It is this negotiated contract that forms the basis of all work. Any alterations or additions are subject to renegotiation of cost and even the unchanged work contracts are subject to cost changes. Most contracts are on a cost-plus-fee basis. Based on an expected resource requirement, the work is begun. If it takes less manpower and/or material, certain credits are given. The far more frequent case is that when it costs more the bills have to be paid. However, there are two principal restraints on such runaway costs for well-defined work. First, governmental monitors and evaluators must have the experience and judgment to recognize the true requirements and distinguish them from deliberate understatements made in competition for the job. The second check is the fee system: a certain fraction of the contract, typically 5 to 10 percent, can be paid to the contractor based on government evaluation of performance. That 10 percent supplement can mean a doubling of the total profit, and so provides a strong incentive to moderate demands and to stick to the original plan. All of this means paperwork, committees, reviews, and visits. It can be a bewildering world of telephone calls, technical people speaking in acronyms instead of language, bean counters, and isolated examples of technical and managerial brilliance. The system works; it is the job that NASA does better than probably any other organization in the world.

5.1 *The Optical Train and Its Support*

The optical characteristics of the Ritchey-Chrétien system adopted for Space Telescope are given in Table 1. This design represents the best compromise between our specific constraints and needs. The primary mirror is of egg-crate construction with a thin front and back face that was thermally slumped to the correct figure by heating while it was lying on a male mandrel. Composed of ultra-low expansion glass and manufactured by the Corning Company, this design represents the best of modern mirror blank technology (Figure 12).

Because of the critical nature of this work, we are actually creating two mirrors. The prime contractor is Perkin-Elmer, the other is the Eastman-Kodak Company of Rochester, New York. They are using very different

Figure 12 The primary mirror blank of the Space Telescope is built of thin sections of ultra-low expansion glass fused together to form a lightweight structure.

Table 1 Optical system characteristics

Aperture: 2.4 m
Focal ratio: 24
Obscuration area: 14%
Mirror spacing: 4.9 m
Front of primary to focus: 1.5 m
Coating: Al + Mg F_2

polishing techniques. Eastman-Kodak is using a Draper machine, with traditional large tools. Perkin-Elmer is using a computer-controlled polishing machine with very small tools. In each case the results are tested interferometrically. At the time this manuscript was submitted, both mirrors were better than $\lambda/50$, rapidly approaching flight quality.

The graphite-epoxy truss structure has already been delivered to Perkin-Elmer. The thermal stability tests performed indicate that our positional goal of 1-micron constancy will be reached (Figure 13).

The optical system will move about slightly during the launch of the Space Telescope. Because of the stringent alignment needs, we have de-

Figure 13 The graphite-epoxy truss structure that holds the primary and secondary mirrors in alignment is shown immediately prior to shipment.

signed the secondary mirror to be adjustable about six degrees of freedom. The alignment is sensed by three Optical Control Sensors which operate as shearing interferometers on the image of bright stars to give a signal to the ground where the adjustment commands are generated and sent to Space Telescope. The most frequent test will be for best focus, also performed by the Optical Control Subsystem, which should not be necessary more than every few days.

Every possible effort is being made to be sure that the Space Telescope primary mirrors are free of gravitational effects in manufacture; however, just in case this is not happening, a set of 24 actuators will be mounted to the back of the primary to distort its figure to the desired shape once it is in orbit. The figure of the mirror will again be determined by the Optical Control Subsystem and the actual analysis and commanding done from the ground. This is not a system that we expect to use, but is there as insurance.

5.2 *Fine Guidance System*

The Fine Guidance System utilizes the astigmatic images at the edge of the field of view to provide the guidance signal for maintaining stability to 0.007 arcseconds (rms). Perkin-Elmer is developing three identical units. The number of units is essentially determined by the need to have two units operating at all times (one to provide the x,y signal and the other to give a roll orientation signal) yet have an acceptable level of reliability. The field of view of the Fine Guidance System is determined by the sensitivity of their units and the number of candidate guide stars. The performance requirement is that adequate guidestars be available for 85 percent of random fields located at the galactic poles.

Guide stars are selected beforehand. Usually these will be known in position on an absolute reference frame to about 1–2 arcseconds accuracy and about $\frac{1}{2}$–1 arcseconds with respect to the object to be observed. Often these positions will be available from existing photographic sky surveys. For wide field scientific instruments, like the Wide Field Camera and the Faint Object Camera, these positions are adequate. For narrow entrance aperture instruments, like the Faint Object Spectrograph and the High Resolution Spectrograph, it will sometimes be necessary to use the Wide Field Camera to determine precise positions immediately before the observation.

The Perkin-Elmer design is shown schematically in Figure 14. The field of view of the interferometer is scanned across the focal surface by means of rotating transfer prisms. The collimated beam is then fed to the Koester's prism which provides a set of interferometer fringes to the pairs of detecting photomultipliers. Those detectors do not scan fringes; they

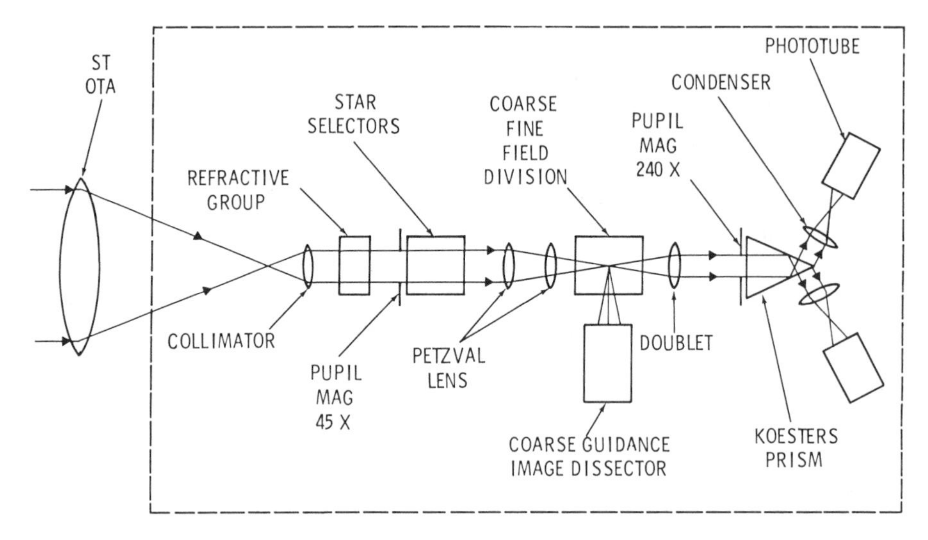

Figure 14 A schematic representation of the Fine Guidance Sensor is shown. The function is described in the text.

look at the total intensity signal, which varies as the star moves. The system sensitivity is such that it should meet our performance specification for stars brighter than 13.5 magnitude. This is the most complex, challenging system of the entire observatory.

5.3 *Pointing and Control System*

The pointing and control system keeps the Space Telescope from becoming lost and does the coarse pointing. The motion of Space Telescope is always controlled by reaction wheels. The control signals come from a variety of sources. The initial reference signal comes from a set of six rate gyroscopes, so called because their position is not so precise, but the rate of change of orientation can be accurately determined and used to calculate just where one is actually pointed. The positions of the gyroscopes are periodically updated by signals from the Fixed Head Star Trackers and the Fine Guidance System. At the beginning of an acquisition of a new object to be observed, the Fine Guidance System signal from the previous source is used to update the gyroscopes. Then the telescope is slewed towards the new position at a rate that reaches about 15 degrees/minute. The Fixed Head Star Tracker, which is a simple telescope plus photomultiplier system, then determines how close one came to the predicted position, using the signal from a few (3) bright stars. The combined signal from the gyroscopes and the Fixed Head Star Trackers will normally give a pointing accuracy of about one arcminute, well within the field of view of the Fine Guidance System detectors.

The Fine Guidance System detector, which has an entrance aperture of 2 arcseconds, then begins a spiral search for the guide stars (one each in two systems). Upon detecting a star, it measures its brightness and if it is within 0.3^m of that predicted, it stops and the second system continues until it has found a star of the correct brightness range. The final confirmation is made by the relative position of the two guide star candidates, which is a much more stringent test than brightness. If the positions do not match, then the search continues until all combinations within 3 arcminutes are exhausted, at which time it tries acquiring a backup candidate (if available) or tells the telescope that it is ready to go on to the next observation.

Narrow field of view scientific instruments require additional help, unless the relative positions of the science sources and the guide stars are known to about 0.01 arcseconds. Usually this will not be the case unless special astrometric observations have been made from the ground or unless the Space Telescope has observed the object before. Three choices then apply. One can use a signal from the scientific instrument to calculate on the ground where the telescope should be pointed; one can use a similar signal to tell the spacecraft where to point (without ground contact); or one can use the Wide Field Camera or the Faint Object Camera to assist in the final acquisition. The last mode means that the guide stars are acquired in the same Fine Guidance System units that will be used for the ultimate instrument (e.g. the Faint Object Spectrograph) except that the object being studied first falls onto a camera (e.g. the Wide Field Camera). A quick exposure is made and transmitted to the ground. The observer then determines exactly where he wants the Telescope to be pointed and sends a set of commands to the spacecraft, which automatically makes the final adjustments. Obviously this last approach requires good communication with the observing scientist and an adequate number of guide stars to allow such a double acquisition.

It has been said that this whole process is like threading a needle on the other side of the world. It is actually more analogous to say that it is like asking someone else to drive a car across 3,000 miles and park within six inches in a parking spot that you knew about already.

5.4 *Data Handling and Communications*

The Tracking and Data Relay Satellite System defines the constraints placed on our communications systems. Within these constraints we have designed the command, the engineering status, and the scientific data systems.

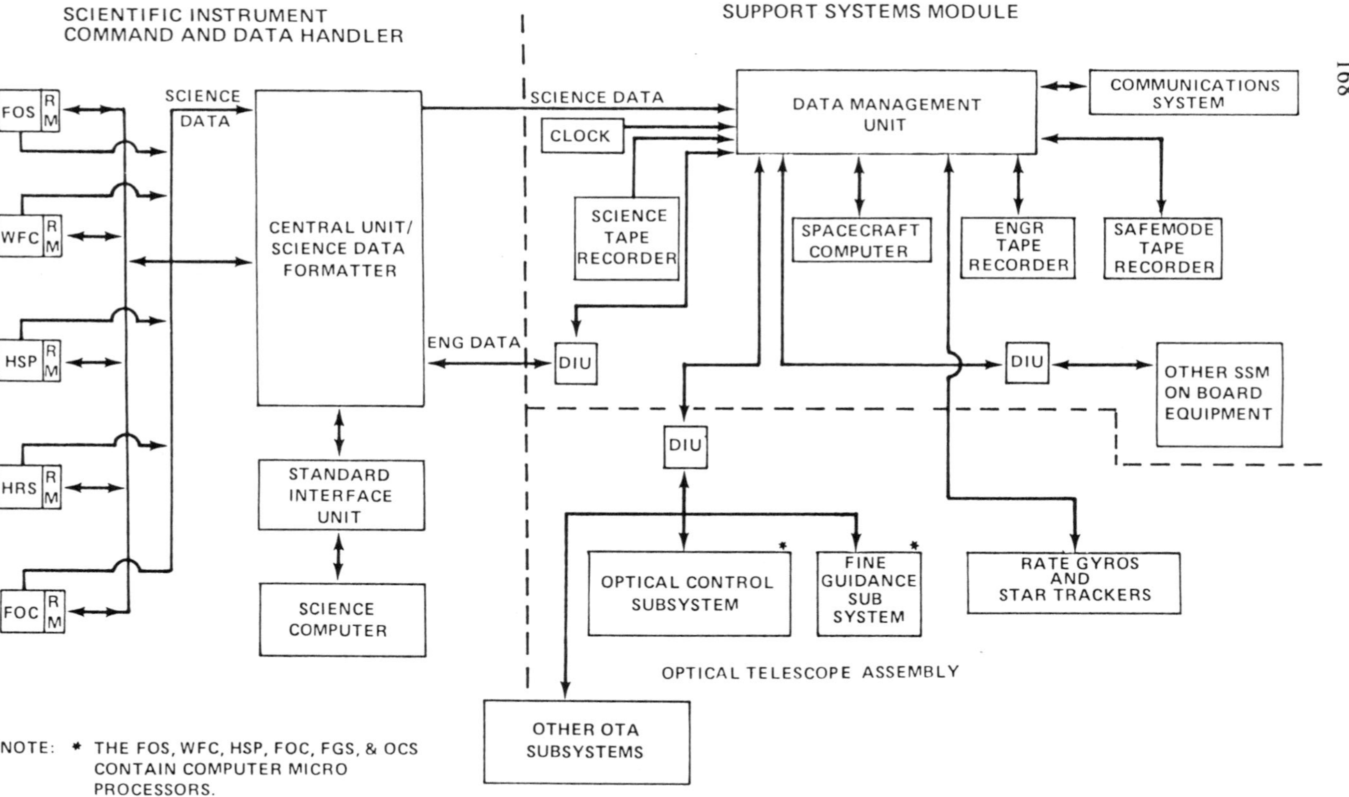

Figure 15 The on-board data management system for Space Telescope uses many components, even in this condensed representation. The acronyms at left are for the five Scientific Instruments; RM means remote module and DIU means Data Interface Unit.

The low data rate (4 kps) is always available through the low gain antennas. This allows communication with the Space Telescope regardless of the orientation and receipt of the basic engineering (health of spacecraft) and low rate science data, restricted only by the visibility factors of the Tracking and Data Relay Satellites. Such a capability is important in routine monitoring of the status of the Space Telescope and also when a malfunction occurs which causes the spacecraft to automatically go into one of its intentional "safe" modes.

The high data rate system is available to accept data from the tape recorders and to support real time involvement of the user with the telescope. All data from the scientific instruments come to the Scientific Instrument Command and Data Handling Computer, an NSSC-I computer system being developed for Space Telescope by International Business Machines Corporation. The data are then sent either directly to the transmitters or any of three tape recorders, all of which are part of the spacecraft system (Figure 15). Each tape recorder can handle 10^9 bits, so that multiple exposures from the Wide Field Camera can be stored (this instrument is the largest data producer by a wide margin). The existence of multiple units allows more data to be stored than we expect to generate between data dumps, so the primary advantage of having them is that there is functional redundancy as the inevitable failures occur.

5.5 *The Power System and Thermal Control*

The heart of the power system is the rollup solar array being developed by the European Space Agency for the Space Telescope. This array of solar cells will provide about 4700 watts average orbital power at deployment and at least 4000 watts after two years in operation. An energy storage system is necessary due to the day/night nature of Space Telescope's low Earth orbit. This is provided by a set of five batteries. By restricting discharge of each battery to not more than 20 percent, we can expect most of these batteries to be operational between our nominal $2\frac{1}{2}$ year maintenance cycles (they are replaceable on orbit). The power budget is distributed as shown in Figure 16. Most of the energy goes to the operation and control of the spacecraft and the scientific instruments. However, an important fraction is used for thermal control.

As the Space Telescope moves in and out of sunlight and earthlight on each Earth orbit, it encounters a dramatically varying radiation energy load which cannot be allowed to cause degradation in the Space Telescope performance. Electronic components operate correctly only over a certain design range of temperatures, lubricants on moving parts (gyroscopes, grating, and filter wheels, etc.) can become too viscous or too thin, and

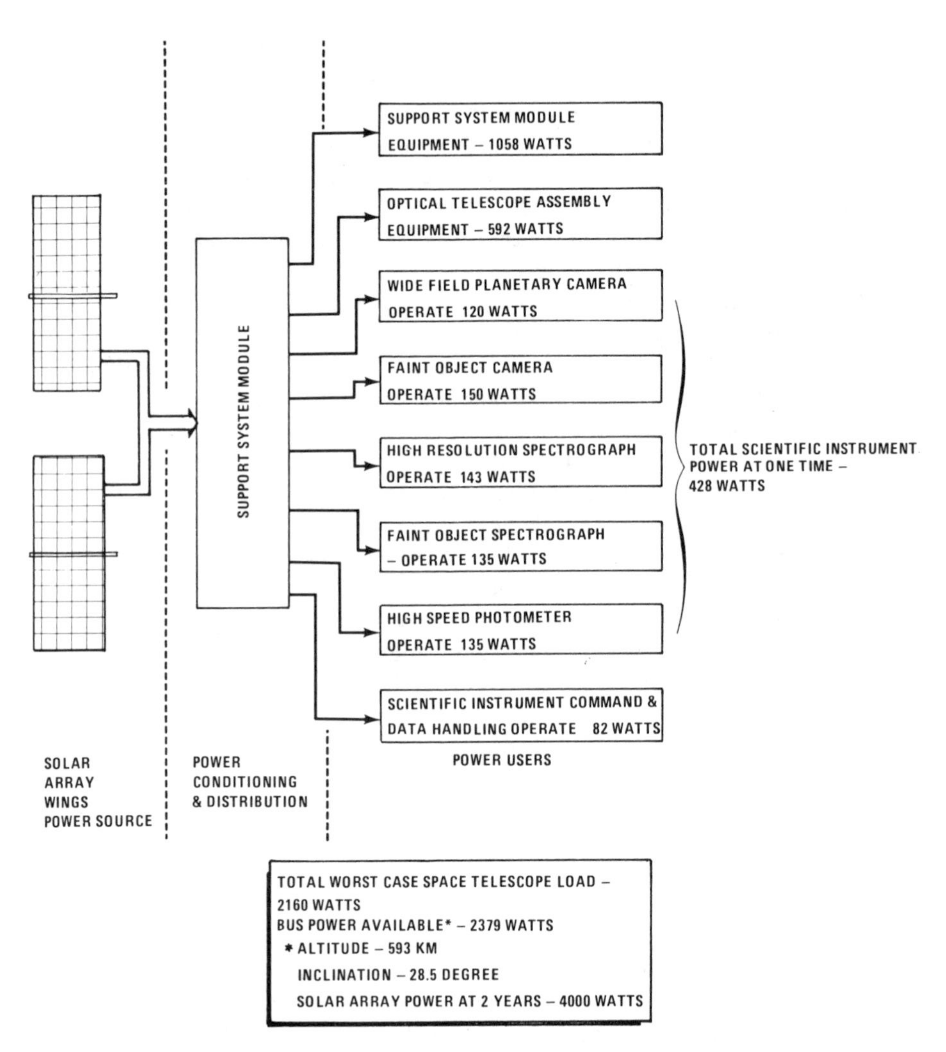

Figure 16 This Space Telescope power system diagram shows both the flow of electrical energy and the power budget allocation. The design of this portion of the spacecraft is closely related to the thermal design.

structures (including the mirrors) can distort. As much as possible thermal control of the Space Telescope is done by means of passive techniques and, in particular, by the use of appropriate amounts of insulation, taking into account the correct thermal time constants for components. A few critical components cannot be controlled passively and these are designed to have thermostatic heaters. Some of these are as follows: The gyroscopes

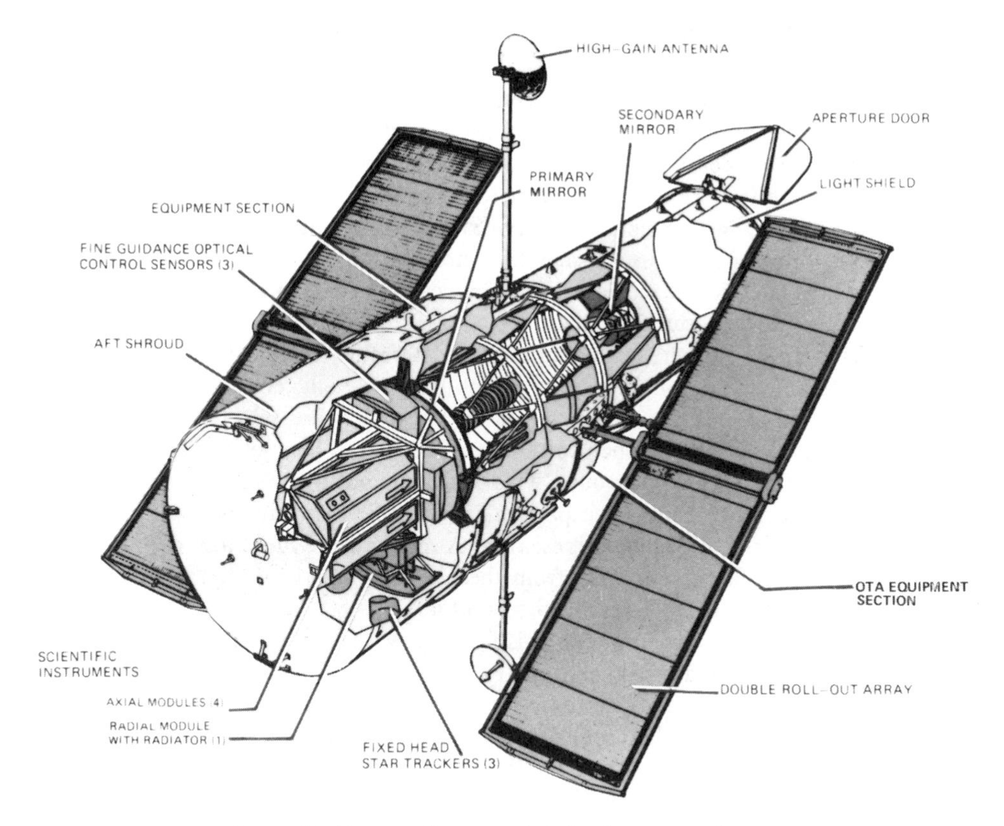

Figure 17 The overall spacecraft design in cutaway view.

demand a high and precisely controlled operating temperature at the rotor. The primary and secondary mirrors will be kept close to the optical shop temperature at which they were manufactured (20°C), which both minimizes figure distortion and keeps any contaminating gases (which are cold) from sticking on their surfaces. Finally, the graphite-epoxy mirror truss and the frame that holds the scientific instruments, the Optical Control Sensors, and the Fine Guidance Sensors will also be thermally controlled by thermostats and heaters. Much of this is probably not essential for routine observations, but does give us insurance in reaching the performance limits.

All of the considerations of this section lend to the integrated spacecraft design shown in Figure 17.

6. THE SCIENTIFIC INSTRUMENTS

Once the Space Telescope is in orbit the users and the results they obtain will be determined by the scientific instruments being used. My emphasis thus far has been on the telescope itself and how we get to that happy state where it works and can be relied on. Now I would like to consider the scientific instruments, the devices that turn the image formed by Space Telescope into useful scientific information.

It was mentioned earlier that many candidate scientific instruments were considered during Space Telescope's preliminary design period (Phase B). These were evaluated both to determine feasible designs and to determine what requirements the scientific instruments could place on the observatory. The scientists involved in Phase B were selected competitively in response to an open call for participation. The intent was to organize them into scientific groups. This was done, but the teams thus formed quickly began to emphasize particular scientific instruments that would best do their own particular research. Almost all the ideas and certainly all of the requirements came from these teams, but the design work was carried out by Space Telescope contractors.

The scientists involved with this activity were divided into six instrument definition teams for astrometry, data management, infrared observations, faint object spectroscopy, high resolution spectroscopy, and imaging optics. The data management team had no particular instrument in mind but it was already clear that data handling would have a major impact on all instruments. Each team worked hard to generate realistic but ambitious performance goals, and then helped the contractors to turn these ideas into designs. The two spectroscopy teams and the infrared team both came out with simple designs for one instrument. The imaging optics team first moved towards a multipurpose camera, then split the functions as it became obvious that the European Space Agency was interested in providing a camera. The astrometry group did produce a design for an astrometric measuring machine; however, when the limitations of the instrument were recognized and the potential of the Fine Guidance System for doing astrometry was quantified, they shifted their emphasis to optimizing the Fine Guidance System design.

The principal determinant of the designs was the nature of the detector. The well-established front runner was the SEC Orthicon which had been under development for several years at Princeton University. This TV-type detector was the best detector then available, although the problems associated with it of low linear resolution and limited dynamic range were well known. ESA felt that it had a better approach for its instrument, based on its own experience.

Each instrument definition team had its own priorities and emphasis and the total complement of possible instruments far exceeded the space, weight, and money available. For this reason it became necessary to make some form of selection within the whole set. The (Large) Space Telescope Management and Operations Working Group was the natural body to do this, involving as it did both the team leaders and other senior scientists.

The Working Group identified a set of three core scientific instruments, the definition of which was that they were the most important scientific instruments and that the Space Telescope would not be a full observatory without them. This set of core scientific instruments included a Wide Field Camera, a Faint Object Spectrograph, and a High Resolution Camera (presumed to be provided by the Europeans). The Working Group then recommended that the Space Telescope not be flown without these instruments, sidestepping the hypothetical situation where one of the instruments might fail immediately before launch when delay costs are extremely high.

All of these studies and recommendations were considered in structuring the hardware phase of the program. It was decided that entirely new science teams would be selected and that each science team would be responsible for its own scientific instrument. This was done through a solicitation for proposals called an Announcement of Opportunity, which delineated the opportunities and constraints. The respondents were expected to describe the science that would be done and the instrument that would be developed to do it.

NASA Headquarters, which controls the Announcement of Opportunity process, was faced with a dilemma over the strong recommendation for the set of three core scientific instruments. If three of the five instrument bays were already claimed by this core of instruments, only two additional instruments could be selected for the observatory (the ESA's Faint Object Camera was already guaranteed space on the flight). The problem was one of relative treatment (financially) of the other two instruments that might be selected. NASA decided that Wide Field Camera and Faint Object Spectrograph teams would selected if acceptable proposals were submitted and that the other bays would be filled on an open competition basis; however, after selection all were to be treated equally. Once the proposals were received, they were technically evaluated for feasibility, cost, and impact on the observatory design. The technical reports then became background information for the peer groups formed to evaluate the science of the proposals. The peer groups were organized into specialty areas, and then their recommendations were considered together by a subset of the reviewers who were free of conflicts of interest. It was very hard to form this last group as most of the major astronomical groups in the United States had submitted proposals.

Selection was made of four scientific instruments, of United States participants on the European Space Agency Faint Object Camera Science Team, and of an Astrometry Science Team. A Data and Operations Team

Table 2 Space Telescope Science Working Group scientists

Scientist	Institution	Responsibility
John N. Bahcall	Institute for Advanced Study	Interdisciplinary Scientist
Robert C. Bless	University of Wisconsin	Principal Investigator—High Speed Photometer
John C. Brandt	Goddard Space Flight Center	Principal Investigator—High Resolution Spectrograph
John J. Caldwell	State University of New York, Stony Brook	Interdisciplinary Scientist
William G. Fastie	The Johns Hopkins University	Telescope Scientist
Edward J. Groth	Princeton University	Data and Operations Team Leader
Richard J. Harms	University of California, San Diego	Principal Investigator—Faint Object Spectrograph
William H. Jefferys	University of Texas, Austin	Astrometry Science Team Leader
David L. Lambert	University of Texas, Austin	Interdisciplinary Scientist
David S. Leckrone	Goddard Space Flight Center	Scientific Instruments Scientist
Malcolm S. Longair	The Royal Observatory, Edinburgh	Interdisciplinary Scientist
F. Duccio Macchetto	European Space Agency, Noordwijk, The Netherlands	Faint Object Camera Project Scientist
C. R. O'Dell	Marshall Space Flight Center	Project Scientist
Nancy G. Roman	NASA Headquarters	Program Scientist (retired)
Daniel J. Schroeder	Beloit College, Beloit, Wisconsin	Telescope Scientist
H. C. van de Hulst	Huygens Laboratory, Leiden, The Netherlands	Faint Object Camera Science Team Leader
Robert W. Hobbs	Goddard Space Flight Center	Data and Operations Scientist
Edward J. Weiler	NASA Headquarters	Program Scientist
James A. Westphal	California Institute of Technology	Principal Investigator—Wide Field Camera

leader was also selected, to head a team composed of representatives of each science team (Table 2). The instruments chosen were the Wide Field Camera, the Faint Object Spectrograph, the High Resolution Spectrograph, and the High Speed Photometer (Leckrone 1980).

To the surprise of many none of the Phase B team leaders were selected except for the Data Management Team leader who became principal investigator for the High Speed Photometer, which was not studied in Phase B. Moreover, none of the detectors studied in Phase B was selected to be in actual flight! There is something to be learned from this, for the quality of scientists and engineers was no less in the Phase B selection than it was in the final selection. The long interval between the two calls for participation was the important factor. The Phase B proposers saw a solicitation in 1973 and responded with ideas that existed then, especially for detectors. As the Phase B activity was prolonged, they were unable to stay abreast of the newest developments, or at least they were unable to drop commitments to old ones. This meant that the newest technology and the freshest ideas were selected (in 1977) and it was hard to be part of that. If the Phase B activity had been as short as was originally planned, things might have been very different.

6.1 *The Wide Field Camera*

The Wide Field Camera, which is sometimes called the Wide Field/Planetary Camera, is being developed by James Westphal at the California Institute of Technology. It will provide a 1600 × 1600 pixel view of the center of the Space Telescope focal surface at both the $f/12.9$ (2.67 × 2.67 arcminutes) and $f/30$ (68.7 × 68.7 arcseconds) focal ratios. The detectors are 800 × 800 element charge-coupled devices, which are thermoelectrically cooled to about −95°C. The instrument is mounted in the radial scientific instrument bay and it has an external radiator which extends beyond the spacecraft wall (Figure 18).

The field at both focal ratios is fed by a rotating pyramid which also divides the field into four equal sections, one for each detector. Each detector at both focal ratios has its own set of relay optics; therefore, there are eight optics-detector chains (Figure 19). $f/12.9$ intentionally undersamples the stellar image, each element of the detector being 0.10 arcseconds, but gains in the field of view. $\phi/30$ will reach fainter objects and produce better resolution, over a smaller field of view.

The heart of the instrument is a charge-coupled device detector. These devices have very low thermal and readout noise and extended wavelength coverage. The red end response is well beyond 1 micron and is determined by the natural response of silicon sensors. The ultraviolet response is extended by coating the front of the detectors with Coronene, a transparent

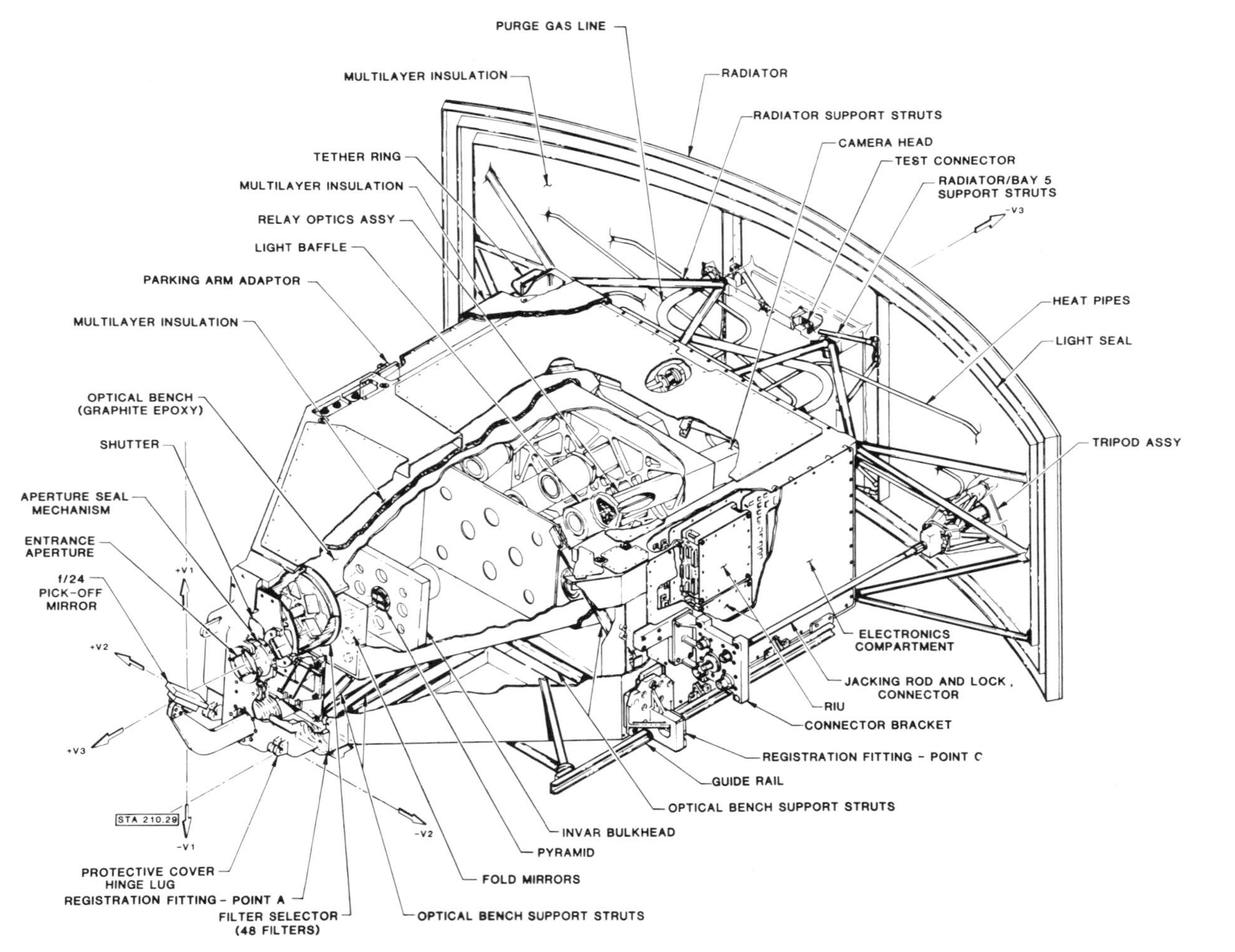

Figure 18 The Wide Field Camera in cutaway view. The arrows refer to the spacecraft reference axes, $+V1$ being forward (out the tube) and $-V3$ being the cold side of the spacecraft.

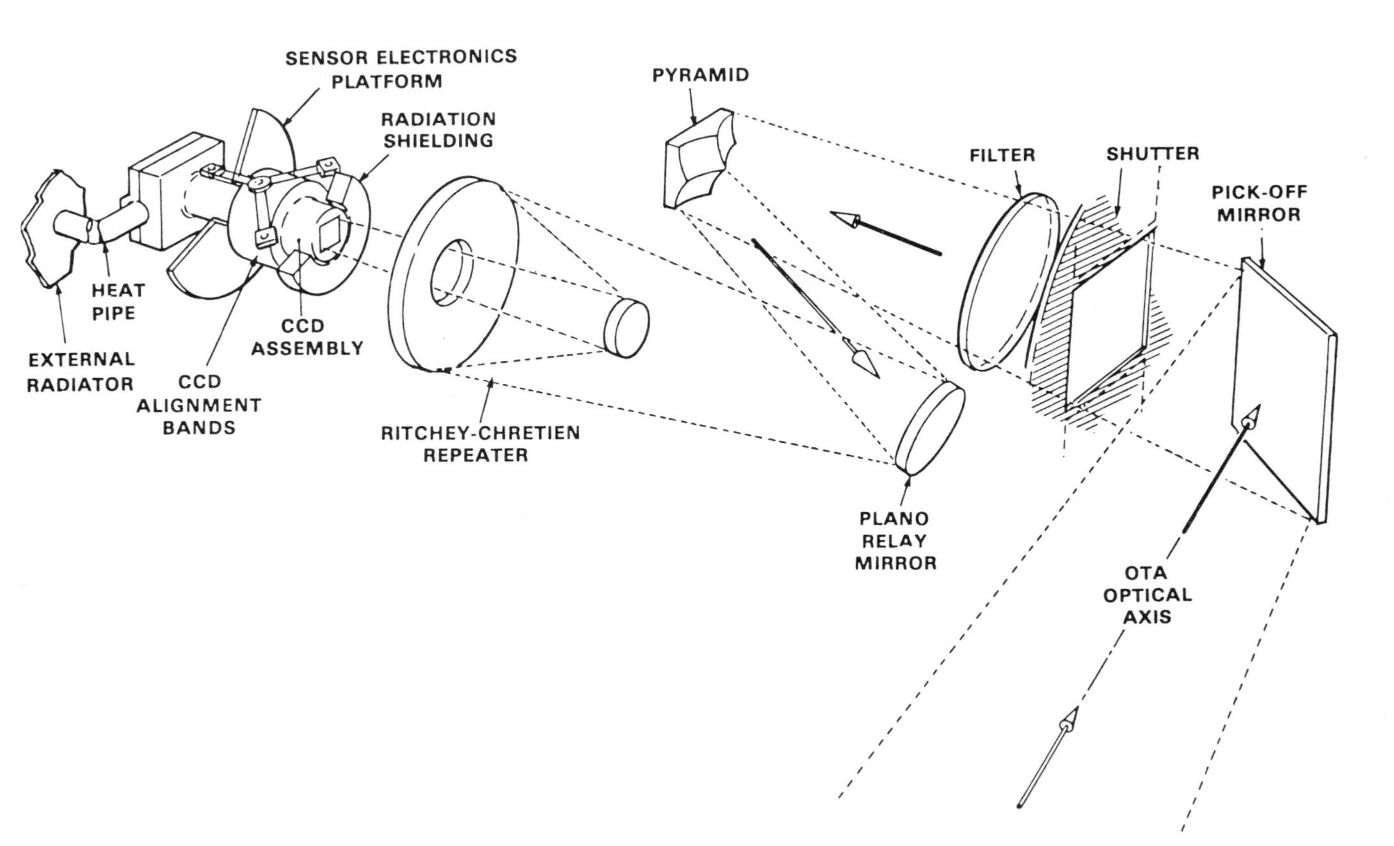

Figure 19 This schematic view of the Wide Field Camera demonstrates the division of the allocated field and how it is imaged onto the charge-coupled device (CCD) detectors.

material that efficiently converts high energy photons to visible light, which the silicon can detect.

The Wide Field Camera is a powerful scientific instrument, which can be operated while other instruments are also in use. It will be able to perform objective prism spectroscopy and polarimetry, in addition to filtered imaging.

6.2 *The Faint Object Camera*

The Faint Object Camera, which is being provided by the European Space Agency, is probably the most versatile instrument on the Space Telescope (Macchetto et al. 1980). Its primary roles are imaging narrow fields at long focal ratios, $f/48$ (22×22 arcseconds) and $f/96$ (11×11 arcseconds). As special features, it can also image very narrow fields at $f/288$ and perform two-dimensional slit spectroscopy, objective prism spectroscopy, and polarimetry (Figure 20).

The detector is a three-stage image intensifier feeding an electron-bombarded-silicon TV tube. Unlike the charge-coupled device detector of the Wide Field Camera, this detector counts photons one at a time. Each photoelectron leaving the initial photocathode becomes a large burst of photons reaching the TV sensor. The location of each burst is determined, one by one, and recorded, and the image is built up bit by bit. This approach means that a maximum photon rate is quickly reached although its ability to measure single photon events is optimum. The initial photocathode is a bialkali, which is sensitive from the telescope's ultraviolet cutoff to about 6000 Å.

The longer focal ratios of the Faint Object Camera mean that each image formed by the telescope is sampled by several pixels. This allows images to be obtained at the highest possible resolution and with the maximum contrast against the background sky. This makes this instrument the one that will be used whenever the highest resolution is necessary. The $f/288$ mode will allow a coronagraphic search around bright stars for faint companions (planets?) and resolution of the ultraviolet images into speckles. The potential for speckle imaging is very great, for the theoretical diffraction limit in the ultraviolet is much smaller than that in the visual and the time rate of change of the speckles should be very slow, so that the technique can be applied to even faint sources.

Due to its greater scale, the Faint Object Camera will be able to reach fainter sources than the Wide Field Camera. However, past about 5500 Å the charge-coupled device detectors of the Wide Field Camera are much more sensitive than the bialkali detectors of the Faint Object Camera. These two instruments are complementary; there is no real redundancy of

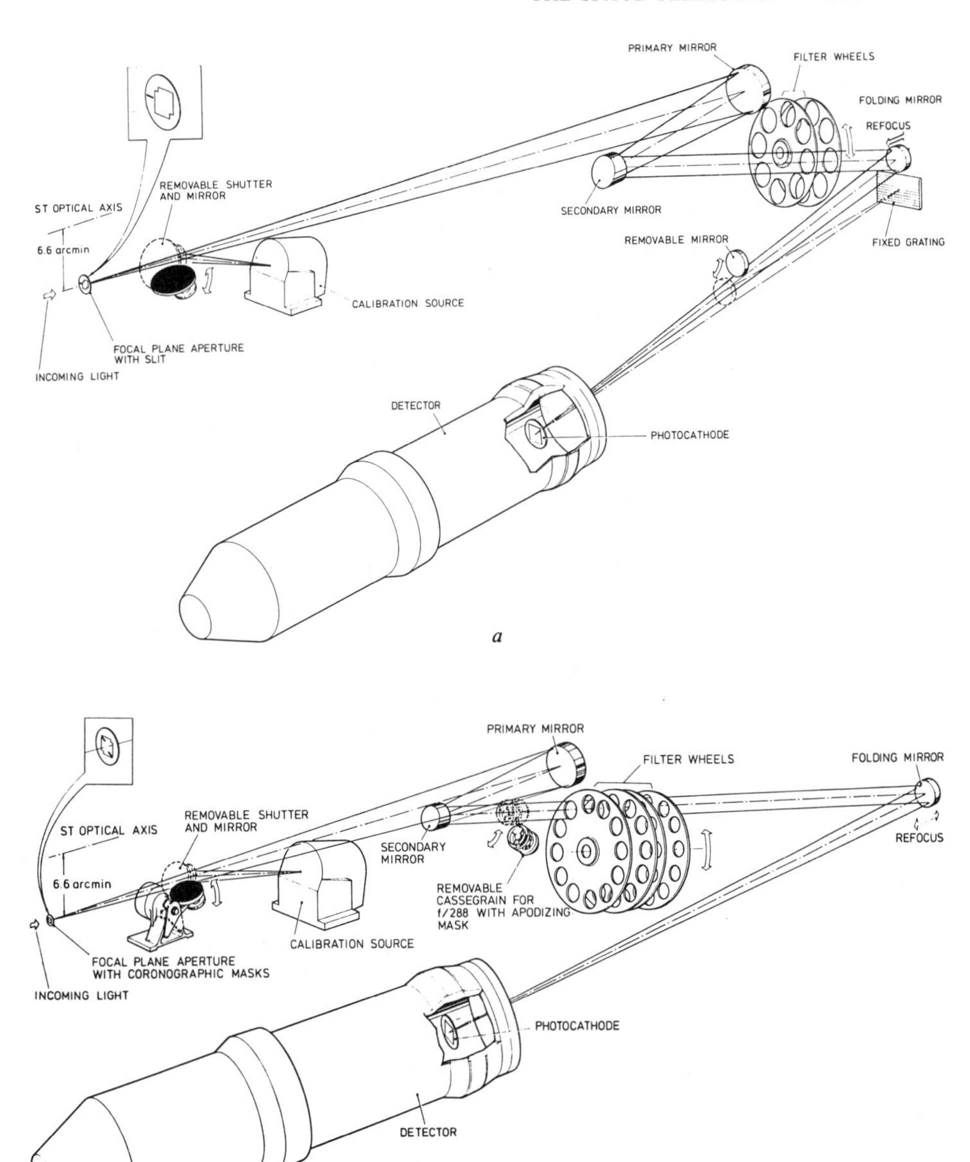

Figure 20 (*a*) This schematic of the Faint Object Camera shows the portion giving an *f*/48 focal ratio, with its optional features. (*b*) This schematic shows the separate *f*/96 system and its options.

capabilities, but they come close enough to give alternative choices if one of the instruments should fail entirely.

6.3 *The Faint Object Spectrograph*

The Faint Object Spectrograph will provide spectral resolutions of about 10^2 and 10^3 on objects as small as the image given by the Space Telescope. This means that very faint sources and compact sources within bright, extended sources can be measured very well. The optical design was selected for its simplicity and efficiency (Figure 21). The linear polarization analyzer can be inserted into the light path to permit spectropolarimetry; otherwise it is a very simple instrument. The most unusual feature of the design is the use of a grazing-incidence mirror which deflects the diverging light beam after it passes through the entrance aperture. This feature allows the entrance apertures to be close to Space Telescope's crowded optical axis, but at the same time to have large gratings and detectors (Figure 22).

The two detectors are both 512-channel digicons, differing only in their wavelength response. These are arrays of 512 silicon detectors that are bombarded with energy-intensified electrons from either trialkali or bialkali photocathodes. This arrangement gives a wavelength response from the reflective cutoff of the Space Telescope to about 7000 Å. Sky signal subtraction is done by deflection of the electron beam from above and below the object spectrum. This will be a highly sensitive system, similar to some already in operation on the ground, and it will be one of the most frequently scheduled instruments on the Space Telescope. It is

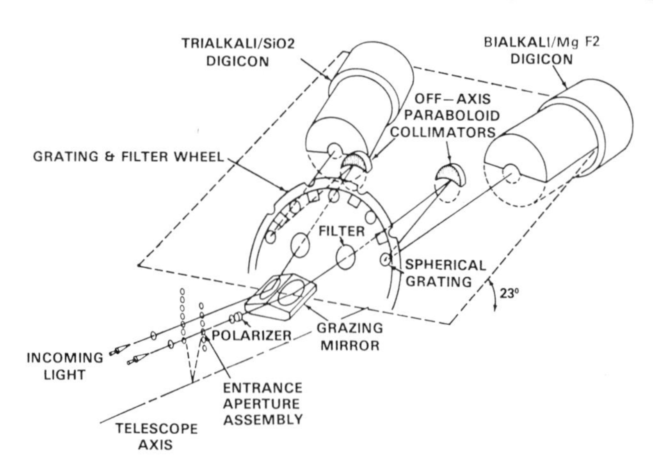

Figure 21 This schematic of the Faint Object Spectrograph shows the basic optical layout.

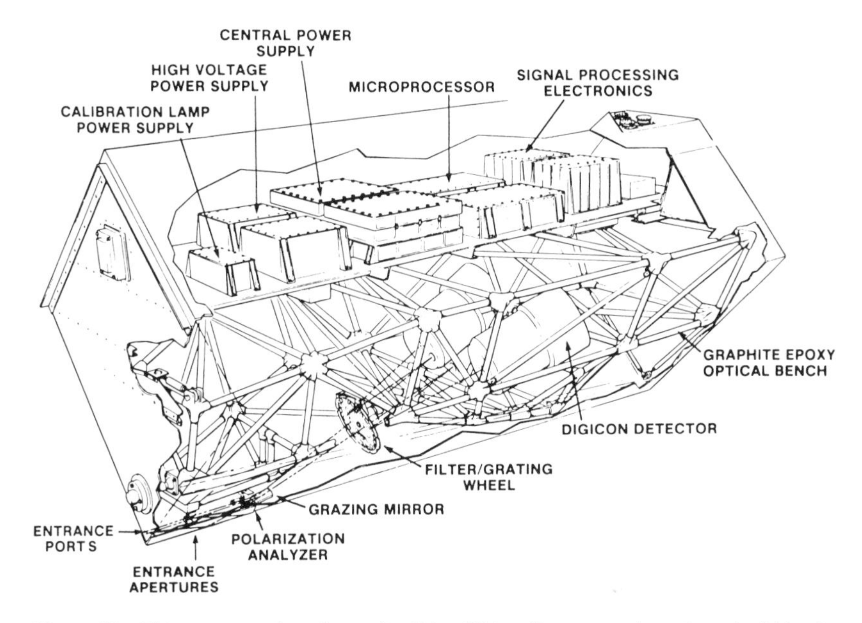

Figure 22 This cutaway view shows the Faint Object Spectrograph packaged within the standard axial scientific instrument box.

being developed at the University of California–San Diego by Richard J. Harms and his team.

6.4 *The High Resolution Spectrograph*

The High Resolution Spectrograph has a role very different from that of the Faint Object Spectrograph in that its goal is to provide uniquely high resolution spectroscopy in the ultraviolet. It has three resolution ranges: 1.2×10^5, 2×10^4, and 2×10^3. The longest wavelengths detected will be about 3200 Å and the shortest will be well below Lyα, as one of its detectors will have a LiF faceplate, allowing full use of the light transmitted by the MgF-overcoated aluminum optics.

The grating turret contains both first-order gratings for the lower resolutions and an echelle grating for the highest resolution. The detectors are again 512-channel Digicons, with slightly longer (perpendicular to the dispersion) elements than those in the Faint Object Spectrograph. The low-order spectra extend well beyond the Digicon face, so that the gratings must be tilted to select the position to be observed. The echelle grating has shorter but multiple orders, which are separated by a cross-dispersion grating (Figure 23). The echelle order selection is done by

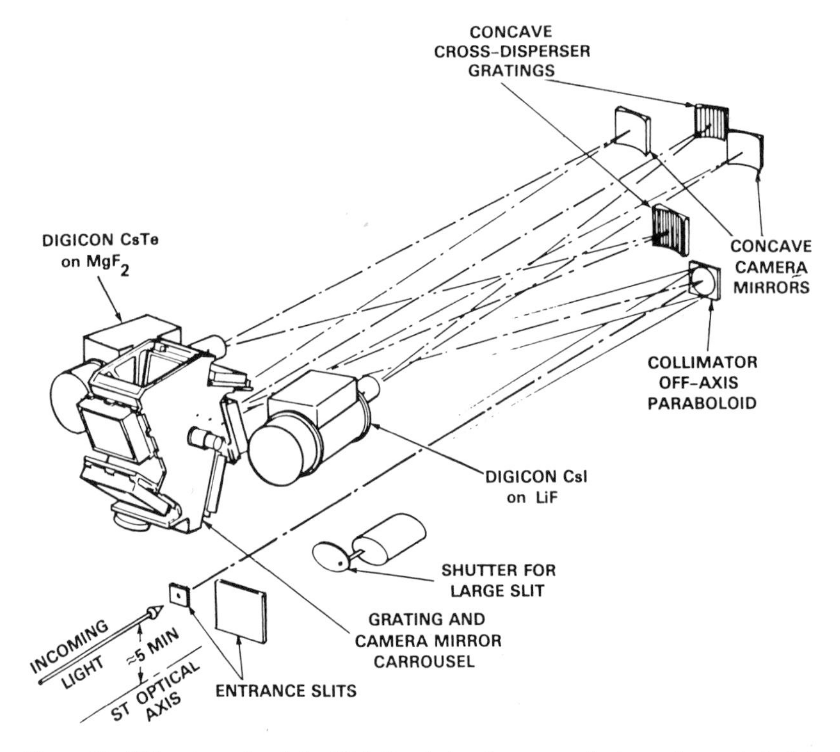

Figure 23 This schematic of the High Resolution Spectrograph shows the method of designing a versatile high resolution spectrograph into a minimum space with the greatest possible efficiency.

magnetic deflection within the Digicon. A mirror-only optical path will permit the imaging of a narrow field of view onto the Digicon as an aid to the acquisition of the source for narrow entrance aperture use. The High Resolution Spectrograph is being developed at the Goddard Space Flight Center by John C. Brandt and his team.

6.5 *The High Speed Photometer*

The High Speed Photometer is the surprise instrument of the hardware phase selection. Originally proposed as a piggyback add-on to another (unspecified) instrument by R. C. Bless of the University of Wisconsin, it has been given its own modularized instrument bay. It is conceptually simple, since it has no moving mechanical parts. There are four image dissectors and one photomultiplier (Figure 24). Each image dissector photocathode is preceded by an aperture mask (to narrow the field of

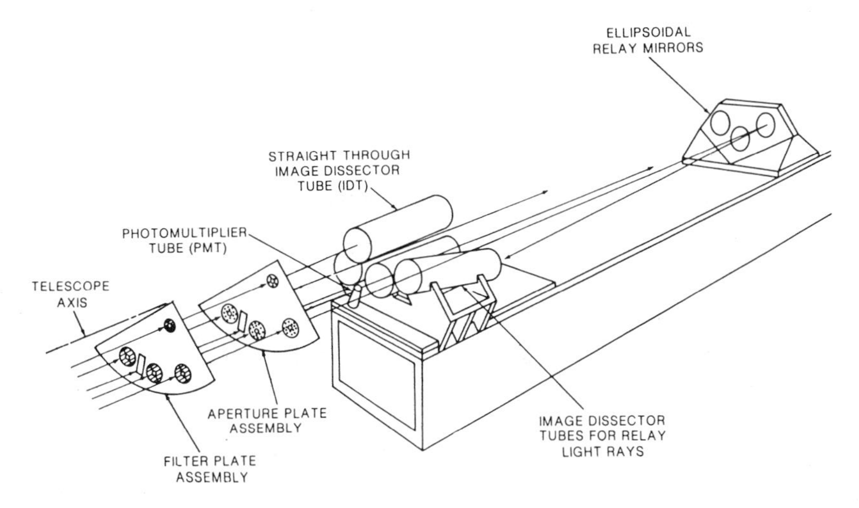

Figure 24 The High Speed Photometer is the simplest of the scientific instruments. There are no moving parts (except electrons).

view) and a set of small filters; therefore, each tube is illuminated by multiple aperture/filter combinations. The object to be studied is located onto the right spot by the telescope pointing system and the dissector tube is commanded to accept electrons from the correct spot. The photomultiplier is fed by a similar chain. The overall spectral response covers the range 1150–6500 Å, and the instrument can also perform ultraviolet linear polarimetry. This scientific instrument will have the high efficiency characteristic of photometers and it will also allow very high time resolution to be carried out down to 16 μs allowing the Space Telescope to exploit the noise-free transmission of bright object images, from stars and planets.

6.6 *The Fine Guidance Sensors*

The Fine Guidance Sensors can also serve to perform astrometric measurements and, in a functional sense, make up a sixth scientific instrument. This capability arises from the fact that the problems of pointing and guiding are so demanding that astrometric accuracy is required. Only two Fine Guidance Sensors are necessary for stabilizing the Space Telescope, leaving the third unit free for astrometry. With a total field area of 69 arcmin^2, the problems are somewhat constrained, but there will be many parallaxes and proper motions to be determined. The design requires that a given Fine Guidance Sensor be able to determine positions of two stars with an accuracy of 2 milliarcsec down to 17th magnitude at a rate of 10 stars in 10 minutes. It should be possible to determine accurate positions of stars as faint as 20th magnitude. Stars of 4th magnitude can also be measured using neutral filters. Further, it is expected that we can use the information as a guide star is scanned by the Fine Guidance Sensor to determine stellar duplicity. W. H. Jefferys of the University of Texas–Austin leads the astrometry science team.

7. THE OPERATIONS SYSTEM

How we actually operate and use the Space Telescope is a problem of comparable complexity to the actual construction. In many ways it is even more difficult to do, for there are more interfaces, the problems are often those of policy and attitude, and everyone is aware that we shall be changing procedures once the spacecraft is up and its in-orbit characteristics are actually known. The basic features of the operations system have been defined and the initial contracts have been let. However, it is quite appropriate that this area of work will be the last part of the Space Telescope development to be accomplished.

Two basic factors determine the approach to Space Telescope operations—the degree of spacecraft automation and the method of scheduling—and, as we shall see, these are related. Automation is a necessity to some degree, for we expect to have direct access to the Space Telescope only about 20 minutes of each orbit because of oversubscription of the relay satellites. Without some automation we would be constrained to changing positions, grating settings, etc. only 20 percent of the time. This would mean the loss of considerable observing time. The question then becomes: How much automation is necessary? To maximize the scientific return, the system should be able to operate on its own for 24 hours. In order to allow the greatest flexibility, the operations system must allow the ground-based observer to participate in as much as 20 percent of the object acquisitions and to see part of the data in real time, in order to monitor observations. This method is now used to some extent in the operation of the VLA, but the versatility and complexity of the Space Telescope is even greater. Large ground-based optical telescopes have evolved towards semi-automation in order to improve efficiency. However, in this respect the Space Telescope will be much more sophisticated.

The method of scheduling is a sensitive issue because it touches on both the tradition of a science and the complex problems of scientific instrument and spacecraft operation. Traditionally, the ground-based astronomer has had in the past what NASA calls "block scheduling." This means that one scientist is given a full night, several consecutive nights, or even an entire half lunation, on a large telescope. He uses the time efficiently or inefficiently, changes auxiliary instruments at his own discretion (in theory), and makes his own program modification as the data comes in. This system has begun to break down as telescopes have become useful throughout much of the 24-hour cycle, transportation to remote sites has become less time-consuming, and the rate of data collection has increased. The Space Telecope will have continuous observing potential, severe constraints on observational procedures, and heavy demand for its use. For these reasons, we shall use "integrated scheduling." Integrated scheduling is a process by which various observing requirements are interleaved in the most efficient manner. This does not preclude observer involvement, but it does mean that how the Space Telescope is used will be well defined beforehand with just enough flexibility to permit changes in the program demanding immediate scientific judgment and observation adjustments.

7.1 *The Science Institute*

The science done with Space Telescope will be controlled through a dedicated facility which lies outside of the NASA organization and is directly responsive to the community of current and potential users of the observatory. This approach goes back to the original formal recommendations for the Space Telescope (National Academy of Sciences 1962), although the issue of operations was not addressed for many years and this recommendation was forgotten. As operations planning began, the idea was spontaneously revived, and it has been enthusiastically supported by American scientists.

The intent is to put the final responsibility and authority for the scientific use of the Space Telescope with the users themselves. This has been done through competitive selection of a consortium of universities to establish a Space Telescope Science Institute. This is the same basic approach used with success for operation of other large national facilities, both in astronomy and high energy physics. Important differences exist for Space Telescope. Space Telescope requires aerospace technology and NASA already exists as the permanent civilian space development organization. If the Science Institute were to attempt the development of the Space Telescope, it would have to temporarily duplicate major parts of an existing body. Therefore, the Space Telescope is being developed and maintained by NASA, while the Science Institute concentrates on the science being done with it. Similarly, NASA already has an elaborate operations system which can easily accommodate the Space Telescope without the development of enormous communications and spacecraft control systems. We expect that the problems of the science and the interaction with NASA will be challenge enough, besides being economical.

The principal functions of the Science Institute will be selecting observers, scheduling, conducting observations, arranging parallel observations on other instruments, data processing, data handling, and research by the resident scientists. Observer selection is a major task, as we expect that demand will far exceed supply. Scientific peer review and what is possible will be important considerations. Scheduling involves making observing plans of progressively greater refinement culminating in daily lists of observing requirements that can be satisfied by NASA's spacecraft operations element. Conducting observations will mean the monitoring of the automated Space Telescope operations, directly interacting with NASA in difficult object acquisitions, monitoring the scientific data flow, and making appropriate program adjustments. Parallel observations will exploit the opportunity to operate more than one scientific instrument at a time. In some cases this will be done to accommodate specific, approved observing programs. In other cases, it will simply mean that we shall be gathering as much scientific data as possible, for the immediate use of any scientist. Data processing is the manipulation of data obtained with the Space Telescope (the scientific instruments and the Fine Guidance Sensors) and turning the raw data (received within 24 hours of the observation) into scientifically useful data that has been calibrated and put into a useful, understandable format. A data analysis capability will also exist for use of the scientific staff of the Science Institute and guest researchers. Data archiving involves the storage of all scientific data in a useful form, an especially important function as all Space Telescope data will be available to any legitimate user one year after the calibrated data is provided to the original observer.

This facility will be the same size as other ground-based astronomy national laboratories, and will be operated by the Association of Universities for Research in Astronomy at The Johns Hopkins University in Baltimore, Maryland. The challenge of developing this new facility is enormous; the scientific potential is even greater.

7.2 *Scheduling Constraints*

The ground-based optical astronomer is accustomed to scheduling his telescope use according to the natural constraints of day/night, lunation, seasonal changes of darkness, rotation of the Earth through the night, etc. The Space Telescope will need to adhere to a similar set of constraints. The Space Telescope's low Earth orbit means that it will be passing from sunlight to darkness about every 100 minutes and will at any time have only slightly more than 2π steradians of the sky available to it. The additional constraints of viewing angles of at least 50° from the Sun, 70° from the bright Earth, and 15° from the moon are needed if the minimum straylight is to be found. As well as this, a further constraint is the passage through the high radiation environment of the South Atlantic Anomaly (where data gathering is not possible for all of the present scientific instruments).

A characteristic snapshot in time of the available sky is shown in Figure 25. This is a projection of the parts of the sky that may be observed with low straylight and shows how long this condition will apply. For observation where the sky background will not be reached or where the zodiacal light is bright, the observation times will be much longer and the observing zone will be extended. An interesting feature of this figure is that there are two continuous observing zones at the poles of the Space Telescope's orbital plane.

This picture continuously changes as the Earth orbits the Sun and the orbit precesses. The Earth's motion causes the center of the optimum observing zone to move across 360° in a year, just as it causes there to be summer and winter constellations on Earth. The orbit's precession is a smaller effect and is a motion of about 7°/day of the ascending node; therefore, the continuous viewing zones move also.

In addition to these constraints imposed by the orbit, there are additional features of the spacecraft which control what can be done. The

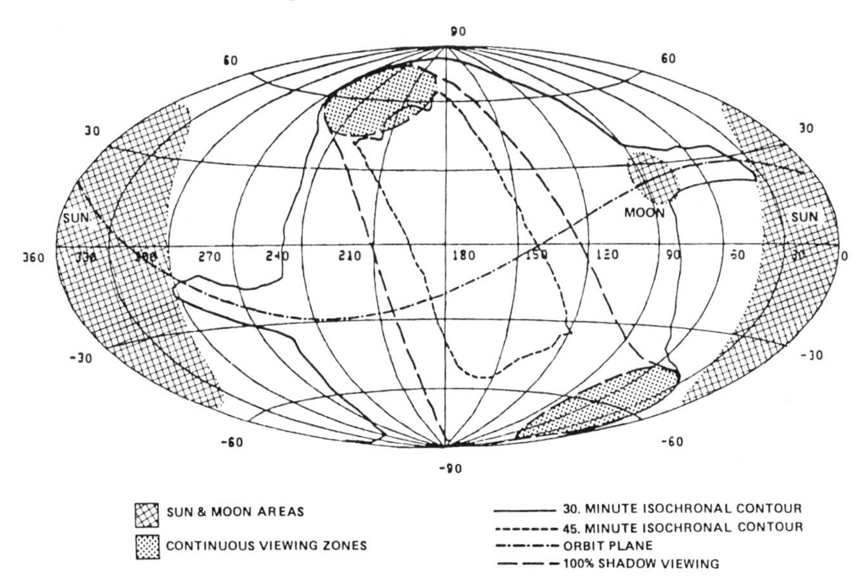

Figure 25 On any given orbit, the Space Telescope can view various parts of the celestial sphere for varying times. The isochronal contours show the boundaries where full straylight suppression can be reached. Certain regions are excluded due to proximity to the Sun and moon, while other parts can be observed throughout the orbit.

most important of these spacecraft constraints are the object acquisition times and the cycle times for the various scientific instruments. In an earlier section, I described how the Space Telescope was slewed across the sky and how the acquisition of a new object was done. If this could be done instantaneously, we could picture the Space Telescope flipping about the sky on every orbit, not wasting a minute; this is not the case. Relocation to a new object takes about two minutes, if the object is close by in position. If the object is far away, then the slewing time becomes important. The slewing rate maximum of 15 degrees/minute is reached by an acceleration of about 0.048 degrees/minute^2 and the telescope is decelerated at the same rate. This means that grouping of the sources to be observed is a natural goal in optimizing observing efficiency. The other major constraint, cycle time, arises from the fact that the thermal energy dissipation and electrical power systems do not permit continuous operation of all five scientific instruments. This has meant that three modes of readiness are planned: "operation," "standby," and "safe," the last only occurring at the time of a major spacecraft problem. Our operational guideline is that the Wide Field Camera shall always be operational, even when one of the axial scientific instruments is operational, and that the others will be on standby. In some bases, the thermal/power constraints will also allow other instruments to be in the operation mode, doing parallel observations or calibrations; unfortunately, this is not a common condition. This constraint means that composition of the observing schedule must also favor the use of a given scientific instrument for multiple observations. This is frequently in conflict with the desire to choose sources that are either near one another or lie in a smooth progression across the sky.

A characteristic observing schedule has been prepared to determine how to best plan the Space Telescope observations. This schedule includes use of all of the scientific instruments and the astrometry function of the Fine Guidance Sensors in about the proportion expected. This is meant to be a characteristic and realistic observing schedule and it has gone through several iterations as the operating characteristics of the observatory become more clear. The conclusion is that scientific observations, including calibrations on celestial objects, are made about 40 percent of the time. A larger fraction would be obtained if we simply tried to maximize time of use rather than carry out specific programs. The small duty cycle seems surprising to many astronomers, but it is comparable with that of ground-based optical telescopes. When one considers that an average night is only 10 hours long and that the best sites are usable as far as weather is concerned about 75 percent of the time, the duty cycle is only

31 percent, without counting the reduced flexibility due to scheduling around the full moon, etc.

7.3 *Spacecraft Operations*

The Space Telescope will be controlled from the Goddard Space Flight Center, which has the responsibility for operation of all near-Earth, unmanned satellites. This control center interacts with the spacecraft through the Tracking and Data Relay Satellite System and with the Science Institute through direct lines, satellites, and the Science Institute's local representatives (see Figure 26). Once the Space Telescope is launched, it becomes the responsibility of the Operations Control Center. The developers of the observatory and the scientific instruments apply the information on how the system should work, and play a supporting role in the event of anomalies. The complexity of this system is illustrated in Figure 27 where we see a schematic diagram of the flow of commands and data.

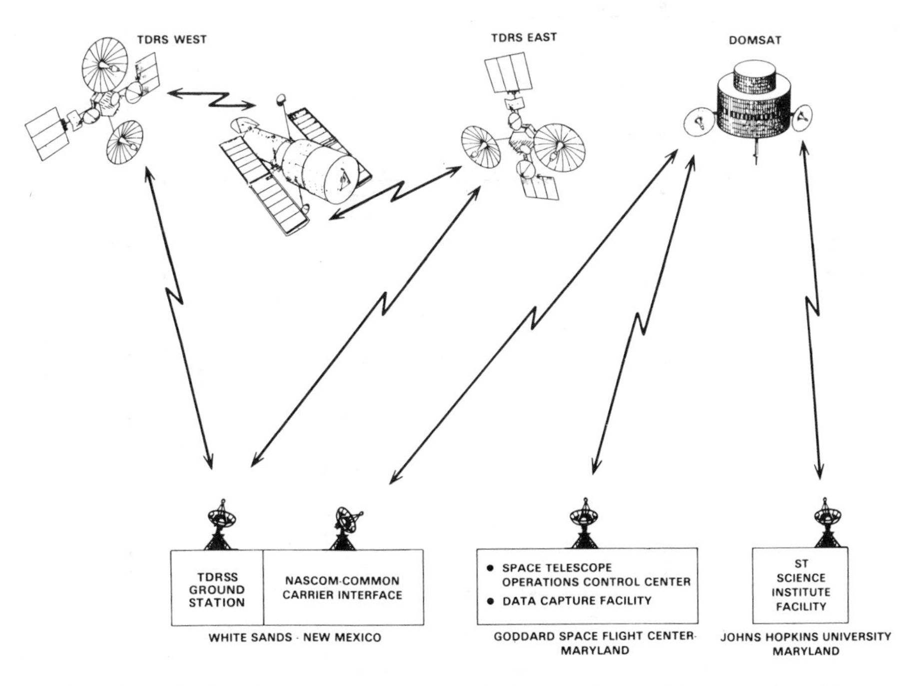

Figure 26 The data flow of Space Telescope includes several ground stations and satellites to provide absolute control of the spacecraft and near real-time availability of a portion of the scientific data.

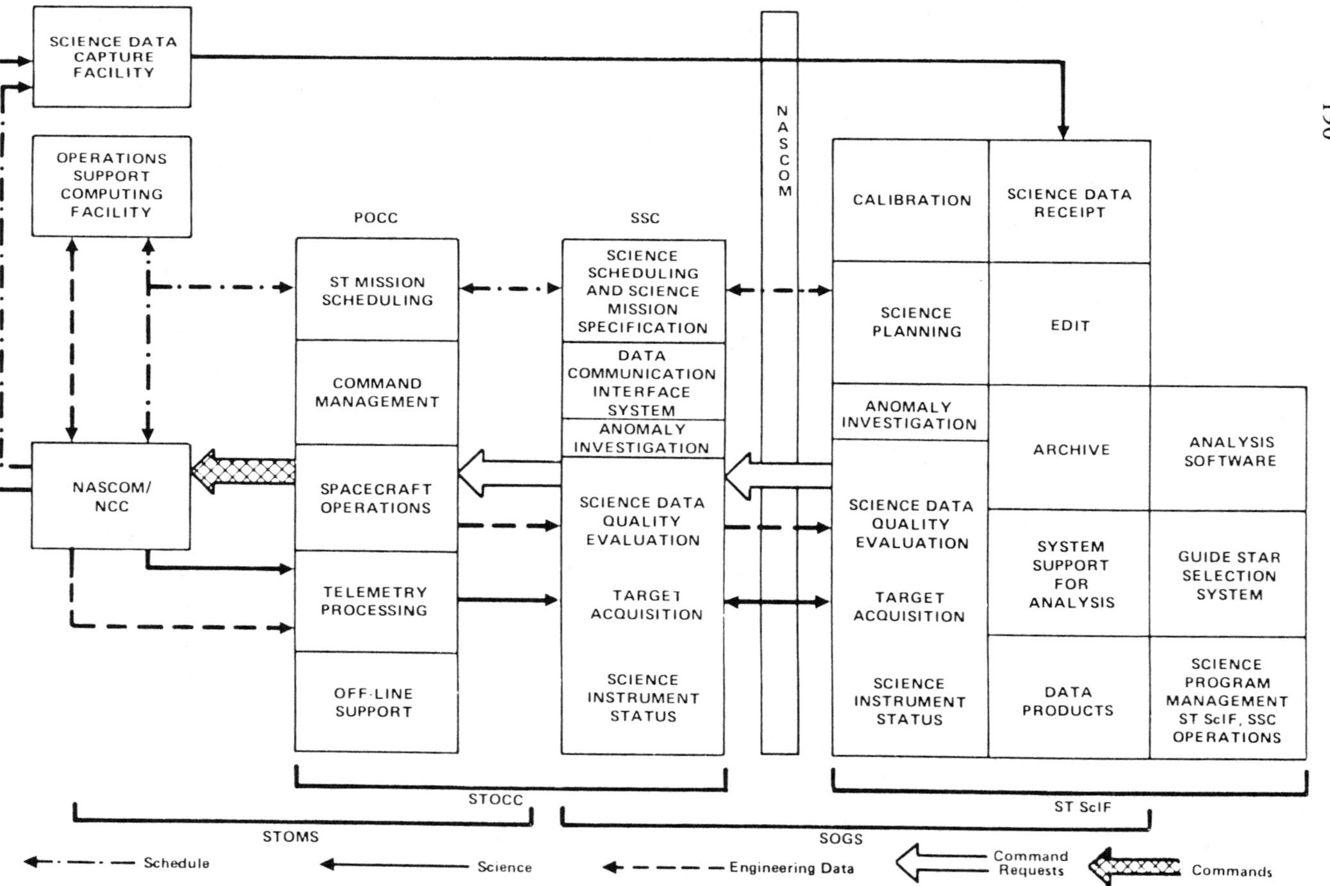

Figure 27 This simplified diagram shows the ground system necessary for Space Telescope operations. The acronyms refer to various organizational elements, including the Space Telescope Science Institute Facility (ST ScIF) and the Space Telescope Operations Control Center (STOCC).

The initial period of operations will be a verification period, when the various parts of the system are checked. Then operations begin. The first scientific users will be the participants in Phase C/D, and they will have exclusive access to the telescope for two months. After that interval, their share drops to 50 percent for six months, then 25 percent for 12 months, 10 percent for 10 months, and then no guaranteed time. Throughout this period, the European Space Agency member nation scientists will get at least 15 percent of the observing time, in return for ESA's contribution to the program. The rest of the observing time will go to guest investigators, whose selection will begin about one year prior to launch.

Maintenance and refurbishment will occur as required. It is expected that the first orbital visit will occur $2\frac{1}{2}$ years after launch, and then the entire spacecraft will be returned to Earth for a one-year refurbishment after five years in orbit. Upon relaunch, the cycle will be repeated. The Space Telescope has many backup systems, which should extend the period between visits. However, the continuous pressure to reduce program costs has meant that risk is present, and this introduces a very real uncertainty into long range planning.

8. SELECTING THE SCIENCE FOR THE SPACE TELESCOPE

The purpose of the Space Telescope program is to do science, not just to build an observatory and an operations system. In this final section we consider the scientific problems. We describe the procedures that will be used to identify the science to be done and what areas of research are presently considered most important.

We can describe the process of observer selection at this time because it will closely resemble the original method of selection of scientific participants—hence the general policy was determined early. When the Announcement of Opportunity for development phase scientist participation was issued, respondents were asked to describe how they would use the Space Telescope. The quality of the proposed investigation was a major factor in the selection. Scientists selected had a specific observing program in mind when their proposals were made in the spring of 1977. The observing programs were based on certain expected characteristics of the Space Telescope and on what the scientific knowledge and gaps were at that time. The nature of the observatory has changed slightly over the development period, and the Scientific Instruments are now well defined and much more versatile than envisioned when the first Announcement of Opportunity was released. Our knowledge of astronomical problems has already changed considerably since 1977, and will change still more before launch in late 1984. The procedure for planning the scientific programs of the Phase C/D participants will be to call for final proposals about one year prior to launch. These proposals would be expected to be within the framework of the originally selected proposals, e.g. a scientist selected originally to study planets could not use his guaranteed access time to work on clusters of galaxies. The Science Institute will then determine the basic programs to be carried out by the development scientists and issue a call for proposals for use of the remainder of the time available. Proposals for work already approved will not be accepted for this additional observation time but there are plenty of other problems to fill out Space Telescope's observing schedule.

Selection will be carried out by the Science Institute within broad policy guidelines prepared by NASA. These guidelines are expected to always emphasize the maximum scientific return from Space Telescope, but they may sometimes include emphasis on NASA-related activities. A good example of this would be observation of a comet that a NASA probe is intercepting or a planet that will be encountered. A more difficult issue will probably be the overabundance of good proposals addressing the handful of natural problems that will exist. It will be a matter of Science Institute policy to resolve this issue.

Two types of Space Telescope user will be selected by the Science Institute, general observers and archival users. General observers will request specific observing programs and work with the Science Institute in their execution. The archival user will get data from the calibrated data stored in the Science Institute archives. Some of this will be fresh data generated by Science Institute staff members using instruments in parallel or at times that would otherwise have been unscheduled. The other archived data will become available to any legitimate scientific user one year after the calibrated data is provided to the original observer. The European Space Agency has already begun establishing a special Space Telescope European Science Facility to be a center for European Space Agency member nation use of the open-archive scientific information.

The more difficult problem is to describe the nature of the scientific programs that will be carried out between 1985 and 2000, the expected operation dates of the Space Telescope. There is little reason to expect that this can be done any more accurately now than a similar prediction could have been made for the use of the 200-inch telescope at Palomar Mountain in 1935. The first attempt at this is the set of proposals that were approved as part of the development phase selection. Because it was known that these would never be used in the form originally submitted, they have not been compiled and published. A revised form of these proposals will be incorporated in the first call for general observers issued prior to the initial launch. A more timely set of proposed uses are found in the proceedings of two meetings. The European Space Agency sponsored a workshop in February 1979 (Macchetto, Pacini & Tarenghi 1979) to share European (primarily, not exclusively) ideas on the best and most practical uses of Space Telescope. Princeton University hosted an International Astronomical Union Colloquium on Scientific Research with the Space Telescope in August 1979 (Longair & Warner 1979).

What became clear in the proposals, the workshop, and the colloquium is that demand will far exceed availability of observing time. As a result, it is clear that Space Telescope must be used only for unique problems, i.e. problems that only the Space Telescope can be expected to resolve. The problems must be not only important and practical, they must also be intractable to other available observing methods. They must be problems that exploit the three major advantages of Space Telescope—steady images some 10 times better than we obtain on the ground, ability to detect point sources 50 times fainter than large ground telescopes, and continuous wide wavelength coverage.

The Space Telescope will be the most powerful optical observatory ever built. It is a challenge to the engineers and scientists who build it. It will be a greater challenge to the users, many of whom are now still in elementary school. There is every reasonable expectation that it will profoundly change our view of the universe.

References

Danielson, R. E., Tomasko, M. G., Savage, B. D. 1972. *Astrophys. J.* 178:887

Leckrone, D. S. 1980. *Publ. Astron. Soc. Pac.* 92:5

Longair, M. S., Warner, J. W. 1979. *Scientific Research with the Space Telescope. NASA CP-2111*

Macchetto, F. D. et al. 1980. *The Faint Object Camera for the Space Telescope. ESA SP-1028,* Oct. 1980

Macchetto, F. D., Pacini, F., Tarenghi, M. 1979. *Astronomical Uses of the Space Telescope.* European Space Agency, Feb. 1979

Meinel, A. 1960. *Telescopes,* ed. G. P. Kuiper, B. M. Middlehurst, p. 25. Univ. Chicago Press

National Academy of Sciences. 1962. *A Review of Space Research. Publ. 1079,* Chap. 2, Astronomy. Natl. Acad. Sci., Natl. Res. Council

National Academy of Sciences. 1966. *Space Research: Directions for the Future. Publ. 1403,* Part 2, Astronomy and Astrophysics. Natl. Acad. Sci., Natl. Res. Council

National Academy of Sciences. 1969. *Scientific Uses of the Large Space Telescope.* Space Sci. Board, Natl. Acad. Sci., Natl. Res. Council

National Academy of Sciences. 1972. *Astronomy and Astrophysics for the 1970's.* Astron. Survey Comm., Natl. Acad. Sci., Natl. Res. Council

NASA. 1972. *The Scientific Results from the Orbiting Astronomical Observatory (OAO-2). NASA SP-310*

Oberth, H. 1923. *Die Rakete zu den Planeträumen.* Munich: R. Oldenbourg

Spitzer, L. 1946. *Astronomical Advantages of an Extra-Terrestrial Observatory. Project RAND Report,* Douglas Aircraft Co., Sept. 1

Spitzer, L. 1968. *Scientific Uses of the Large Space Telescope.* Publication of the Natl. Acad. Sci., Washington, DC

Spitzer, L. 1979. *Q. J. R. Astron. Soc.* 20:29

Spitzer, L., Jenkins, E. B. 1975. *Ann. Rev. Astron. Astrophys.* 13:133

Telescopes for the 1980s, pp. 195–278

THE EINSTEIN OBSERVATORY AND FUTURE X-RAY TELESCOPES

Riccardo Giacconi, Paul Gorenstein, Stephen S. Murray, Ethan Schreier, Fred Seward, Harvey Tananbaum, Wallace H. Tucker, and Leon Van Speybroeck

Harvard/Smithsonian Center for Astrophysics,
Cambridge, Massachusetts 02138

1. INTRODUCTION

A complete history of the events that led to the development and flight of the Einstein Observatory (Figure 1) should proceed on at least three different lines of approach: 1. a description of the development of the technology, 2. an account of the struggle to gain acceptance by the scientific community and the funding agencies of the need for such a mission, and 3. a chronicle of the vicissitudes that occurred during the execution of the program and that ultimately determined the characteristics of the observatory in orbit.

The distinguishing aspect of the Einstein Observatory is its use of grazing-incidence optics (see Section 1.1). The introduction of grazing-incidence optics to extrasolar X-ray astronomy has not been easy. This is to be contrasted with the vigorous support the National Aeronautics and Space Administration (NASA) has given to high energy astronomy in general and to solar X-ray astronomy in particular. The support given by the Agency to missions such as the Small Astronomy Satellites (SAS-1 [Uhuru], SAS-2, and SAS-3), the Orbiting Solar Observatory, the Apollo Telescope Mount on Skylab, and, most recently, the High Energy Astronomy Observatory (HEAO) has given X-ray astronomers in the United States the opportunity to be at the forefront of research. The obstacles encountered in the introduction of X-ray telescopes may be those experienced in the adoption of any new technology, particularly when, as in this

8243-2902/81/0820-0195$02.00

Figure 1 Artist's conception of the Einstein Observatory (courtesy of TRW, Inc.).

case, this new technology brings about a significant sociological change. The flight of the "Einstein Observatory" marked the beginning of a departure from the manner in which X-ray investigations are carried out, from individual experiments to the shared use of large facilities. We hope that what can be learned from our narrative may be helpful in planning the development and operation of the new X-ray telescope facilities of the 1980s.

1.1 *Technological Developments*

Grazing-incidence telescopes, used in X-ray astronomy, focus X-rays by making use of the fact that they behave like light rays if they strike surfaces at a shallow enough ("grazing") angle. The great advantage of using grazing-incidence focusing optics for extrasolar X-ray astronomy was evident in the early 1960s and was first pointed out by Riccardo

Giacconi and Bruno Rossi (1960), who called attention to the large collection area, high angular resolution, and large signal-to-noise ratio that could be obtained by use of nested grazing-incidence paraboloid collectors. They also pointed out the possible extension of this technique to true image-forming optics, as was originally suggested by Wolter (1952) for microscopes.

The design of grazing-incidence optics had been of interest to X-ray microscopists for decades (Kirkpatrick & Baez 1948). All such designs were based on the effect, measured by Compton in 1922 (Compton & Allison 1935), that X-rays can be reflected with high efficiency (about 50%) by polished surfaces provided that the glancing angle of the incident radiation to the surface is small enough ($\theta \lesssim 1°$ for $\lambda > 10$ Å). Among the different approaches that had been studied for microscopes, Wolter's designs appeared the most interesting for adaptation to telescopes because they allowed a large collecting area and high angular resolution over a large field. However, they required the use of two reflections from surfaces that were conic sections of revolution (Figure 2). Since the technology for polishing such surfaces to the high degree of precision required did not exist, none of Wolter's designs was ever translated into practice. The precision requirement is set by the fact that surface irregularities must be kept to a few angstroms to avoid scattering which degrades efficiency and resolution. Giacconi and Rossi thought that the larger physical size of the optics required for telescopes would make the achievement of the required precision easier.

Giacconi and his coworkers at American Science & Engineering, Inc. (AS&E) initiated, under NASA sponsorship (Contract NAS5-660), a program of experimental investigations which succeeded by July 1961

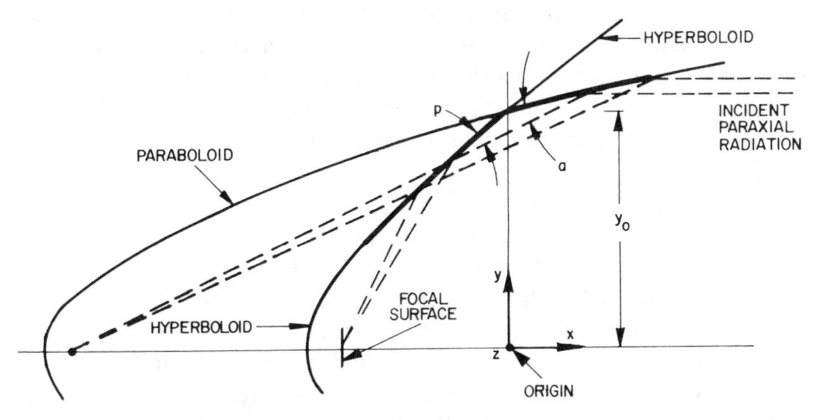

Figure 2 Ray tracing of the optical system on which the design of X-ray telescopes is based.

in producing an X-ray collector of about 1 cm^2 area, grazing angle of 2°, consisting of a glass cone with an optically polished interior surface (ASE-142). Then by January 1963 they constructed an aplanatic telescope for soft X-rays (Giacconi et al. 1965a) with 2 cm^2 area, 1 arcminute of resolution in 40 arcminutes field of view, focal length 64 cm, and 5% efficiency at 44 Å. The telescope was built by machining and polishing the interior surfaces of an aluminum tube. The surface was coated with evaporated gold. Although very far from theoretical efficiency, the successful laboratory demonstrations of these prototypes, obtained with relatively primitive manufacturing means, were extremely encouraging.

Even at this early state of development it was clear that grazing-incidence telescopes would be quite useful in the X-ray study of celestial objects, particularly the Sun, since they were 100 times more sensitive than the X-ray pinhole cameras then being used.

The experimental research program that had been carried out was funded at a very modest level. No funding was available to build a telescope with optically polished surfaces, nor to explore the preparation of surfaces beyond the state of the art. Although some new techniques for fabrication were developed, such as epoxy and electro-deposited replicas of polished mandrels, the main aim of the program was simply to develop working prototypes suitable for space applications. Further advances in the technique could come only after acceptance of the concept for flight missions by NASA, since much more money is normally made available for experiment under these conditions than for laboratory research programs.

1.2 *Use of X-Ray Telescopes in Solar Astronomy*

X-ray emission from the Sun had been studied intensively by Herbert Friedman and his colleagues at the Naval Research Laboratory (NRL) since 1948. They had shown that the sources of X radiation tended to be localized and of small angular diameter (Blake et al. 1963). However, pinhole camera techniques if designed for 1-arcminute resolution were barely capable of obtaining one picture during an entire rocket flight. Features of much smaller angular size could well exist but they could be detected only by use of grazing-incidence telescopes with photographic image recording as was proposed by the AS&E group.

John Lindsay, of Goddard Space Flight Center (GSFC), had been involved in studies of the Sun for many years, first as a member of the NRL group and later at NASA's GSFC. He had furnished the initial stimulus for the development of the Orbiting Solar Observatory series of satellites and was technical monitor for the NASA contract that permitted the AS&E group to start the development of X-ray telescopes. He was personally interested and enthusiastic about using X-ray telescopes for the study of the Sun, and was most helpful in having the techniques adopted as part of the NASA program of solar investigations.

NASA granted Giacconi and Harmon of AS&E (ASE-106, 1961; ASE-311, 1962) a contract for development of an experiment to be carried in the pointed section of the Orbiting Solar Observatory (OSO-D) spacecraft (Contract NAS5-3569). A collaborative program between the AS&E and GSFC groups also began. It succeeded in obtaining on October 15, 1963, and in a reflight on March 17, 1965, the first X-ray pictures of a celestial object (the Sun) with grazing-incidence optics from a rocket-borne instrument (Giacconi et al. 1965b). The telescopes developed for this flight were of electroformed nickel. They had a diameter of 7.6 cm, a focal length of 83.6 cm, and a collecting area of 1.6 cm^2. The telescope efficiency at 8 Å was of 15% and the angular resolution was better than 1′.

A much more ambitious experiment was also planned as part of the Lindsay-Giacconi collaboration for the Advanced-OSO spacecraft and later for the Apollo Telescope Mount (Lindsay & Giacconi 1963; ASE-623, 1964). Work was initiated in 1964 on a telescope of 26-cm diameter, 200-cm focal length, and 38-cm^2 collection area which would yield 5-arc-second resolution. The development of an Advanced OSO spacecraft was cancelled by NASA in 1965 and NASA began studies for incorporation of the solar experiments in a manned vehicle. In September 1965 John Lindsay died, and the AS&E–GSFC collaboration was dissolved. The seeds of this work, however, came to fruition in the two experiments, S-054 (by Giacconi and his collaborators at AS&E) and S-056 (by Underwood and his collaborators at NASA/GSFC) which were flown as part of the Apollo Telescope Mount mission in 1973.

The development of X-ray telescopes had proceeded by this time to the achievement of 20-arcsecond pictures of the Sun from rocket-borne telescopes (Vaiana et al. 1973, 1968), and to the achievement of 5-arcsecond resolution in the 31-cm diameter, nested mirrors of 213-cm focal length produced by the AS&E group for the Apollo Telescope Mount (Figure 3). [A useful summary of these developments is given in an article in *Space Science Reviews* (Giacconi et al. 1969).] The debt that these achievements owe to John Lindsay's advocacy and interest and to the support given by Harold Glaser of the Solar Physics Branch of NASA Headquarters cannot be overestimated.

1.3 *Use of X-Ray Telescopes in Extrasolar Astronomy*

In contrast to the rapid adoption of the use of grazing-incidence telescopes for solar physics experiments, the introduction of this technique to extra-

Figure 3 The nested mirror configuration used in the ATM Solar X-ray Experiment S-054. The mirror material is beryllium; the reflecting surfaces are Kanagen coated.

solar X-ray astronomy was quite slow. Proposals to develop a small telescope (10-cm^2 area) for use in rockets to study celestial sources were submitted to NASA as early as 1961, and proposals for larger telescopes (up to 1.2-m diameter) to be used on satellites were submitted as early as 1963, all without success.

Perhaps the success of conventional techniques using mechanically collimated proportional counters in the discovery of X-ray stars of unexpectedly high luminosity (Giacconi et al. 1969) and the abundant scientific returns achieved in the early 1960s using these techniques in rockets made the adoption of telescopes for the study of extrasolar sources seem less urgent. Considerable gains in sensitivity could be expected by using these techniques on satellites such as Uhuru and even larger ones that were being proposed, for instance, by H. Friedman.

Some substantial technical questions also remained. First, the grazing-incidence telescopes used in solar astronomy were not well suited for stellar research. This was due to the following technical considerations: two types of imperfections can occur on the surface of grazing-incidence mirrors—1. deviation from the desired overall figure (slope errors), and 2. microscopic surface roughness. The first quantity establishes the half width of the point response function and controls the angular resolution that can be achieved; the second determines the amount of scattering loss of the incident radiation out of the reflected beam, and establishes the fraction of incident beam that falls within the resolution element. In a high flux situation, such as that encountered in solar observations, loss of power from the reflected beam is relatively unimportant, since one can still achieve high resolution images. In the study of extrasolar sources, where the incident fluxes at earth are more than a million times less intense than from the Sun, the efficient collection of photons is imperative. To do this, the surface finish of X-ray reflection surfaces had to be improved from the typical values of about 50–100 Å roughness to values better than 30 Å, as was finally achieved in the Einstein Observatory. To improve the collecting area, nested mirror configurations had to be developed and the concommitant problems of co-alignment solved. Finally, in a satellite application such as on the Orbiting Astronomical Observatory (OAO), one could not use film as the detecting medium; electronically read out X-ray imaging cameras capable of high resolution and single photon detection had to be developed.

Lack of support for a laboratory program devoted to the development of these techniques for stellar research meant that such developments could only be bootlegged as part of the development of solar telescopes. In fact, in the S-054 Apollo Telescope Mount program a 2-nested-mirror configuration was used, although it was not strictly necessary, and first generation X-ray television cameras were developed although film was used as the flight detector. The first fused silica telescopes polished to optical precision were also fabricated (Mangus & Underwood 1969) as part of the solar program.

The lack of support for use of focusing optics by the scientific community was a more serious problem for the advancement of the art. Opinions expressed at the time by advisory committee members ranged from stressing the need to accomplish all-sky surveys with conventional detectors prior to embarking on the use of high resolution optics to the flat statement by some theoretical astrophysicists that since the X-ray production processes would give rise in the main to diffuse sources, there was no need then or in the future for high angular resolution telescopes in X-ray astronomy observations. In the mid-1960s the Orbiting Astronomical Observatory (OAO) was the only satellite platform that had the pointing capability and the data and control system required for an X-ray telescope, but the inclusion of X-ray experiments on the OAO might have interfered with the observational programs in UV astronomy which space astronomers had been working on since the 1950s. The only X-ray collector based on grazing incidence ever included in the OAO was a small pickaback experiment by R. L. F. Boyd and collaborators from the Mullard Research Laboratory in England (Bowles et al. 1974). Because the mirror area was only about 10 cm^2 and the instrument lacked imaging capabilities, it was scarcely more sensitive than the satellite-borne conventional detectors being used at the time of the flight in 1972.

The Woods Hole Summer Study of the Space Science Board of the National Academy of Sciences held in June and July, 1965, provided the first opportunity to fully discuss the scientific advantages of the use of focusing optics in extrasolar X-ray astronomy. The X-ray and Gamma-ray Astronomy Panel included, among others, H. Friedman, Chairman, G. Clark, P. Fisher, R. Giacconi, W. L. Kraushaar, R. Novick, and B. Rossi.

For the first time the limitations imposed by background noise and source confusion on the development of more sensitive detectors of the conventional type were clarified in open debate. Among the major conclusions in its report the panel mentioned that grazing-incidence reflection telescopes "appear to be the only tools capable of many of the refined observations that will be needed beyond the early exploratory stage." The superior angular resolution of the imaging optics, the sensitivity to the detection of very weak sources, and the possibility of using the telescope in conjunction with polarimeters and dispersive spectrometers were recognized. The panel recommended that "a program of X-ray astronomy using total reflection telescopes should be started at the earliest possible time and pushed vigorously to exploit its ultimate capability." Further, the use of an OAO dedicated to X-ray astronomy was recommended by the panel (NAS Publication 1403, 1966).

In addition, there emerged at this meeting a consensus among cosmic-ray physicists who needed heavy payloads, X-ray astronomers interested in the utilization of extremely large-area detectors, and those hoping to use X-ray telescopes, that a large but inexpensive spacecraft, a "super-Explorer," could accommodate all of these requirements. In particular, the point was made that X-ray telescopes required only modest (arcminute) pointing since image reconstruction to high accuracy (about 1 arcsecond) could be done post facto on the ground because of the small number of impinging photons. Therefore, one could conceive of a series of identical and relatively inexpensive spacecrafts which could carry out both survey and roughly pointed missions. This germinal idea later led to the High Energy Astronomy Observatory (HEAO) series of satellites.

In a June 1967 letter to Homer E. Newell, Associate Administrator for

Space Science and Application, which was widely circulated within NASA, Giacconi reiterated the need for an OAO-sized telescope (30-cm diameter, 300-cm focal length) mission to be followed by a more ambitious mission (1-m diameter, 10-m focal length telescope) in the 1970s. In that letter Giacconi proposed that the smaller telescope be implemented as an experiment under a single principal investigator. He recommended that the larger telescope be implemented as a multi-experimenter facility, and offered the services of the AS&E group as the lead group, to act as principal investigator providing the mirror optics and X-ray camera, and integrating other instruments, such as spectrometers and polarimeters, into the payload.

Dr. Newell responded in August 1967, expressing interest in proceeding with the smaller telescope mission, which would be suitable for flight on an Apollo Telescope Mount, but expressing NASA's intention to consider even this small version a *facility,* rather than an *experiment.* At his direction a meeting was held on October 9–10, 1967, under the chairmanship of Dr. Nancy Roman and many of the active US X-ray astronomers participated. The conclusion was that "a large grazing incidence parabolic focusing X-ray telescope is necessary for future high energy space astronomy programs."

As a result of the interest expressed by several groups in participating in such a mission, and of the explicit recommendation of the Astronomy Missions Board (NASA SP-213, July 1969) that a telescope mission be carried out, an Announcement of Opportunity was circulated by NASA (Memo Ch. 6 to NHB8030.1A) on February 16, 1968. The announcement read in part: "NASA is currently planning the use of a parabolic focusing telescope for X-ray astronomy. This instrument is to be considered a facility rather than an experiment. At present, we will welcome proposals for experiments to be performed at the focus of such an instrument, together with the necessary auxiliary instrumentation, but not proposals for the basic telescopes."

As a result of this solicitation several groups, including those of Giacconi at AS&E, George Clark at Massachusetts Institute of Technology, Robert Novick at Columbia University, and Elihu Boldt at Goddard Space Flight Center, were selected as potential investigators and were asked to assist Marshall Space Flight Center in defining the mission.

A first meeting with the interested scientists was held at Marshall Space Flight Center on March 19, 1969, and chaired by Ernst Stuhlinger, Associate Director of Science of the Center. A series of meetings then ensued, which made clear the fact that focusing X-ray optics had not yet become conventional and that it was not possible to state the requirements for the observatory without knowing in greater detail what the experiments were attempting to do. Scientists from the various organizations that had been invited to participate in this effort decided, therefore, to hold a series of meetings at their own institutions, in order to define the scientific objectives of the mission, the experiments necessary to carry out these objectives, and then to derive the necessary telescope design parameters. The results of these meetings were presented to NASA on August 1, 1969, in a document entitled "Preliminary Study: Telescopes and Scientific Subsystems for a High Energy Astronomy Observatory."

What had evolved and was described in the report was an integrated facility incorporating two X-ray telescopes [one with low resolution (10″), high energy response, large area (5000 cm^2) and one with high resolution (2″), lower energy response, smaller area (1000 cm^2)], focal plane instruments, and support subsystems designed to perform sophisticated observations in X-ray astronomy. It was stated in this report: "The characteristics of the telescopes and experiments are so intimately related that we conclude that the specifications for the optics must be ultimately set by the users. In addition, the users must retain technical control of the implementation and testing of the devices." The scientists recommended that the payload be included in the NASA "super-Explorer" series to be launched with a Titan III-C vehicle.

Following this preliminary study the scientists of the four institutions involved decided that, in order to preserve the important aspect of an integrated observatory design, they would respond to the expected Announcement for Opportunity of NASA as a Consortium with a single proposal and with a single principal investigator. A Letter of Understanding between the four institutions, which established the ground rules for writing the proposals, carrying out the design for the experiments, setting the objectives for the mission, controlling the changes in the hardware, retrieving and utilizing the data, was written on January 27, 1970, and signed by Boldt, Clark, Giacconi, Gursky, and Novick, each representing the X-ray astronomy groups at their institutions. It was felt that naming a single principal investigator supported by a Scientific Steering Committee would insure the competent and recognized scientific leadership necessary for the success of the mission.

1.4 *The HEAO Program*

In response to NASA Memorandum Change 27, NHB8030.1A of March 3, 1970, a proposal for a Large Orbiting X-ray Telescope was submitted by the four institutions (AS&E, Columbia, Goddard, and MIT) in May 1970. Giacconi was the designated principal investigator. Notwithstanding the problems raised for NASA in entrusting to a single principal investigator an experimental payload requiring an entire spacecraft of substantial complexity and cost, the technical proposal was well received by the reviewing committee and NASA management in October 1970, and the Large Orbiting X-ray Telescope was assigned the third flight of the HEAO series of four spacecrafts.

At the outset the Consortium had designed the observatory as a single integrated payload, which would be built and operated by the Consortium on behalf of NASA as a national facility available to all interested astronomers. The proposers expressed willingness to waive their rights as principal investigators to exclusive use of the data for a period of one year, and offered to share observing time with guest observers as soon as the observatory became operational.

Work started on this project in 1971. Dr. Fred Speer of Marshall Space Flight Center was designated HEAO Program Manager. Particular emphasis was placed on such crucial elements as the 1.2-meter-diameter high resolution optics and the imaging cameras. In January 1973, however, major changes took place in the HEAO program. Because the cost of the program for the four missions had escalated to about $450 million, and because of other budgetary problems, NASA was forced to reduce the budget for the entire program by one half. This implied scaled-down spacecrafts and three, rather than four, launches. In a series of meetings at NASA Headquarters, the program was restructured accordingly. The Large Orbiting X-ray Telescope lost one of its telescopes (the 5000 cm^2, low resolution one) and its 1.2-m-diameter, 6-m focal length high resolution telescope shrank to 0.6 meters in diameter and 3.5 meters in focal length. The only consolation was that in the revised program the telescope was assigned to the second, rather than the third, flight.

A new special review by the Office of Space Science occurred on September 13–14, 1973. In a proposal by a Consortium made up of the Harvard/Smithsonian Center for Astrophysics (where Giacconi and his group had moved from AS&E), Columbia, Goddard, and MIT, the new observatory HEAO-B was described. Its validity in view of the new findings in X-ray astronomy, particularly the Uhuru discoveries, was reassessed. Its role as a national facility with substantial guest investigator participation was reaffirmed. Technical advances that had taken place in the intervening period, primarily the ability to polish mirrors to very high accuracy, partly compensated for the loss of area and allowed the main scientific objectives stated for the Large Orbiting X-ray Telescope to still be carried out in HEAO-B at the expense of longer observing times.

The most serious restriction in the revised program was the imposition of a fixed life on all three HEAO spacecrafts. This was an unusual procedure for NASA since most spacecraft, although designed for a minimum specified life, were usually allowed to operate, if producing valuable science, until they failed. For instance, Uhuru, designed to operate for six months, continued its operation for years. The new policy was adopted by NASA's Office of Space Science as a result of considerable pressure from the Office of Management and Budget in an effort to control overall program costs by limiting post-flight mission operations and data analysis costs to fixed amounts. Most scientists involved in the program were convinced that this policy would eventually be abandoned.

A five-year struggle ensued to retain a maximum of the scientific content of the program within the boundaries set by the limited resources and the cost increases due to inflation and subcontractor overruns. We had to accept a reduction in the complement of instruments, a reduced performance specification for the telescope optics, and a minimization of test and calibration programs. An early victim of this approach was Columbia University's polarimeter, which was removed from the payload at the time of program revision. The high efficiency mirror and the All-Sky Monitor of the Goddard Space Flight Center suffered the same fate.

In the years 1973–1975 many serious deletions occurred in the program. The single test mirror, which was intended to test the fabrication techniques, was first delayed and then eliminated, causing the fabrication of the flight optics to proceed without any X-ray testing until the final system calibration. The two-chamber Imaging Proportional Counter (IPC) became a one-chamber IPC, and its gas was changed to argon-CO_2 to allow use of only one gas reservoir in common with the Focal Plane Crystal Spectrometer. This modification caused an endless string of problems from which the mission is still suffering. The Monitor Proportional Counter, a 1° field-of-view detector, was intended to have two units: one co-aligned with the telescope and monitoring the source; the other offset and monitoring the background. One unit (the background detector) was deleted, greatly reducing the scientific utility of the detector. The replacement of a 10-position filter wheel by a 2-position filter utilizing the same mechanism as the grating spectrometer greatly reduced the filter spectroscopy that could be done. Actually, in flight, one of the two filters failed and could no longer be used.

In addition, a number of prototypes, spare components, and other hardware items was reduced. The calibration program at the 1000-foot X-ray facility in Huntsville was shortened from the required six months to one month. HEAO-B scientists compensated for this in part by working, in shifts, 23.5 hours per day during that month.

Most serious, however, was that under these constraints we could not implement a magnetic torquing system for momentum dump. At issue, of course, was the extension of the mission from a one-year program to an open-ended one. HEAO-A and -B experimenters pleaded on many occasions with NASA Headquarters to reconsider this approach. In a letter dated June 20, 1974, to Dr. J. E. Naugle, Acting Associate Administrator for Space Science, Giacconi pointed out that "the HEAO-A and -B missions should be considered as more than just individual experiments. They are the only opportunities now existing for significant X-ray observations. They represent a unique national resource in astronomy. It is clear that they would continue to give important results for at least several years." He stressed, in the same letter, that "X-ray astronomy deserves and needs the commitment of resources that will establish permanent X-ray observatories in space."

The fact that these requests could not be granted essentially determined that we would have no major US X-ray facility in orbit after the demise of the HEAO-A and -B spacecrafts in late 1981 or early 1982 until the flight of the Advanced X-ray Astrophysics Facility (AXAF) in the late 1980s.

Using all their ingenuity the scientists of HEAO-B (Einstein Observatory), together with several Marshall Space Flight Center engineers, conceived and implemented a momentum management program, which minimizes the use of gas by selecting optimum times to observe specific targets, by deleting targets in unfavorable celestial positions which are not unique, and by pairing targets with off-setting momentum buildup when feasible. Although we have ameliorated the situation with respect to the consumable gas, thereby extending the mission from the original one year to the currently planned two and a half years, the low orbital altitude, the finite amount of gas, and possibly the low reliability gyros will cause the mission to terminate sometime in 1981.

1.5 *Scientific Objectives of the Einstein Observatory*

In spite of this rather severe obstacle course, the Einstein Observatory is still the most powerful X-ray astronomy mission ever flown. It has achieved a sensitivity to point sources of 10^{-14} erg cm^{-2} s^{-1} in the 1–3 keV band, a flux 500 times smaller than previously detected and about 10 million times smaller than that of the first extrasolar source Sco X-1 detected in 1962. The original observational objectives set in 1970 for the Large Orbiting X-ray Telescope (LOXT) have in general been met by the actual achievements of the Einstein Observatory, and will be summarized in the remainder of this article.

For a detailed description of the LOXT scientific objectives, which is too long to reprint here, we refer the reader to the Consortium Proposal of May 1970 (ASE-2410-III, pp. 2.0–2.4). It is of historical interest because it correctly anticipated many research areas in X-ray astronomy which later proved fruitful in the Einstein mission.

2. DESIGN PHILOSOPHY OF THE OBSERVATORY

In an undertaking as complex as the design and development of the Einstein Observatory an overall design philosophy is required. This philosophy includes both formal plans and statements and empirical approaches to problems. The Einstein experience can be used to illustrate three principles of the design philosophy which can be applied to future programs: the unified scientific approach, the establishment of spacecraft requirements at lowest acceptable levels that still allow operating flexibility, and the idea of "soft" rather than "hard" failure.

2.1 *Unified Scientific Approach*

The decision, described above, by AS&E, MIT, Goddard, and Columbia to submit a single proposal to build the experiment was the starting point for a unified scientific approach to the Einstein Observatory. Given the complexity of the undertaking, the limited relevant experience, and the high degree of cost control required, a unified approach was necessary to deal with the many problems that arose, and to minimize the negative scientific impact of the numerous decisions and trade-offs required in the course of the project.

The absence of a single imaging detector with all desired properties (high spatial resolution, large field of view, high quantum efficiency, and good spectral resolution) or a single spectrometer (high energy resolution and high efficiency) meant that a unified scientific approach was necessary in order to develop a complement of focal plane instruments capable of achieving common observational objectives. The Consortium scientists agreed that, in addition to obtaining data from the instruments for which their individual institutions had responsibility, all scientists would share in the data from all instruments.

2.2 *Spacecraft Requirements*

The general concept of the Einstein Observatory is similar to that of a ground-based one: a mirror is relatively rigidly connected to its focal plane where various instruments may be placed. This system can be pointed towards objects of interest. The mirror itself, however, was to be radically different. In addition to the design and development of this mirror (see Section 3), major areas of technical concern when the observatory was proposed were the pointing accuracy and stability that would be required and the reliability of the individual instruments. In the course of the project, we worked out specific technical requirements in these areas. These requirements were established at the lowest acceptable level that still allowed operating flexibility.

One important decision along these lines involved whether or not to process data on board the spacecraft. The energy of X-ray photons enables them to be detected individually with little background; furthermore, the counting rates for most objects are low, so it is often feasible to transmit the characteristics of individual photons to the ground rather than accumulating them in different categories on board the spacecraft. We decided to limit the Observatory to instruments in which individual photons are detected and to exclude those that integrate over extended periods, either in the detector or in on-board memory elements. The ability to transmit all of the data to the ground for reduction and analysis proved especially important when we recognized a problem in reconstructing high resolution images about one year after launch. The problem was traced to aspect distortions caused by residual magnetic fields; the effect can be removed by calibration of the star trackers coupled with knowledge of the earth's magnetic field. Had we processed data on board and not transmitted the raw data to the ground, this post facto correction would not have been possible, and several arcseconds degradation would have been unavoidable.

We also found that it was possible to relax the pointing accuracy and stability requirements significantly, from approximately the telescope angular resolution to a fraction of the size of the useful field of view. For stability, it is sufficient that acceptably small motion occur during the time required for an aspect determination rather than over the much longer time required for a complete observation. The sky coordinates eventually assigned to each photon are calculated from its detector coordinates and the pointing direction measured at the time of the event.

The final pointing tolerance allows the optical axis to be pointed anywhere within 0.5 arcminutes of the desired direction, and the motion of the optical axis to be anything less than one arcsecond during a second of time. The fields of view of all detectors are large enough to accept these errors, and the aspect system response is sufficiently fast that the allowed motion does not degrade resolution. The actual spacecraft performance is somewhat better than the allowed tolerance, and the relatively modest requirements were important in constraining overall costs.

This principle of establishing spacecraft requirements at the lowest acceptable levels should be an important part of any design philosophy since it minimizes new developments and associated costs. The price paid is more complicated reduction and analysis on the ground, but as we have seen above this also may allow the removal of previously unanticipated errors. The lowest acceptable level must still provide sufficient operating flexibility if one is to avoid pitfalls such as discussed in the section on observatory operations.

2.3 *Soft Failure*

Another principle of the Einstein design philosophy is that of "soft" failure. Whenever possible, redundancy in the design is such that a single failure cannot cause the loss of a substantial portion of the mission's scientific objectives. Examples are the use of three (rather than two) star trackers approximately aligned with the X-ray telescope axis, the design of a thermally matched X-ray optical train to guard against degradation or failure of the active control system, and the incorporation of redundant components in the design of the observatory.

Three sets of data are used to determine the aspect solution. The absolute direction and moderate frequency error terms are determined from star tracker measurements. The spacecraft includes three star trackers, each with a 2° by 2° field of view, sensitive to approximately 9th magnitude stars, and digitized to about 0.2 arcseconds. The fields of view are displaced along a diagonal, one being coaligned with the X-ray optical axis (x-axis) and the other two being approximately $2\sqrt{2}$ degrees away. This arrangement gives a probability of at least 95% of finding suitable guide stars for a target anywhere in the sky. Originally there were to be only two aspect-determination trackers, with two additional trackers of more modest capability looking in the y-z plane to be used by the spacecraft for attitude control. When we recognized that this control could be carried out with the slightly offset x-looking trackers, NASA and TRW (which was responsible for spacecraft design and construction) agreed to drop the two y-z trackers in favor of the third x-tracker. This redundancy has proven crucial with the early failure of one tracker.

The individual star tracker measurements required 0.32 seconds integration and have rms errors of 1.5 arcseconds for 9th magnitude stars. The prelaunch estimate of the systematic part of the error, due to residual magnetic fields, uncalibrated distortions, etc., was about 0.75 arcseconds, and the accuracy of the aspect solution could approach this value by using gyroscope signals to form a local aspect solution which effectively averages over a sequence of star tracker readings. The gyroscope errors are less than 0.3 arcseconds in a 10-second period, and so, for example, the rms error averaged over 10 seconds would be reduced to about 0.8 arcseconds. In fact, the performance in orbit showed magnetic field effects larger than expected, as mentioned above. The later corrections for these effects contributed an additional 1-arcsecond systematic error.

The lowest frequency error terms are due to incorrect positioning of the instruments in the focal plane, to thermal distortions, and to long-term structural changes. These effects are sensed with a fiducial light system, as shown schematically in Figure 4. Each instrument carries fiducial

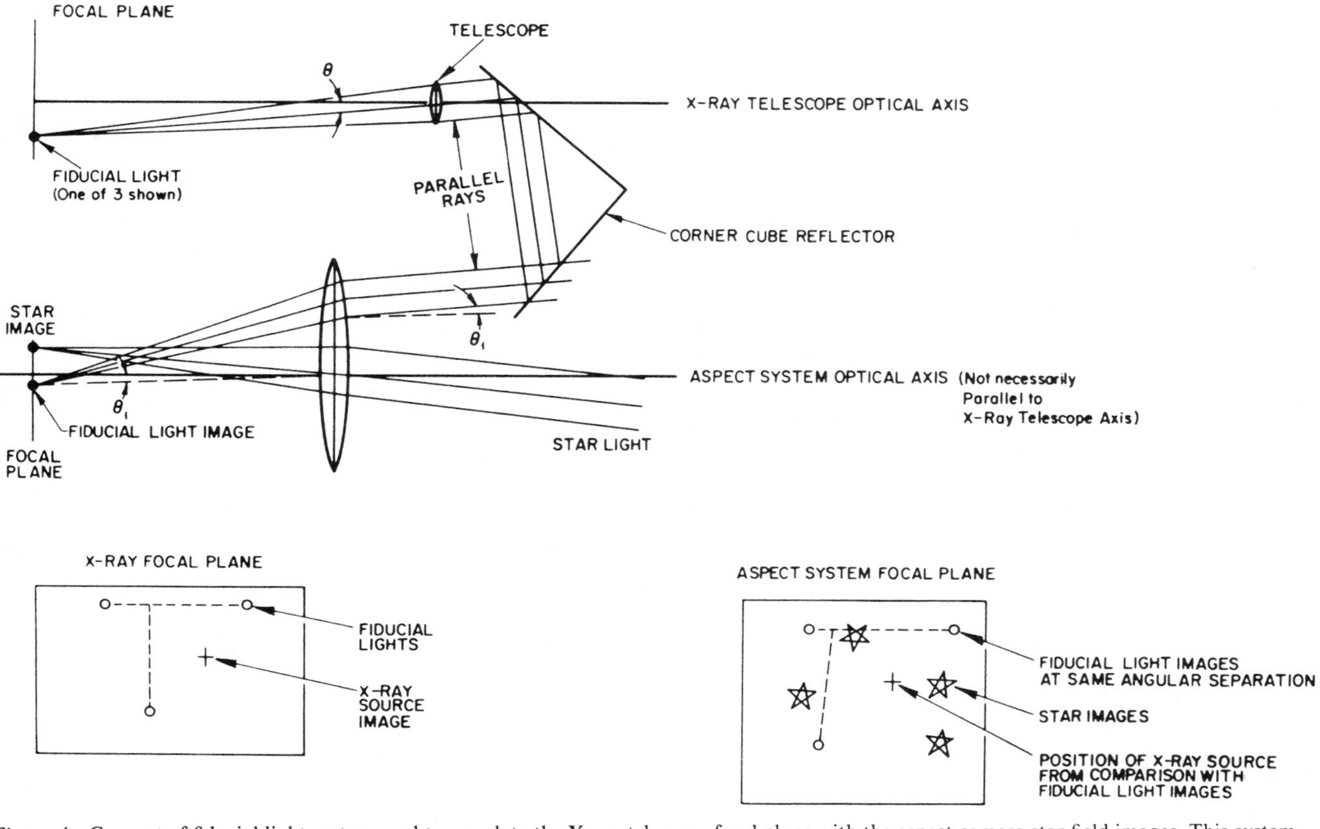

Figure 4 Concept of fiducial light system used to correlate the X-ray telescope focal plane with the aspect camera star field images. This system provides the data required to correct for small motions of the X-ray focal plane (from ASE 2410-III).

lights which are collimated by a lens near the nullpoint of the mirror, reflected by a corner cube, and imaged by the star trackers. Improper instrument positioning, bending of the optical bench, or motion of the aspect sensor optical axes can be detected and corrected using these signals. The only absolute dimension that must be maintained between components is the separation of the mirror and focal plane; this is accomplished with a low expansion graphite-epoxy optical bench designed and built by the Convair Division of the General Dynamics Corporation.

The use of materials with low coefficient of thermal expansion is common throughout the mirror assembly, the optical bench, and the focal plane housing and transport assembly. An active thermal control system is used to keep the mirror assembly temperature at 70° ± 2° F in order to minimize loss of resolution due to thermally induced deformation. At the same time the use of low expansion materials with matched thermal coefficients precludes the necessity of a more extensive active control system and also provides assurance of reasonable performance even in the event of failure of the active control system since image degradation is caused in this design only by temperature differential rather than by changes in the average temperature.

The final example of the principle of soft failure is the use of redundant focal plane detectors to prevent loss of scientific data-gathering capability because of the failure of a single detector. Also, critical spacecraft and experiment support systems are all redundant. The cost of building, testing, and integrating a second system is usually significantly less than that of the first, which also involves design costs. Since the Observatory is not serviceable after launch, the cost of the redundant systems is an acceptable price for avoiding loss of mission. Future space-borne astronomical observatories will probably be serviceable using the Space Transportation System (Shuttle) and, hence, the cost trade-offs between on-board redundancy and unscheduled servicing will have to be examined carefully for various different subsystems.

3. MIRRORS

Because the high resolution mirror assembly had to be both designed and developed, it was the most uncertain essential component of the Einstein Observatory. It was also most likely to dominate the construction schedule of the Observatory and therefore it received more attention than any other single item. The largest previous high resolution X-ray mirror was used in the AS&E S-054 experiment on the Skylab Apollo Telescope Mount; this mirror was about one tenth the size of the eventual Einstein high resolution mirror assembly. The S-054 mirror response included a narrow central peak, FWHM about 3 arcseconds, but large scatter—50% of the incident energy being outside of a 48″ radius at 1.7 keV and only 10% within a 5″ radius (Vaiana et al. 1977). Reduction of scatter by achieving smoother surfaces was and continues to be the major problem in X-ray telescope fabrication.

Several preliminary projects were carried out prior to the mirror development. Ray tracing programs were developed which gave theoretical mirror response functions; these were used to develop empirical relationships between resolution and optical design parameters (Van Speybroeck & Chase 1972, Chase & Van Speybroeck 1973), and to perform design trade-off studies. The theoretical effective area of the final design is shown in Figure 5*a*, and the resolution in Figure 5*b*. An experimental program to measure the scatter from polished flats also was conducted using a method suggested by Zehnpfennig and described by Van Speybroeck (1973). The object of the scattering measurements was to choose a satisfactory material for the mirror substrate. Though this program did not include sufficient samples to establish statistical distributions it did demonstrate that 1. fused silica flats could be polished to yield little scatter outside of a one-arcsecond aperture, and 2. evaporated nickel or platinum, which yield better energy response than the bare fused silica, did not introduce significant scatter.

We thus knew that there were no fundamental barriers to achieving a one-arcsecond mirror, although, of course, it is much easier to polish flats than the conics of revolution of X-ray telescopes. There were other substrate materials and polishing techniques tested, but, in general, the number of samples was small and satisfactory materials or techniques may have been rejected because of a single defective example. Our objective, however, was to choose a possible material and technique, and this was accomplished. We also were guided by the experience obtained in developing smooth surfaces for laser flats and other applications—particularly the work of H. Bennett, J. Bennett, and associates.

The choice of a fused silica substrate caused substantial engineering problems. The S-054 mirrors, for example, had been made of beryllium, a single, very stiff material. Fused silica has poorer stiffness characteristics and must be supported by a cell of other materials. It has a low coefficient of thermal expansion, which must be matched by the support cell. A mirror assembly must be assembled and tested on the ground in a gravity environment, and then used in orbit. W. Antrim, R. Hall, and other members of the AS&E mechanical engineering department created the final design after many iterations, including important contributions from scientists then at AS&E and engineers at the Perkin-Elmer Corporation. The cell had to support the mirror with small distortions for

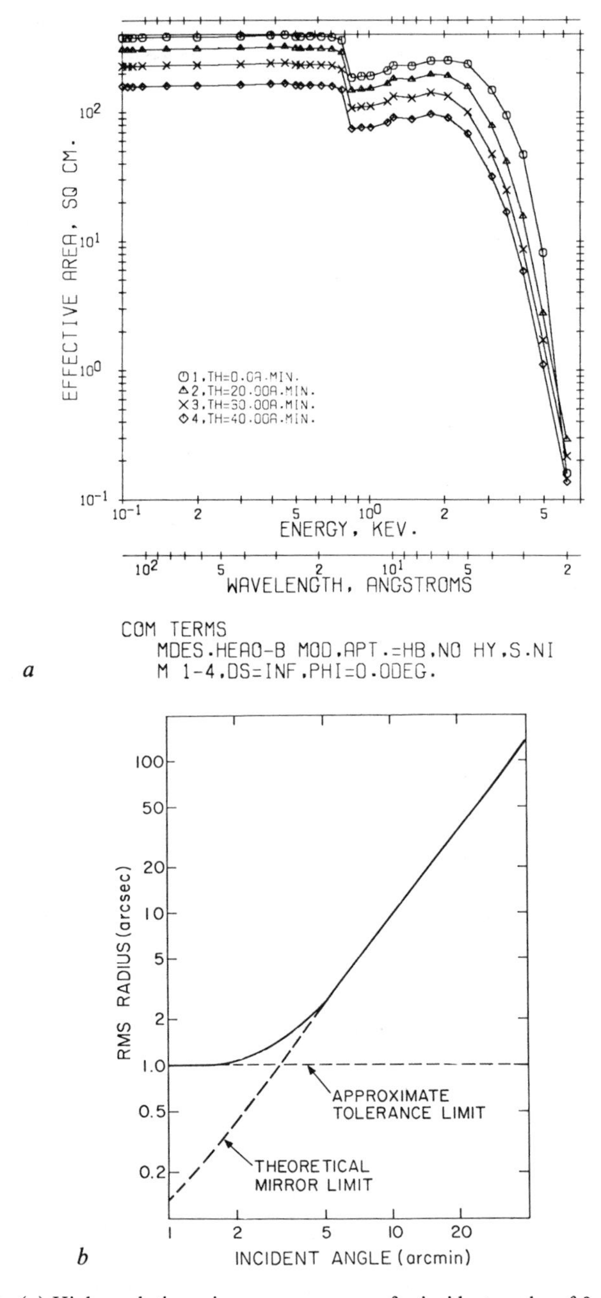

Figure 5 (*a*) High-resolution mirror area vs energy for incident angles of 0, 20, 30, and 40 arcminutes. (*b*) Mirror theoretical resolution vs incident angle.

ground testing, protect it in flight, be thermally and temporally stable, be non-contaminating, and, if possible, allow disassembly and reassembly without loss of alignment. This was achieved with an invar and graphite-epoxy structure as shown in Figure 6. There are four concentric confocal mirror pairs; each ray strikes a paraboloid and then a hyperboloid. The focal length is 3.44 meters, the diameters of the paraboloid optical surfaces vary between 0.34 and 0.58 meters. The reflecting surfaces are fused silica or quartz with evaporated nickel coatings; these are supported by an invar center mounting plate and invar end rings; the latter are supported from the center mounting plate by graphite-epoxy cylinders. The fused silica mirrors are bonded to the support cell by means of flanges which provide tangential support but are deliberately weakened in the radial direction to avoid coupling radial forces into the mirror elements.

The fact that the fused silica mirror elements were not self-supporting greatly complicated the design task. The mirror elements could not be polished in the final support cell, and therefore had to be placed in the cell without introducing distortions. The original concept involved polishing the glass elements in fabrication fixtures and then transferring the mirror elements to the final support cell with the glass always being supported. This approach did not work.

The fused quartz material for the mirror walls was provided by Heraeus Schott Quarzschmelze of Hanau, Germany. The approximately

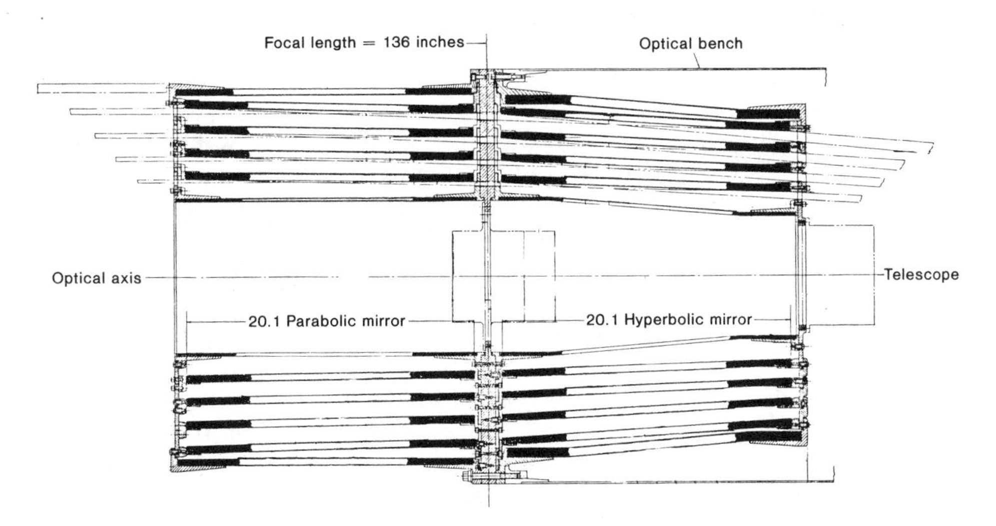

Figure 6 Mirror mechanical design. The eight fused quartz mirrors are supported by invar flanges; the end flanges are supported by graphite-epoxy cylinders. The assembly is mounted from the center invar plate.

conical blanks were made by fusing barrel stave–shaped elements in a centrifuge. This involved procedure required almost one year for completion.

The mirror was fabricated at Perkin-Elmer by a team headed by P. Young. The Smithsonian Astrophysical Observatory, AS&E, and NASA–Marshall Space Flight Center groups continued to work closely with Perkin-Elmer Corporation during the fabrication of the mirror assembly. It is greatly to Perkin-Elmer's credit, and especially to those directly involved in the project, that this degree of cooperation was achieved. It was beneficial to the program, and yet it had to be a burden upon the Perkin-Elmer staff to operate with such constant customer presence.

Young (1979), Mathur et al. (1979), and Ledger (1979) describe the fabrication of the mirrors. The mirror elements were first diamond ground to approximate shape. Measurements taken on the larger elements at this time were not reproducible because of inelastic deformations of the rigid fabrication fixtures which induced changes in the shapes of the optical elements. We were much more successful after abandoning the rigid fixtures and depending upon loose abrasive grinding to achieve the final shapes. This is a fundamentally more satisfactory approach: the only shape that can be reproduced at the desired tolerances is the free one; any fixture-constrained shape is essentially unreproducible. However, great care is required to achieve the "free" state, and the elements ultimately had to be supported in mercury while being measured.

The basic polishing of the mirror was accomplished horizontally using a submerged process similar to the "bowl-feed" method described by Dietz & Bennett (1966), but adapted to the X-ray telescope geometry. The process was iterative: a correction was made, the surface was measured, and another correction was made. The most difficult tolerances were the axial slope and surface roughness. Most tolerances were achieved.

The mirror fabrication presented difficult problems beyond those normally encountered in programs of this size because of our lack of experience with X-ray optics. The people making the mirror were not able to test it with X rays, and, consequently, all specifications were indirect—essentially manufacturing tolerances rather than performance requirements. Their own extensive experience with visible light optics was not transferable to X-ray optics, which have an unfamiliar shape, and whose eventual performance depends critically upon parameters that often are not specified for visible light optics. Finally, the theory of scattering from rough surfaces in the X-ray wavelength was poorly developed at the time and not supported by experimental data. Consequently, in many instances, decisions were based on judgment rather than on solid theoretical or experimental grounds. We knew that a small degradation of the mirror surface finish, for example, could result in a large loss of performance, but we could not make quantitative estimates of this loss. We, therefore, strove to carry each process to the point where little further improvement seemed possible without major changes in the production procedures. This naturally led to conflicts with the NASA staff responsible for the project budget.

Although the mirror represented less than 5% of the cost of the Einstein Observatory, it had a large potential cost impact if its delivery was delayed. There also was a psychological impact—as long as the mirror was not delivered, all other contributors to the project felt under less pressure to complete their tasks. The actual extra time required to approach the limits imposed by the fabrication procedures and equipment being used typically was measured in hours or days, and appeared to us as an inconsequential fraction of a ten-year project. The NASA program office feared that longer times would be required, and were very conscious of the cost of each day of the total program. They, therefore, were inclined to take a lenient view of the specifications and to relax them if possible. T. Kirchner of AS&E and P. Young of Perkin-Elmer expended great effort rearranging the schedule to provide extra hours for critical procedures with minimum impact on the final delivery date. They also kept the NASA program office informed of the state of the project. Not withstanding these efforts, the NASA program office did direct Perkin-Elmer to stop polishing the inner elements before the surface roughness specification was achieved, in part because they were behind schedule and thought that the as yet untried coating and alignment procedures would require longer than the time scheduled for them. These concerns, although understandable, proved to be unjustified.

The program had originally included an X-ray test mirror, which was intended to provide learning experience and to allow X-ray testing before the final polishing of the flight mirrors. Unfortunately, this mirror was delayed until it could not be tested before the final polishing of the flight mirror and had to be dropped from the program. This test mirror might have given both scientists and administrators a better basis for decisions had it been completed.

The mirrors were coated with chromium and nickel after polishing (nickel being the reflecting surface). The equipment used for coating the mirrors (Mathur et al. 1979) fit inside the conical elements to enable coating at normal incidence, which resulted in the least additional surface roughness. X-ray measurements performed on witness samples, which were coated prior to and concurrently with the X-ray mirrors, showed no additional scattering due to the coating.

The mirror elements were aligned optically and bonded to the support cell. Each mirror element was supported during this procedure by 32 calibrated forces placed so that the mirrors essentially floated and could assume their unconstrained, gravity-free state; this quite successful approach was devised by Perkin-Elmer. The mirrors were positioned by a minimum set of fixed attachments designed not to overconstrain, and, therefore, possibly deform the mirror. The sensing of the proper optical alignment was not difficult, but thermal distortions of the support fixtures during bonding caused small alignment errors.

The X-ray calibration of the mirror system is described below. The central part of the system response is about 3.5 arcseconds FWHM independent of incident energy; the mirror is responsible for most of this width, but the detector and test setup also contribute somewhat. This performance was expected from the measured axial slope, roundness, and alignment errors. The mirror has an effective focal length of 3.44 m; this determines the focal plane scale of 1 arcminute per mm.

The mirror has survived a few months' storage, a one-year integration period, X-ray testing, launch, and almost two years in orbit without degradation. The performance in flight is indistinguishable from that measured on the ground.

4. FOCAL PLANE INSTRUMENTATION

Four focal plane detectors are available to investigators utilizing the Einstein Observatory (Giacconi et al. 1979b). Two of these, the High Resolution Imager (HRI) and Imaging Proportional Counter (IPC), as their names suggest, provide spatial resolution and are selected for studies requiring imaging. The HRI offers the highest spatial resolution, in fact, better than that of the telescope itself. However, it offers no energy resolution, its field of view is small, and its quantum efficiency is low (except at the longest wavelengths) compared to the IPC. The IPC has better overall quantum efficiency, provides a modest degree of energy resolution based on its pulse amplitude, and has good particle background rejection capability. Its angular resolution, typically 1.5′, is only moderate. The other two focal plane instruments are devoted to spectroscopy. The Solid State Spectrometer is a non-dispersing detector with good quantum efficiency and low noise above 0.8 keV. The highest energy resolution is obtained using the Focal Plane Crystal Spectrometer. For sources that are sufficiently strong, dispersive X-ray spectroscopy with the Focal Plane

Crystal Spectrometer provides the best opportunity to identify and study line components in the X-ray spectra.

Two auxiliary instruments located forward of the focal plane are available to augment the capabilities of the observatory. One is an Objective Grating Spectrometer whose dispersed spectrum is viewed by the HRI. The entire spectrum from point sources can be observed below 2 keV with the Objective Grating Spectrometer. For strong point sources, it allows one to detect the most prominent lines and to obtain the shape of the continuum. The other auxiliary instrument consists of two filters (beryllium and aluminum). Either one can be inserted forward of any of the focal plane instruments to remove soft X rays and to modify the detected spectrum in a manner that enhances the ability to obtain broad-band spectral information with the imaging detectors. In practice, the filters have not been utilized during the first twenty months of observatory operations in order to minimize the operation of a complex mechanical system

Table 1 Characteristics of Einstein Observatory instruments

Instrument	Field of view	Spatial resolution (50% diameter) within field of view	Effective area
High Resolution Imager (HRI)	25′ diameter	2″ within 5′ of axis (determined by mirror response)	$20\ cm^2$ at 0.25 keV $10\ cm^2$ at 1 keV $5\ cm^2$ at 2 keV
Imaging Proportional Counter (IPC)	60′ × 60′ (with loss of four 3′ wide strips)	1.5′ at 1.5 keV	$100\ cm^2$ at 0.25 keV $110\ cm^2$ at 1.5 keV $30\ cm^2$ at 3 keV[c]
Solid State Spectrometer[a]	6′ diameter		$110\ cm^2$ at 1.5 keV $30\ cm^2$ at 3 keV[c]
Focal Plane Crystal Spectrometer	6′ diameter 1′ × 20′ 3′ × 30′		$1\ cm^2$ at 0.25 keV 0.1–1 cm^2 at $>$ 0.5 keV
Objective Grating Spectrometer (used with HRI)			$< 1\ cm^2$
Monitor Proportional Counter[b]	1.5° × 1.5°		$667\ cm^2$

[a]Operations terminated September 1979 with exhaustion of coolant.
[b]Outside focal plane.
[c]Limited by mirror > 2 keV.

Table 1 (*continued*)

Instrument	Energy resolution ($E/\Delta E$, FWHM)	Time resolution	Principal uses
High Resolution Imager (HRI)		8 μsec	High resolution images. Highest sensitivity in long exposures.
Imaging Proportional Counter (IPC)	1 at 0.25 keV 1.3 at 1.5 keV 2 at 3 keV	63	Images of diffuse sources. Fluxes and spectra of sources. Highest sensitivity in short exposures.
Solid State Spectrometer[a]	14 at 1.5 keV 20 at 3 keV	2–5000	Nondispersive spectroscopy of sources.
Focal Plane Crystal Spectrometer	50–100 at < 0.4 keV 100–1000 at > 0.4 keV	8	Highest spectral resolution, measurement of line profiles.
Objective Grating Spectrometer (used with HRI)	50		High spectral resolution, measurements of entire spectrum of point sources. Measurements of continuum.
Monitor Proportional Counter[b]	5 at 6 keV	10 (TIP mode)	Time variations. Extension of flux measurements beyond 3 KeV.

following a partial failure of the Be filter mechanism during checkout in the early days of the mission.

Situated outside of the focal plane, the Monitor Proportional Counter has a viewing direction that is parallel to the telescope axis. It operates continuously and provides information on the intensity or spectrum of strong sources in the energy range beyond the high energy cut-off of the grazing-incidence optics. It is also used to measure changes in these quantities with time, both for investigations concerned primarily with time variations and as a service for observers utilizing the imaging and spectroscopic detectors where information on temporal variations is needed for the interpretation of the data.

Table 1 summarizes the characteristics of each of these instruments. A capsule description of each focal plane detector is given below. The material is an update of the description given by Giacconi et al. (1979), based upon the initial twenty months of Observatory operation. The descriptions reflect the instrument configurations in actual use in the Einstein Observatory.

4.1 *High Resolution Imager*

The High Resolution Imager (HRI) was developed at the Harvard/Smithsonian Center for Astrophysics specifically for the Einstein Observatory. Unlike the other instruments, which are adaptations of devices that have operated in other applications, there was no previously existing prototype. The imager provides 1-arcsecond spatial resolution over the central 25′ region of the focal plane. X-ray events are read out individually with their spatial positions in two dimensions and their time of detection. The image of a source can be reconstructed to a spatial resolution that is comparable to that of the telescope itself, and centroids of the blur circles of point sources are limited only by knowledge of the Observatory aspect as a function of time. Three separate HRIs are included in the Observatory to provide redundancy. A detailed description of the HRI has been given by Henry et al. (1977). Excerpts from their description are given below.

The detector consists of two microchannel plates in a cascade configuration. The front plate has a tube length-to-diameter ratio (L/D) of 80, center-to-center tube spacing of 15 μm, and a bias angle of 0° (the tubes are perpendicular to the face of the plate). The plates are separated by a gap of 38 μm. The second plate has an L/D ratio of 80 and a bias angle of 13°. After another gap there is a crossed grid of wires. These wires are electrically connected to a resistor chain and every eighth wire is connected to a charge-sensitive preamplifier. In back of the crossed grid is a metal reflector plate, which is necessary for obtaining good imaging performance.

The input face of the first microchannel plate is coated with MgF_2 to improve the quantum efficiency, which is a function of incident angle and X-ray energy. In addition, there is a normal nickel electrode coating applied by the plate manufacturer on both front and back faces of each plate. The X-rays enter at angles from 2.5° to 5° with the normal to the front face of the first plate, due to the geometry of the grazing-incidence telescope reflective optics. Of order 10^{-1} of them produce a photoelectron, which creates a cascade resulting in $\sim 10^{7}$–10^{8} electrons leaving the back of the second plate. This cascade emerges from a few tubes at surface (*4*) as illustrated in Figure 7. The electrons have axial energies distributed from about 30 to 300 eV and transverse energies of $\sim$ 30 eV. They are accelerated axially, adding another 250 eV. The charge cloud spreads to a diameter varying between 0.5 and 2 mm in the vicinity of the crossed grid, thus striking several wires. This spreading is essential for position deter-

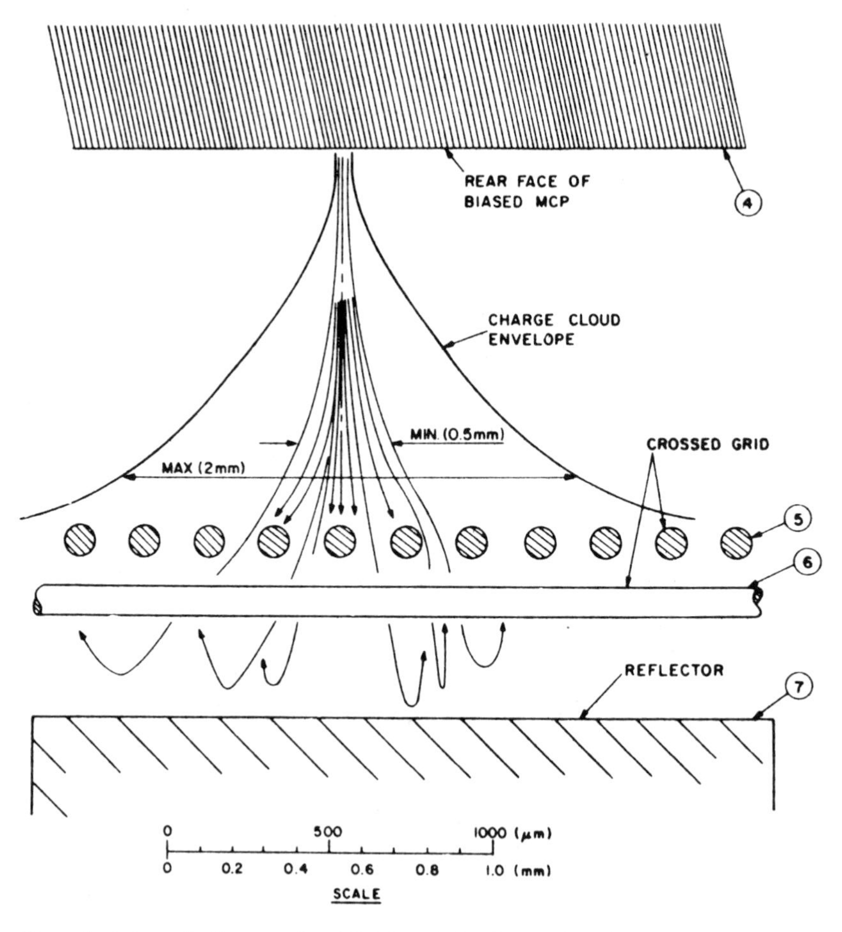

Figure 7 Schematic diagram of the High Resolution Imager (HRI). Electron cloud exiting from rear of microchannel plate (MCP) spreads over crossed grid assembly. Two-dimensional centroid position of charge cloud is determined from charge division.

mination. Since the charge cloud is collected by several wires, the electronic processor can measure the centroid of it to a sufficient accuracy that a resolution of a small fraction of the wire spacing is achieved.

The centroid of each event is measured in two orthogonal directions in the telescope focal plane (parallel to the input microchannel plate face) by collecting half of the charge from each event on each crossed grid axis. The coarse position is measured digitally by determining which of the seventeen preamps per axis has received the largest signal. The local or fine position is measured relative to the coarse position by charge division along the interwire resistor chain. Since the charge cloud only spreads to encompass a few wires, a valid event will be detected by at most three preamps. This enables the electronics to be simplified by summing the outputs of every third preamp. This scheme is partially redundant against single-preamp failure and has a large signal-to-noise margin against RFI and other forms of electrical noise that are encountered in a satellite environment.

The count life of the plates is more than sufficient for the expected life of the Einstein Observatory, since the high resolution detector will only be operated for about 10^7 s (due to the use of other detectors on the focal plane) and the expected counting rate from most known X-ray sources is less than 10 s^{-1}. Furthermore, the jitter in the spacecraft pointing system will move the image through a finite area so that even bright X-ray sources will not burn out individual tubes.

The entire assembly of the microchannel plates, crossed grids, and UV/ion shield is enclosed in a vacuum housing. This permits ground testing of the plates, provides a stable environment, and substantially reduces the launch loads on the parylene UV/ion shield. The vacuum is maintained by a small attached ion pump. In orbit, a vacuum door opens to permit the focused X-ray beam to enter the detector. (This door is the only moving part of the instrument.)

An in-flight calibration system, consisting of a weak diffuse X-ray source mounted in the vacuum door, a focused UV source which projects a calibration pattern onto the detector, and a set of fiducial lights that are projected into the star trackers, is included with each detector. Microchannel plates have a very small sensitivity to far UV. This permits the detection of a UV calibration pattern projected onto the plates using a UV light source and appropriate optics.

As measured in orbit the non–X-ray background rate of the HRI is $\sim 5 \times 10^{-3}$ cts/(arcmin)^2 -s which is about an order of magnitude larger than the diffuse cosmic X-ray contribution (from a region free of strong sources) expected at the focal plane of the telescope. This rate is about a factor of three above laboratory levels and varies with orbital position.

Three components contribute to the non–X-ray rate in approximately equal amounts. The one component observed in the laboratory is temperature insensitive and is probably due to field emission operating on microchannel plate defects such as cracks or whiskers. Another component is due to residual UV sensitivity in an environment of geo-coronal emission. The third component can be attributed to cosmic rays and trapped particles which vary with orbital position. Three identical detectors were flown to provide redundancy.

4.2 *Imaging Proportional Counter*

The Imaging Proportional Counter (IPC) is a position-sensitive proportional counter which provides the Einstein Observatory with good efficiency and nearly full focal plane coverage, plus moderate spatial and

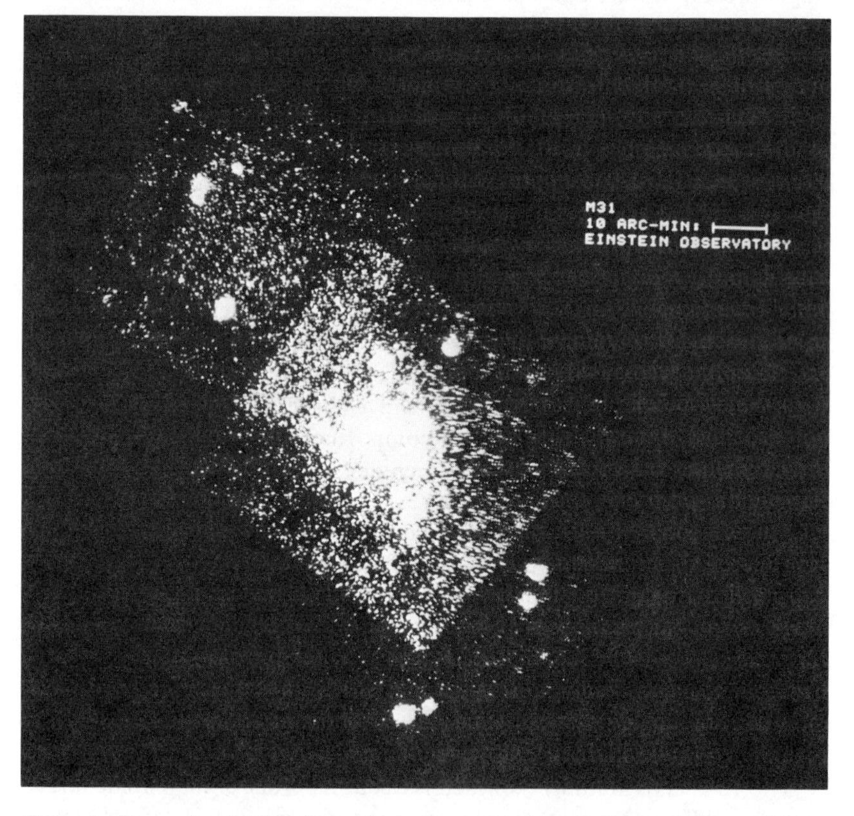

Figure 8 Composite of three observations with the Image Proportional Counter (IPC) of galactic-type X-ray sources in M31.

Figure 9 HRI observation of central portion of M31 showing resolution of extended IPC region into many discrete sources.

spectral resolution. It was developed by the Harvard/Smithsonian Center for Astrophysics. It is generally used to measure the X-ray flux of quasars, active galaxies, and stars and to image extended emission from clusters of galaxies and large-diameter supernova remnants. The large field of view of the IPC is responsible for most of the serendipitous source discoveries of the Einstein Observatory.

The complementary manner in which the HRI and IPC are used is exemplified by the observations of our twin galaxy, Andromeda (M31). Figure 8 is a composite of 3 IPC exposures which use the $60' \times 60'$ field of view to cover the entire galaxy. The spiral structure of M31 can be seen from the distribution of the outermost discrete sources. The central region of the galaxy is not fully resolved in the IPC detector with its $1'.5$ resolution. The HRI data shown in Figure 9 for the center of M31 resolve this central extended IPC region into $\sim$ 30 separate discrete sources. Time variations have been detected in several of these HRI sources, including one identified with the nucleus of M31. Over 80 sources have been detected in the combined IPC and HRI data. The data are being used to study the spatial and luminosity distributions of the X-ray sources and their associations with various galactic features.

The on-axis effective area of the primary detector (IPC) in conjunction with the mirror reaches a maximum of approximately 150 cm^2. The effective area falls off below 1 keV due to the transmission of the detector window which is coated with 2 μm of polypropylene with a thin layer of Lexan plus 0.2 μm dag on the interior to provide conductivity. At high energy the area falls off at 4 keV due to mirror inefficiency.

The counter body houses the electrodes in a density-regulated gas mixture at a pressure of 800 torr, STP. The composition of the gas is 84% argon, 6% xenon, and 10% CO_2 as supplied to the detector. The equilibrium composition of the gas mixture is maintained passively by the gas resupply system which achieves a balance with diffusive losses through a controlled leak. The detector system contains a calibration source in which a fluorescent line of aluminum is produced by α-particle bombardment. The spectral position of the aluminum line, as observed in a set of pulse height channels, is monitored every 12 to 24 hours to correct for changes in gain caused by temperature or gas composition changes.

Early in the operation of the Einstein Observatory, the transparency of the controlled leak decreased to the point where the differential diffusion of the window became an important factor in determining the equilibrium concentration of CO_2. The IPC gain had to be actively controlled by commanding the anode high voltage to the appropriate step to keep it within the desired range. In January 1980, a second controlled leak was placed in parallel with the first. Consequently, the equilibrium CO_2 concentration increased, the detector gain was much less sensitive to temperature, and it was no longer necessary to vary the high voltage to keep the gain variation within acceptable limits.

The IPC electrodes consist of a plane of anode wires (12-μm gold-coated tungsten) centered between two orthogonal planes of cathode wires (75-μm nichrome). Each cathode plane is effectively a single wire wound back and forth at a pitch of 1 mm. A 3-mm spacing separates the anode from each of the cathodes. Avalanches at the anode result in localized induced signals in each cathode that propagate to both ends of the plane. The difference in the rise time between the two signals permits a determination of the X-ray position, a method that was developed originally by Borkowski & Kopp (1972). The diameter of a circle which encompasses

50% of the counts from a point source image is 1.5 mm (1.5′) at 1.5 keV and improves to 1 mm at 3 keV. Below 1.5 keV, the spatial resolution broadens approximately inversely as the square root of the energy. The centroid of the resolution circle can be determined to a precision of 20″ when calibration measurements are used to generate a correction function that compensates for distortions. The 20″ figure represents the one sigma precision for determining positions of point sources.

A modest degree of energy resolution is provided by 32-channel digitization of the pulse amplitude of the anode signal. The intrinsic energy resolution, $E/\Delta E$, where ΔE is the full width at half maximum of the peak due to a line at energy E, is typically 1.3 at $E = 1.5$ keV. The gain has been observed to vary over the active area of the detector with the larger point-to-point gain changes amounting to 10–20%. The greatest challenge in the calibration of the IPC has been to map out the spatial variation in gain so that the spectrum of sources can be obtained accurately to the degree allowed by the energy resolution and count statistics. Almost two years after launch this project is still in progress with the gain of the central region reasonably well mapped by inflight observations of a point source. Calibration of the entire field would require one month of observations.

Background from particles is rejected by a wire plane in the vicinity of the rear wall which provides an anti-coincidence signal. There is further particle background rejection inherent in the differential risetime method of reading out positions. The Lexan coating for the window seems to be effective at eliminating residual UV sensitivity. The counting rate in the IPC from a region free of strong sources is 5×10^{-4} cts/s-(mm)^2 in the energy range 0.25–3 keV. Especially at lower energies, most of these counts can be attributed to the diffuse X-ray background. More details on the detector have been given by Gorenstein et al. (1975), Harvey et al. (1976), and Humphrey et al. (1978).

4.3 *Solid-State Spectrometer*

The Solid-State Spectrometer, developed by the Goddard Space Flight Center, uses a Si(Li) crystal for the observation of X rays in the energy range 0.8–4 keV in a non-imaging mode. As a nondispersive spectrometer, it offers a distinct improvement in energy resolution over equivalent proportional counter techniques. The resolution of the flight system is 160 keV over virtually the entire energy range. Above 1 keV the effective area of the Solid-State Spectrometer is similar to the IPC and much higher than for the Focal Plane Crystal Spectrometer and the Objective Grating

Spectrometer. This makes the Solid-State Spectrometer the instrument of choice for spectral observations of moderate intensity sources.

Each detector consists of a 9-mm diameter (6 mm effective), 3-mm thick Si(Li) chip, which has two concentric grooves machined on the n-side or reverse of the crystal. A ≤ 200 Å thick gold contact is deposited on the front face of each detector to bias the central detector volume inside the innermost groove. This gold layer is clumped and irregularly distributed on a scale of microns. The average transmission for soft X rays through the gold layer is therefore smoothed by mounting the detector slightly out of the telescope focal plane to produce a resultant defocusing to a size of less than 1 mm^2. A 0.1-μm parylene filter with an aluminum coating eliminates sensitivity of the Solid-State Spectrometer to UV and visible light.

Two detector assemblies are mounted in a common solid ammonia/methane cryostat, which maintains the operating temperature of the solid-state devices at approximately 100 K. Redundancy is provided since either of these detectors may be positioned at the telescope focus by command. For additional details of the instrument, see Joyce et al. (1978).

A design problem in the cryogenic cooling system was detected during pre-flight calibration. A time-dependent buildup of ice on the detector surface arose from the cryopumping of water vapor from the insulation material. The detector was periodically "defrosted" by heating to 220 K, but for the first seven months of the mission, observations typically had an accumulated ice buildup in excess of 10^{-4} cm corresponding to an effective low energy cut-off of 0.8 keV. During the final four months of operations, the problem ameliorated to the point where the effective cut-off had fallen to 0.55 keV. The operation of the Solid-State Spectrometer ended after 11 months in orbit when the ammonia/methane cryogen was expended.

The total background in the spectrometer due to all nonsource contributors is 0.28 s^{-1} above 0.5 keV, and 0.19 s^{-1} above 1 keV. Background is not measured simultaneously with each source observation, and must be estimated from data accumulated over source-free exposures at other times. Uncertainties in the background exist at the 10% level, and can compromise the spectral fitting for very weak sources.

4.4 *Focal Plane Crystal Spectrometer*

The Focal Plane Crystal Spectrometer is a curved crystal Bragg spectrometer developed by MIT. X rays from a celestial source are focused by the X-ray telescope and pass through an aperture/filter wheel assembly

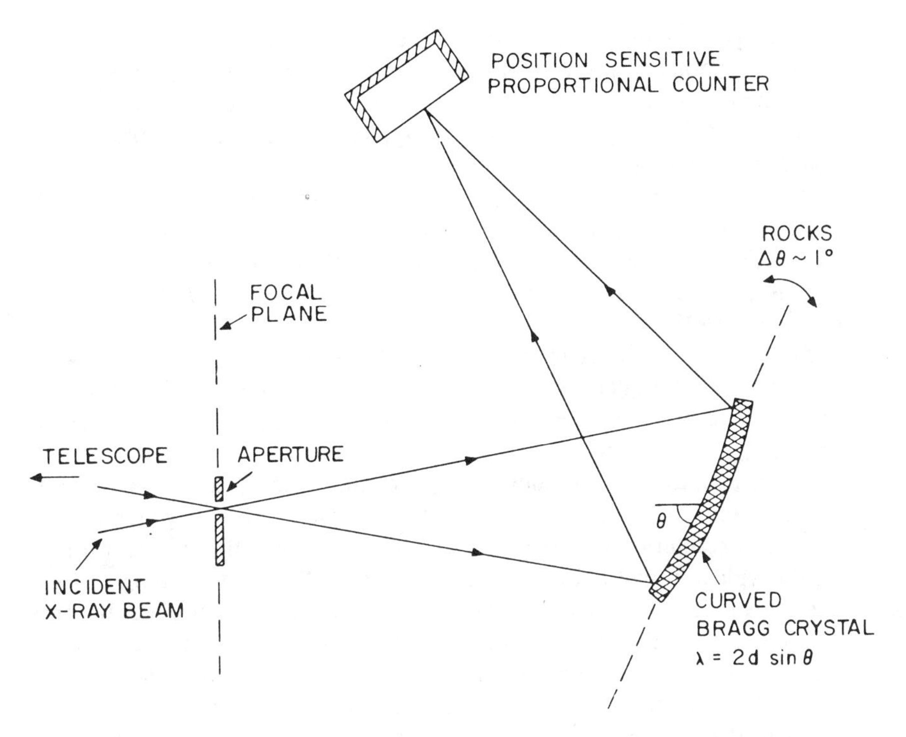

Figure 10 Schematic diagram of the Focal Plane Crystal Spectrometer.

located at the focal plane. They diverge in an annual cone with half-angles of 2.5° to 5° and strike a curved diffractor. X rays satisfying the Bragg condition are reflected by the crystal and detected with a multi-wire proportional counter. This technique results in high spectral resolution, but relatively low efficiency. Figure 10 is a schematic diagram of this spectrometer. Six crystal diffractors are available on a turret mount. The crystals that were used most often are TAP and RAP, which cover the energy bands 0.7–1.2 and 0.5–0.8 keV respectively. These bands include lines from O VII and O VIII.

The relative position of the diffractor and the point in the focal plane from which the X-ray beam diverges is maintained so that each lies on the circumference of a circle (the Rowland circle) with a diameter equal to the radius of curvature of the diffractor lattice. As the Bragg angle is changed, the relative position of the diffractor and focal plane is also changed. When the Focal Plane Crystal Spectrometer is operated as a conventional curved crystal spectrometer, the detector is positioned so that it also lies on the Rowland circle. The diffracted X-ray beam is refocused at this location to form an astigmatic image of the X-ray intensity distribution at the telescope focal plane. The effective detector area, defined post facto on the ground, is typically less than 1 cm^2, so the non–X-ray background is small.

A particular spectral feature is studied by scanning a narrow range of crystal angle (and therefore of wavelength) with all other elements held fixed—the deviation from ideal Rowland circle geometry during such a scan is small. (For more details, see Canizares et al. 1977, Donaghy & Canizares 1978.)

The imaging property of the spectrometer and detector allows study of extended sources such as supernova remnants. These can be observed through one of four ($3' \times 30'$, $2' \times 20'$, $1' \times 20'$, and $6'$ diameter) apertures.

When operated in the conventional manner, the Focal Plane Crystal Spectrometer will achieve its highest sensitivity, but will generally be limited in resolution by the geometric effects of the diffractor curvature. These typically give a Bragg angle spread of 2′–20′, depending on the angle. For extended sources the resolution is further limited by the imaging capabilities of the detector. For the study of point sources it is possible to improve the spectral resolution by departing from the conventional Rowland circle configuration. In this mode the detector is located closer to the diffractor so that the reflected X rays are intercepted before they converge to the usual line image. The dispersion of the curved diffractor is utilized to improve the spectrometer resolution at the expense of a larger effective area at the detector and thus a higher non–X-ray background.

The detector is a multi-anode, position-sensitive proportional counter with resistive anodes. It has $\sim$ 1-mm resolution along a direction parallel to the plane of the spectrometer and 1-cm (wire-to-wire) resolution along the orthogonal direction. Anticoincidence anodes are located on three sides of the active volume and are used to reject non–X-ray events. There are two detectors which share the gas supply with the IPC and have a similar pressure regulation and replenishment system. The instrument covers the range 0.2–3 keV, with a resolving power, $E/\Delta E$, of 50–500. A typical observation involves a scan width of three to five spectral resolution elements centered on some specific spectral feature, and lasts 5000–40,000 s. Figure 11 is the spectrum of Pup A as observed by the Focal Plane Crystal Spectrometer (Winkler et al. 1980). Details of the oxygen line structure, as well as other lines, are seen.

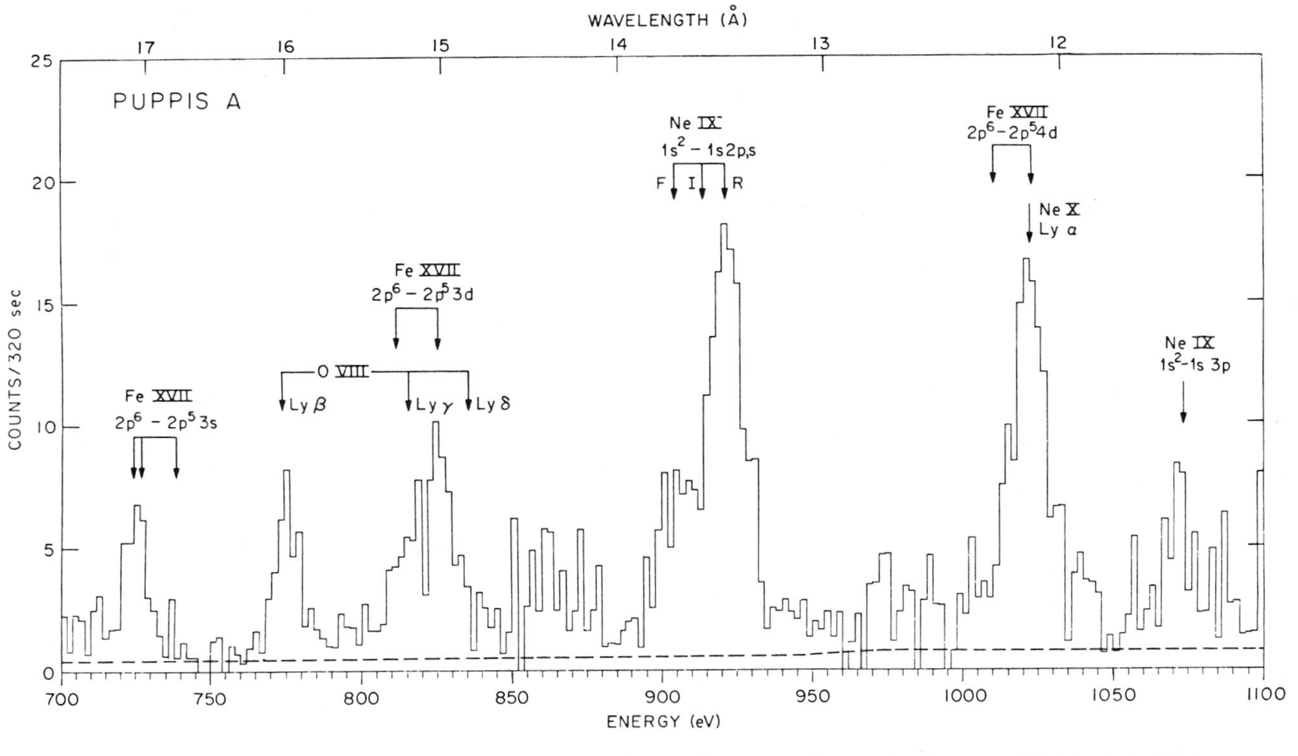

Figure 11 X-ray lines in the supernova remnant Pup A as observed by the Focal Plane Crystal Spectrometer (Winkler et al. 1980).

5. OBSERVATORY OPERATIONS

5.1 *Introduction*

The planning and operation of the Einstein Observatory presented a significant challenge, occurring as it did at a transition point in X-ray and space astronomy from principal investigator experiments to use of national facilities. Although contracted by NASA to be built and managed by a single principal investigator, it was recognized as a potentially powerful observatory by the scientists involved, and it, in fact, became a widely used astronomical facility. The areas of calibration, mission planning, mission operations, and data analysis were greatly increased in scope, though little in budget, from the initial conception, in order to prolong mission life, to allow the participation of hundreds of guest observers, and to meet the general requirements of an ongoing observatory.

A specific center for scientific operations to coordinate the observing program and to carry out the data reduction and analysis was needed to increase the scientific usefulness of the Observatory. An overall systems approach allowed the use of computers for detector development, calibration, and integration; for data simulations and software development; for mission planning and real time operations; for observation accounting; and for data reduction, analysis, and archiving.

5.2 *X-Ray Calibration*

Our first experience in operating the instruments of the Einstein Observatory as an integrated scientific facility took place in August–September 1977 with the X-ray calibration of the experiment. In preparation for this activity, a team of scientists and engineers from the Consortium institutions, the Marshall Space Flight Center, and the industrial subcontractors designed and built a special X-ray test facility at Marshall Space Flight Center.

The need for such a facility was clear as early as 1970 in the Large Orbiting X-Ray Telescope proposal submitted by the Consortium. The primary requirements for the test facility were 1. substantial length with a moderate quality vacuum to permit effective focusing of a point X-ray source at a large distance from the telescope, 2. a vacuum chamber capable of housing the overall experiment with a mechanism for rotating the payload, and 3. an X-ray source capable of producing line radiation of sufficient intensity to be useful in measuring the system effective area and resolution. We recognized that the scope and potential use of such a facility was beyond the means of individual experimental groups, and it was therefore recommended that the facility be built at a NASA center.

While the Einstein Observatory hardware was being developed during 1974–1977, the calibration facility was being built at Marshall. The facility included vacuum pumps, a 1000-ft-long pipe to provide the required separation of source and telescope, a 20-ft-diameter by 40-ft-long vacuum chamber to house the experiment, and a microfocus X-ray source utilizing electrons accelerated onto interchangeable target foils. During the facility design and development, scientists and engineers provided additional details concerning baffle requirements to prevent scattered X rays from reaching the telescope; vacuum requirements at the source chamber, telescope chamber, and in the intervening pipe; alignment requirements including system stability and measurability; thermal control system requirements; cleanliness requirements to avoid contamination of the X-ray telescope surface; and flux monitor requirements to determine the X-ray source output during testing. Figure 12 is a picture of the Einstein Observatory experiment being installed in the calibration facility vacuum chamber.

The scheduling of the calibration testing was a complex task. With tests desired at several wavelengths, many angles of incidence, and at least three temperatures for the 9 detectors comprising the 4 focal plane instruments, we recognized that our ability to complete the tests within the 30 days allotted in the schedule would depend on efficient organization of the individual tests. We knew that changes in table positions to vary the azimuth and/or elevation of the telescope relative to the source were readily accomplished (1 to 2 minutes), while temperature changes required longer times (several hours to 1 day). Instrument reconfigurations and target source changes would require 5 minutes to 1 hour.

With this information as a basis, Leon Van Speybroeck developed a computer program to optimize the calibration testing. The program also generated an accurate estimate of the total time required and a bookkeeping system for following the test. A schedule of 1397 separate tests was developed. The initial plan to run 16 hours per day for 30 days was revised to a plan to run 24 hours per day for 18 days. This was actually accomplished using two 12-hour shifts supported by scientists, engineers, test set operators, and facility personnel. The prime reason for the change to 18 days of 24-hour per day testing was to allow us 12 days for reworking hardware problems and retesting the experiment. This "reserve" capability turned out to be very important as several problems did arise, and the opportunity afforded by this scheme to analyze data from the first run, rework the hardware, and retest the instrument was critical to the success of the mission.

The calibration program included tests to determine the effective area, scale factor, linearity, and full point response function of the X-ray

Figure 12 Photograph of the Einstein (HEAO-B) experiment being installed in the Marshall Space Flight Center calibration facility vacuum chamber.

mirror–HRI system. An example of the X-ray mirror–HRI point response to a 1.5 keV X-ray source observed during this calibration test is shown in Figure 13. The full width at half maximum for these data is 3.8 arc-seconds and 50% of the power is contained in a circle of radius 5.6 arc-

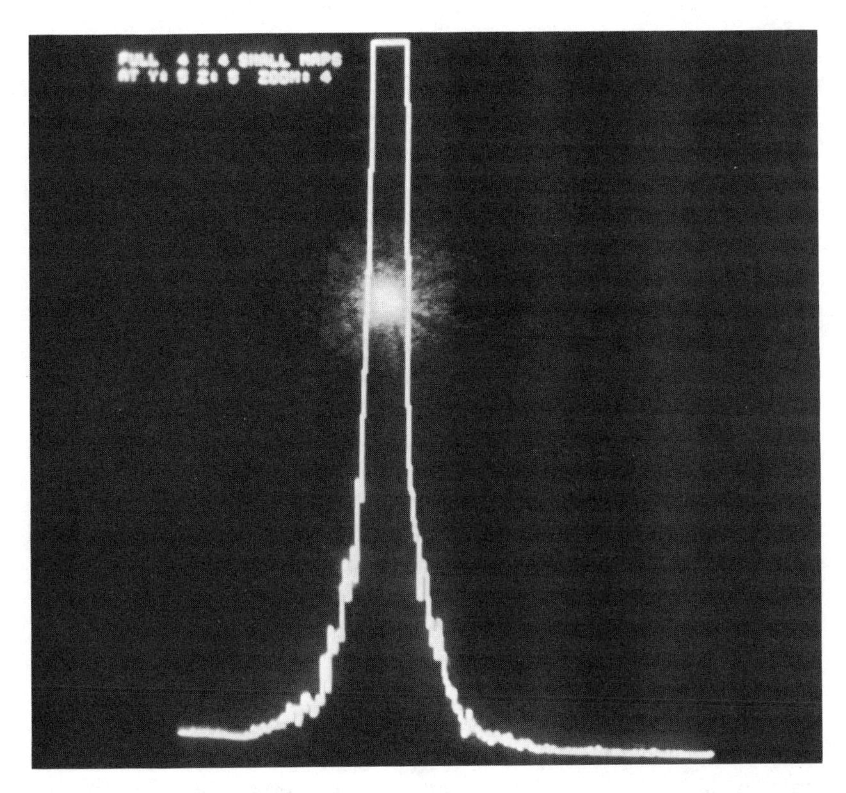

Figure 13 Point response of Einstein X-ray mirror–HRI system obtained at 1.5 keV during calibration test at Marshall Space Flight Center. See text for details.

seconds. The central peak in the figure saturates the selected scale in order to allow the data in the wings to be displayed.

Tests were also made to determine the effective area, scale factor, point response function, gain, energy resolution, and sensitivity to threshold settings of the IPC. Measurements with the three spectrometers were directed primarily to the determination of system efficiences, energy scales, and energy resolutions.

An important aspect of this calibration activity was the accompanying development of the software required to analyze the large volume of data obtained—equivalent to more than one month of operation in orbit. Much of this software formed a basis for that required to reduce and analyze data after the launching of the Observatory. The early availability of this software and the cooperative experience of the Consortium scientific team during calibration contributed significantly to our ability to process data and obtain meaningful scientific results immediately after launch.

5.3 *Mission Operations Planning*

Planning of mission operations began approximately $3\frac{1}{2}$ years before launch, with one scientist (Ethan Schreier) assigned full time to the operations and data systems planning. Functionally, "operations" consist of two general areas—1. generation of the observing program from science planning to detailed scheduling, and 2. real time control and monitoring of the satellite. Pragmatically, it is necessary to define an operating philosophy, specify detailed requirements necessary to implement this philosophy, and develop data bases, software, hardware, and procedures necessary to meet the requirements. In fact, the entire procedure is an iterative one involving tradeoffs in hardware design and limitations, organizational and budgetary constraints, and operations limitations.

Several difficulties rapidly became apparent. The Consortium scientists were able to formulate a technically reasonable observing philosophy but they were not sufficiently familiar with NASA operational restrictions and the details of setting up a control center. NASA personnel were familiar with satellite operations in general, but they did not fully understand the scientific requirements. The flexibility required to optimize scientific output was regarded by NASA as incompatible with the long lead times for scientific planning and strict definitions which they always required. By the time operations planning was well underway, some $2\frac{1}{2}$ years before launch, it was too late and too costly to make significant changes either to the experimental equipment or to the spacecraft hardware. Related to this was the problem involved in gathering the command and telemetry lists and the basic operational data. This difficulty was caused by the late start in operations planning and by the split organizational responsibilities. It is obvious that, in the future, mission operations must be taken into account at an earlier stage, during instrument design and command format selection. As instruments are developed, command and telemetry data bases compatible with control center requirements must be developed. Both NASA and the experimenters must recognize the importance of mission operations right from the start. Adequate manpower and financial resources must be made available early in the program if these activities are to be carried out effectively.

Some problems involved decisions based on the budget. Attempts were made to avoid hardware modifications and, instead, make changes in the operational methods and in software. This was not always successful. A good example is associated with the attempt to extend the life of the mission. The method for getting rid of momentum built up during stable pointings due to gravity gradient torque is through a propellant-based reaction control system. As the scientific importance of the mission as well as the probable delay until the next such mission became more widely recognized, the possibility of adding a magnetic torquing system to prolong the operation was considered. Although this was technically feasible, it was considered to be too expensive and it was therefore rejected. However, it was finally agreed that the life of the mission should be extended and serious studies were made to develop targeting and operational strategies and software to minimize gas consumption. The number of man years expended to do this have cost more than it would have cost us to make hardware modifications to extend the mission. Even now, "momentum management" is still the dominant factor in scheduling observations, and many observing proposals are curtailed because of constraints associated with extending the life of the mission.

The dominant aspect of routine operations involves loading the on-board command memory. This is the result of another trade-off. It was recognized quite early that the Observatory was extremely complex to operate, and the targeting scenarios which later were required by momentum management only exacerbated the situation. The Stored Command Programmer and Attitude Control and Determination System Memory capacities were clearly marginal. However, it was decided that through judicious use, they would be adequate. To make them function efficiently considerable software development has been required, and difficult operational problems must be handled twice a day in loading and verifying the command memory.

While these problems and others show that perhaps some budgetary decisions were not very wise, the solutions were obtained by individuals in NASA, TRW, and the science Consortium, who worked together to overcome the original limitations. This collaborative work was essential for the success of the mission.

5.4 *Scientific Operations*

Two years after the launch of Einstein, the observing program contains approximately 10,000 entries, each one corresponding to a separate scientific observation. The organization of these 10,000 entries began with a Consortium observing program, circulated in preliminary form in May 1977 and formally released in revised form in June 1978 as "The HEAO-B X-Ray Observatory Consortium Observing Program."

As was described in the introduction of this chapter, the Consortium scientists and NASA had agreed in 1978 to institute an immediate Guest Observer Program, thereby waiving the traditional principal investigator's exclusive rights for the first year's data. The Consortium scientists felt it was important to specify their target lists for the 80% of the first year's observing time available to them, in order to provide prospective guest observers enough information for them to propose independent new observations, correlative observations and/or collaborative investigations. While at first sight this initial list of over 3000 targets might have had a discouraging effect on potential guest observers (comments to that effect were, in fact, heard on several occasions), actual experience shows this was not a real problem. In the first two years of the Guest Observer Program over 500 proposals were received, were reviewed by a committee of reviewers, and almost 400 were selected for implementation with over 2000 targets involved. By March 1980 approximately half of these targets had already been observed. The increase of the guest observer fraction of the observing time, intended to reach 50% by April 1981, allows even more programs to be accepted, as well as longer observations of individual targets. The responsibility for overseeing the Guest Observer Program at the Center for Astrophysics was undertaken by Fred Seward.

In support of the Guest Observer Program a Users Committee made up of Consortium, guest observer, and NASA representatives was established to advise the principal investigator who acts as Observatory Director. This Committee meets quarterly to review the overall operation of the Observatory. Instrument performance, target scheduling, and data processing status are reviewed and possible improvements are discussed. Most guest observers using imaging data visit the Center for Astrophysics to work with their data. Typical visits last one week. Guest observers are supported by the scientific personnel and data aides at the Center during their visits. The *Einstein Observatory Users' Manual* provides guidance in using the data processing system.

The listing of the Consortium observing program, and a subsequent volume for the second year of the mission, served several other functions. The listings provided a basis for working out conflicts and eliminating duplications among the separate observing programs of the Consortium institutions. Targets are organized by scientific categories. This facilitates planning of the future observing program and establishing workable scientific data analysis projects. The listings are also used to generate the computerized Einstein Observing Catalog and Accounting (OCA) database, which is the keystone of the mission planning and data archiving activity.

The OCA contains information such as the target name, target right ascension and declination, instrument to be used, desired and achieved observing time, observer's name and institution, and status of the observing request and data reduction activity. Each target is assigned a unique and permanent sequence number when it is entered into the OCA, as well as a record number based on increasing right ascension which does, in fact, change as more targets are entered into the OCA, which is continually updated.

Several factors are involved in the actual scheduling of targets. The first requirement is that a target be within a 30° band perpendicular to the earth-sun line, so that the solar array can generate sufficient power. This means that a target is accessible for at least one month every six months. The next requirement is that the targets be selected so as to minimize the buildup of angular momentum in the reaction wheels and the resulting consumption of reaction control gas. Propellant consumption is minimized by selecting optimum target orientation with respect to the satellite orbit (passive momentum management) and by selecting target sequences that tend to offset perturbing torques (active momentum management). These choices have been quite effective, and current projections are that the reaction control gas, which was originally expected to last 13 months, will actually provide more than two years of useful life in orbit.

Based on the pool of acceptable targets, instrument sequences are determined for one month at a time. An individual instrument may be used for as short a period as one day or as long as three weeks. From this point onward, additional factors must be considered such as the proper allocation of observing time among the individual Consortium institutions and guest observers, as well as the scientific priorities of the individual investigators with multiple targets available. A target list called the Master Observing Program is generated in increments of a few days to cover the one month period. The Master Observing Program, which is oversubscribed by approximately 30%, is forwarded to the Einstein mission planning group at Goddard Space Flight Center with a lead time of one to three weeks. This group, consisting of Consortium, Goddard, and TRW personnel, then generates a Detailed Observing Program. Software at Goddard is used to carry out the target sequencing required for effective momentum management, while also taking into account factors such as earth occultations, data loss during transit of the South Atlantic Anomaly, slewing requirements to change from target to target, and the availability of suitable guide stars for satellite attitude control and determination. The result of this activity is the Detailed Observing Program, which is a specific list of targets and commands to be linked up to the satellite at approximately 12-hour intervals. The on-board Stored Command Programmer and Attitude Control and Determination Computer then use these commands lists and assigned times to execute the desired observing program.

5.5 *Data Processing Facility and System*

Planning for the Einstein scientific data system started at the same time as mission operations planning, some $3\frac{1}{2}$ years before launch. As the general operating philosophy was developed, a preliminary data flow and data analysis system was conceptualized, and the possible choices for a computing facility were studied. It was both natural and useful that these areas fell under the same scientific guidance.

A computer facility was required for several major functions: experiment simulations and calibration data analysis; mission planning and operations (on-going observing program generation, observation accounting, quicklook experiment monitoring); software development; data reduction and analysis; and data archiving and distribution.

The Einstein computer facility at the Center for Astrophysics was, therefore, set up in accordance with certain basic guidelines and objectives. Among these were 1. to allow active participation of the instrument scientists in the development of software (this would require an efficient, interactive, multi-user environment); 2. to coordinate scientific operations for all Observatory users, Consortium scientists as well as guest observers (observing programs would be scheduled only a short time before they were to be executed); 3. to provide a data analysis facility for all users, and a data archive for future use; 4. to develop a useful image processing facility; and 5. to meet these objectives within a low and almost "fixed" budget.

It was apparent that we would have to run the computer as a dedicated facility. Mission planning, operations, and quicklook data activities must have very high priority. Data analysis and distribution for other users and visitors also require high priority. Finally, unexpected problems and requirements might force reallocation of priorities at a future time in the face of a fixed budget. The above considerations led to a system based on dedicated minicomputers operated by the Einstein scientific staff.

The choice of computer was affected by several factors in addition to the basic hardware features. These factors included commonality with laboratory and instrument development hardware and software, experience with computer architecture and hardware, and the availability of a powerful operating system. The first two factors are obvious, allowing sharing and interchange of existing hardware and software and optimization of the computer configuration. The last factor becomes critical for automating a data reduction system which contains many software components, where the processing flow may change depending on the results of previous processing.

The Einstein data processing facility consists of two independent computer systems incorporating Data General Eclipse machines. The hardware is summarized in Table 2.

The "Production" machine is dedicated to on-going mission planning and data reduction. Access to it is limited, and the machine serves as a somewhat automated data pipeline. The "User" machine is used for inter-

Table 2 Einstein computer hardware

Production (data reduction) computer
Eclipse S/230 512 bytes memory
3 disks (200 + 200 + 100 M bytes)
5 tape-drives 800/1600 BPI
1 Versatec Printer/Plotter
6 terminals 9600 baud
1 phoneline 300 baud
1 MCA connection to User computer
User (analysis) computer
Eclipse M600 1024 K bytes memory
3 disks 200 M bytes each
2 tape-drives 800/1600 BPI
1 Varian Printer/Plotter
1 lineprinter 600 Lpm
10 terminals 9600 Baud
MCA connector to Production computer and to display system
Measuring engine (2 axis digitizer)
2 display systems (512 × 640 pixels)

active data analysis and advanced software development. It also supports two image-processing systems (512 × 640 × 8 bit color display) and a measuring engine. Data are routinely transferred between the two machines. Although the computers are operated and configured independently, there is complete compatibility and either machine can be reconfigured to support any given function.

The scientists themselves developed much of the software for data analysis. Thus, those most knowledgeable about the instruments implement the software and are available to modify it in response to actual instrument performance. Significant post-launch modifications have been made by a staff of professional programmers, working with the scientists. It would have been highly beneficial if more of this support had been available before launch.

Implementation of the software system to satisfy the basic requirements and ground rules led to several design features:

1. The comprehensive Observing Catalog and Accounting (OCA) database incorporates a target list with scheduling, analysis, and archiving status. Observing constraints, observing time scheduled, observing time achieved, data processing completed, and location of archived data are automatically updated and made accessible via a catalog editor.

2. Routine reduction is semi-automated via a sophisticated system of software macros. The system correlates the incoming data with the OCA,

executes reduction software appropriate to the data, and maintains a restart capability. The system is routinely operated by data aides, with scientific input required only at the beginning (data processing priority) and the end (data quality). Non-interactive data reduction includes decommutation, aspect analysis, image or spectral file generation, first-order source detection, and data archiving.

3. Basic archival file formats are designed around efficient use of the image processing system taking into account the principles of efficient archiving.

The Einstein Observatory data rate is 6.4 kilobits s^{-1}, of which about 5 kb s^{-1} is scientific data. The source counting rates are typically low, and processing time is closely correlated to observing time. Approximately two thirds of real time on one computer is required for routine data reduction. The remaining capacity of that computer is used for quicklook data processing and mission planning functions. The "User" computer is typically called on to support the analysis activities of some 20 to 30 scientists at a time—resident, visiting, and guest observers. It is often oversubscribed, and a factor of two increase in capacity could easily be justified.

The Einstein experience can be extrapolated reasonably to the expected X-ray astronomy computer needs of the next decade. We can estimate that the data processing requirements of an X-ray astronomy facility for the next decade will be no more than an order-of-magnitude larger.

6. SCIENTIFIC IMPACT

X-ray astronomy has already demonstrated its power as a tool for the investigation of some of the great problems of modern astrophysics: the end points of stellar evolution, the existence of black holes, the nature of the energy sources in the nuclei of active galaxies and quasars, and the question of the large scale structure and evolution of the universe. In some cases, it provides us with a means of studying matter in physical states not directly accessible to optical or radio observations.

With the launching of the Einstein Observatory with its imaging capability and increased sensitivity (up to 500 times that of previous observations) X-ray astronomy has joined the mainstream of astrophysics. Over the course of two years, 4000 target fields have been observed and 8000 individual X-ray sources have been detected. Almost every type of star in the galaxy has been detected; we have X-ray photographs of supernova remnants, of galaxies, and of clusters of galaxies, and we have detected some quasars that may be the most distant objects ever observed by optical telescopes.

In what follows, we sketch the results in some of those areas where the impact of Einstein X-ray observations has been especially significant, and note the progress that can be expected when the Advanced X-ray Facility starts operation in the late eighties.

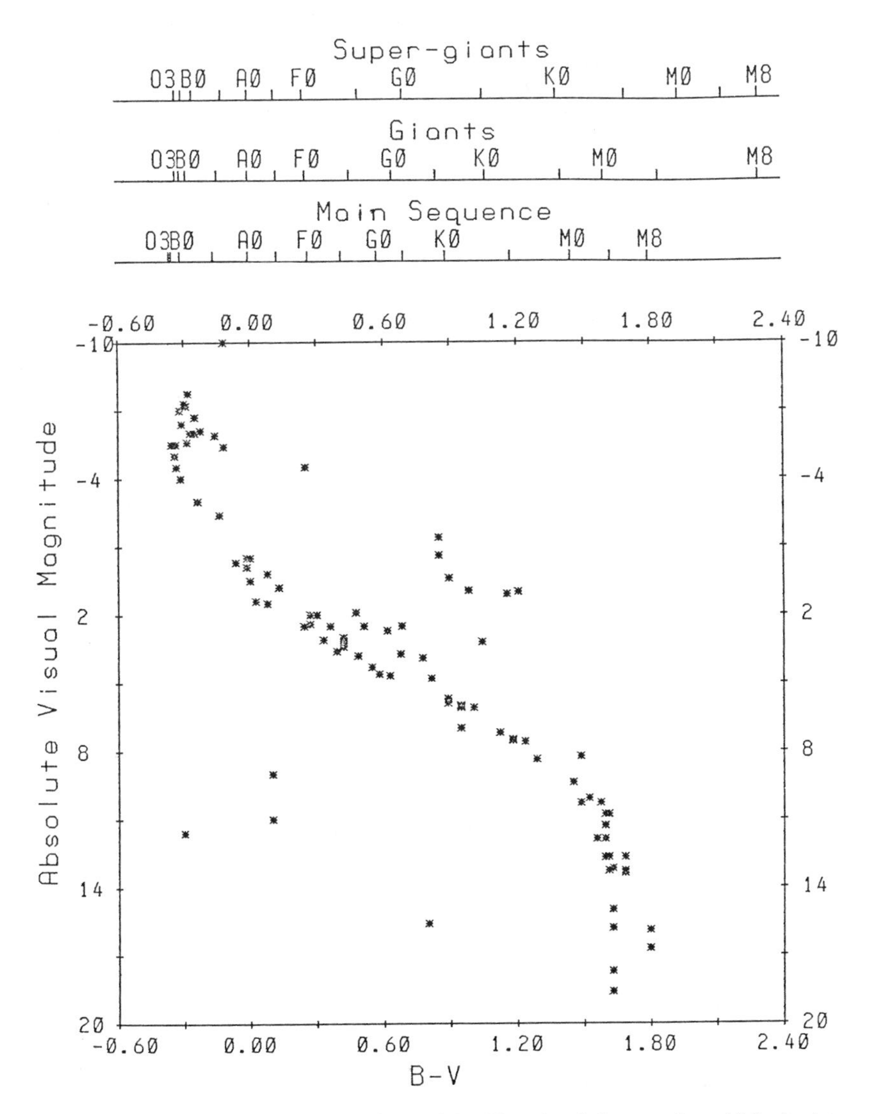

Figure 14 H–R diagram for stars detected by Einstein. Only stars for which absolute visual magnitude could be determined are shown.

6.1 *Stellar Coronae*

Early observations carried out with the Einstein Observatory have detected more than 150 low-luminosity stellar X-ray sources. In many cases, the X-ray positions are accurate to less than 4 arcseconds, and there is little doubt of the identification. In the case of IPC source detections un-

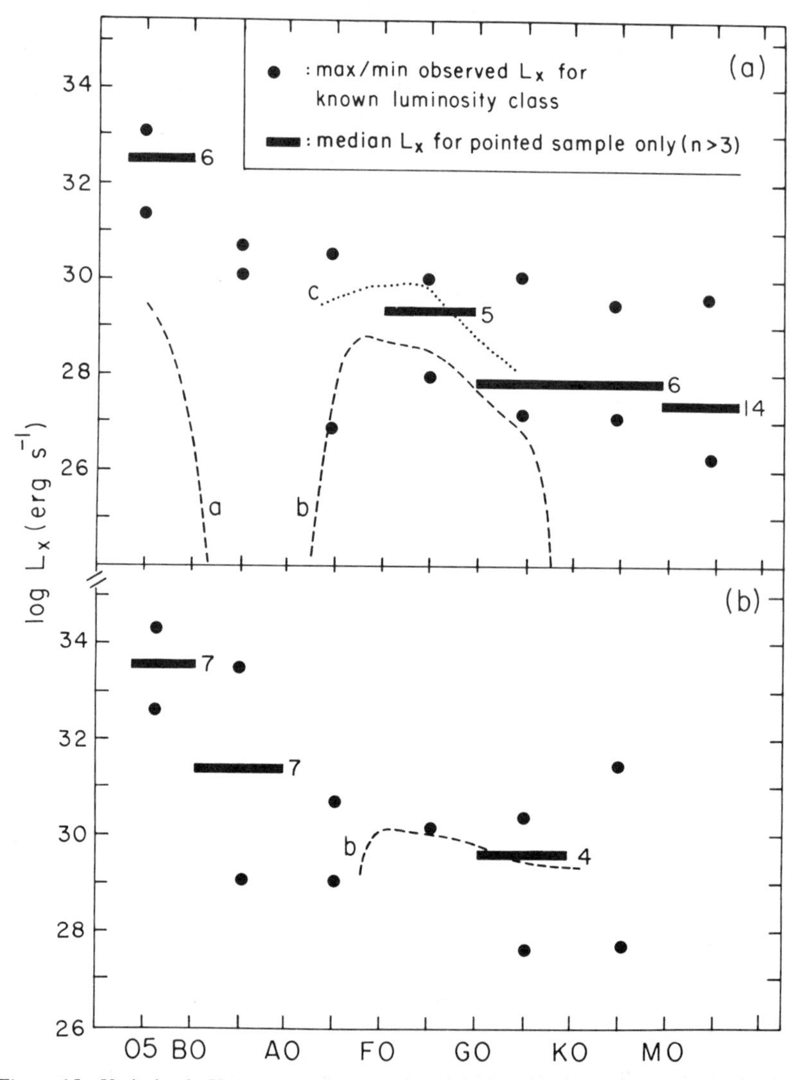

Figure 15 Variation in X-ray luminosity (L_X) vs spectral class of the stars detected by Einstein (as plotted in Figure 1): (*a*) main sequence; (*b*) giants and supergiants. The horizontal bars are the mean values of L_X, the dots are minimum and maximum values.

certainty in position is as large as 60 arcsec, and there is a greater probability for error. With such a large sample, however, a few spurious identifications should not alter the main conclusion that the X-ray sources in stars arise from stellar coronas (Vaiana et al. 1980). Figure 14 is an H-R diagram for all stars detected in a pointed survey of observations of selected nearby stars of known spectral type and luminosity classes. For the few stellar types not yet detected (late-type giants and supergiants) the upper bounds on luminosity are relatively high. Figure 15 summarizes the variation of L_x within each class.

Although the conventional picture relating coronal emission to stellar surface turbulence (via acoustic wave generation and dissipation) is inadequate to explain these observations, a connection between coronal formation and heating and the vigor of surface motions seems unavoidable. We simply do not understand the mechanism providing this coupling. Stellar rotation, convection, and dynamo generation of surface magnetic fields may all be involved. Elucidating the connection between these processes and the generation of X-rays from stellar coronae is one of the principal goals of the future Advanced X-ray Astrophysic Facility (AXAF). Its hundredfold increase in sensitivity over the Einstein Observatory will enable us to increase the number of stellar coronas observed to the thousands and to make detailed spectral studies of hundreds of nearby sources.

X-ray observations in the near future should also give us an understanding of the contribution of stars (particularly late-type dwarfs) to the diffuse emission and mass distribution in star clusters, our Galaxy, and other galaxies. Given present estimates of the Facility's limiting sensitivity, and assuming the mean X-ray luminosities derived from late M dwarfs from Einstein data, the AXAF will be able to probe for dM stars down to $V \cong 22$ mag. The X-ray observations will therefore provide an observational tool for studying the space distribution of these stars comparable in power to the most sensitive ground-based optical telescopes.

6.2 *Strong X-Ray Sources in Binary Stars*

The Einstein Observatory has made an important contribution to our understanding of strong X-ray sources in binary stars. We have detected approximately 100 sources in the Andromeda Galaxy. Since the sources are all at the same distance, we can construct the upper end of the luminosity function, and begin to determine populations and evolutionary trends.

In the future, we can expect tremendous advances in our understanding of the details of these X-ray sources as a result of timing observations with the X-ray sources as a result of timing observations with the X-ray Timing Explorer satellite and refined spectral studies with the AXAF.

Observation of spectral changes in fluorescent iron lines and possibly others as a function of pulse phase will constrain the geometry and physical conditions of the emission zone near the central object, be it a white dwarf, a neutron star, or a black hole. Given the long life time of the AXAF Observatory, it should also be possible to study long-term spectral changes and, thereby, to gain insight into such questions as the free precession of neutron stars and the nature of the mass transfer process from the giant star onto the collapsed companion.

6.3 *Globular Clusters*

A systematic study of globular clusters by the Einstein Observatory has refined the positions of known sources and yielded two new X-ray sources, bringing the total number of sources associated with globular clusters to eight. The accuracy of the preliminary positions measured for these sources is of the order of 1 arcsec. In several cases, the measured X-ray positions are substantially greater than 1 arcsec from the optical center of the cluster. Analysis shows that if the masses of these objects are all the same, then that common mass is certainly greater than the average masses of cluster stars. The likelihood function falls off significantly above $M_x = 5\ M_{\odot}$, providing strong and direct evidence that these sources are not black holes with masses of a hundred or more solar masses (Grindlay et al. 1981, in preparation).

The improvement in positional accuracy that would be achieved by AXAF offers the prospect of greatly sharpening this conclusion. In particular, we will be able to examine the assumption that all of the masses are the same. For example, if several sources were found to be within .05 r_c of the cluster centers, and the rest were outside of, say, 0.5 r_c, where r_c is the core radius, then we could conclude that some cluster sources are very heavy and are probably black holes, while the remainder are less massive and are probably neutron stars that have captured a nuclear burning companion star.

6.4 *Supernova Remnants*

More than twenty supernova remnants have now been detected in X rays, including several in the Large Magellanic Cloud. Einstein studies have produced detailed images of many of these with spatial resolution ranging from one to ten percent of the extent of the remnants. A detailed comparison of the X-ray, optical, and radio structure of individual remnants should clarify our understanding of the shock wave dynamics and the origin of the high-energy electrons responsible for the radio emission. Improved spectral data from the Einstein Solid-State Spectrometer (Becker et al. 1979) have provided important information on the ioniza-

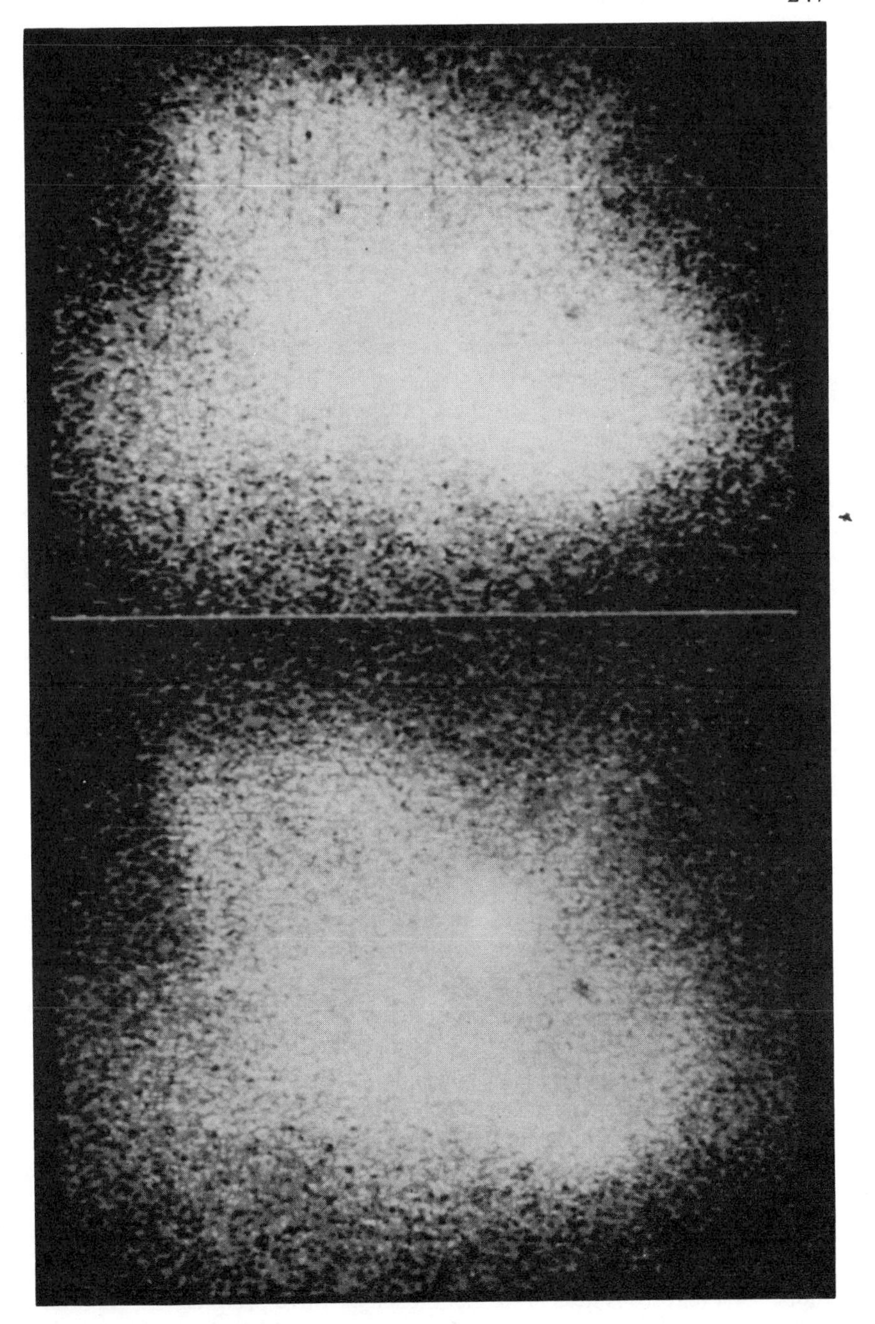

Figure 16 Einstein HRI images of the Crab Nebula. The top is during the pulsar on phase, the bottom during the pulsar off phase.

tion state and abundances of heavier elements in some of the remnants. Two young supernova remnants, the Crab Nebula and Cas A, illustrate the current status of this work, its potential and its limitations.

An X-ray image of the Crab Nebula is shown in Figure 16. The pulsar is clearly detected. By separating the image data into time-phased segments, the contribution of the unpulsed emission can be measured. In particular, at the off-pulse phase an upper limit to any residual blackbody X-ray emission from the surface of the neutron star is obtained. This corresponds to a luminosity of less than 1.3×10^{34} erg s^{-1}, or a surface temperature less than 2.5 million degrees, assuming a radius of 10 km (Harnden 1980).

The extended emission from the Crab Nebula shows no shell structure and is apparently due to synchrotron emission from high-energy electrons generated by the pulsar. If there is any emission from a shock wave around the expanding envelope, its brightness is too low to be detected in the presence of the extended nonthermal radiation.

With AXAF, it should be possible to study the structure of the Crab Nebula as a function of X-ray energies ranging from 0.2–8 keV. This information, together with the radio and optical maps, should allow us to put together a fairly complete picture of the acceleration and diffusion of energetic electrons in the Nebula. The inclusion of an X-ray polarimeter would be especially useful in mapping out the magnetic field structures and in detecting the presence of a nonthermal component in other supernova remnants.

An entirely different structure is apparent in the youngest known galactic supernova remnant, Cas A. The X-ray image shown in Figure 17 illustrates the shell-like appearance of the supernova remnant. In Cas A, the search for a collapsed object results in an upper limit that corresponds to a blackbody temperature of less than 1.5 million degrees for a neutron star of radius 10 km (Murray et al. 1979). Figure 18 shows a composite picture of Cas A in radio, optical, and X-ray wavelengths. Comparisons of these images indicate a general correspondence in the optical and X-ray bands, particularly in regions of high velocity knots, which correspond with the brightest X-ray features. The optical spectra of these knots show strong oxygen and sulfur lines, and the X-ray spectrum of Cas A (Figure 19; Becker et al. 1979) shows strong silicon sulfur and argon lines, consistent with the idea that they contain processed stellar material ejected in the supernova explosion. Both the X-ray and radio maps indicate the presence of a faint halo or outer shell just beyond the region of the bright knots. This feature is interpreted as the shock wave produced in the interstellar medium by the explosion. Observations of a high temperature component to the Cas A X-ray spectrum (Pravdo & Smith 1979) are in agree-

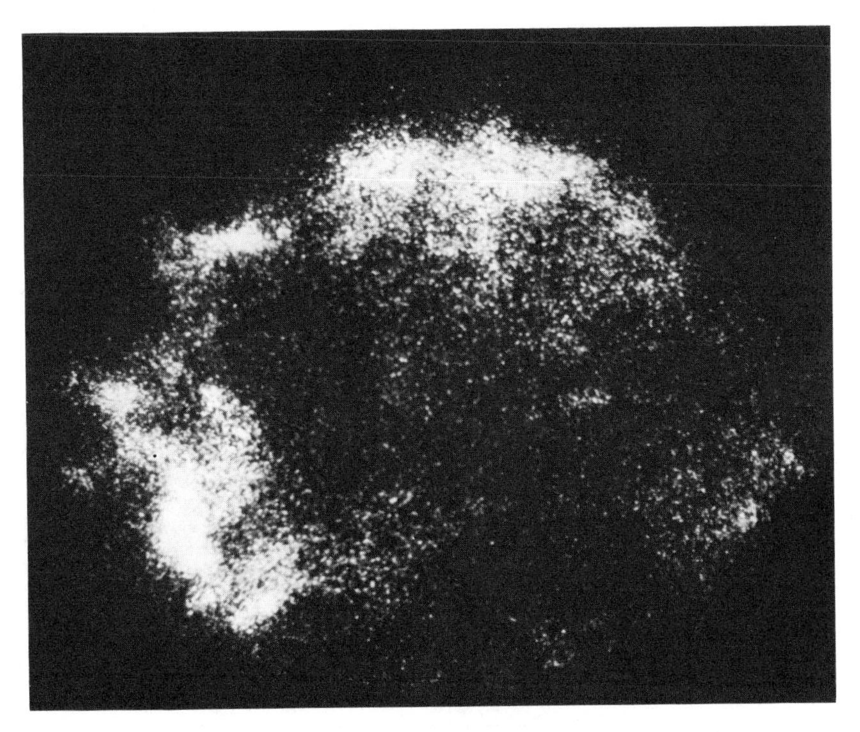

Figure 17 Einstein HRI image of the Cassiopeia A supernova remnant.

ment with this picture. On the assumption of hydrostatic equilibrium between the various temperature components of the gas, Fabian et al. (1980) have estimated the mass of the X-ray–emitting material in Cas A to be about 15 solar masses. This requires that the progenitor star was massive, consistent with the mass required to give the elemental abundances in the optical knots (Arnett 1975, Chevalier & Kirshner 1978, Lamb 1978).

The combined spectral and spatial resolution of AXAF will allow the variation of temperature within supernova shells to be traced, making possible more definitive studies of the energetics and dynamics of the supernova process and its aftermath.

One of the most interesting results that has emerged from the study of supernova remnants by the Einstein Observatory is that only a few of them contain central point-like X-ray sources. Either the formation of neutron stars by supernova explosions is not the general rule, or the neutron stars produced in these events are able to cool more quickly than

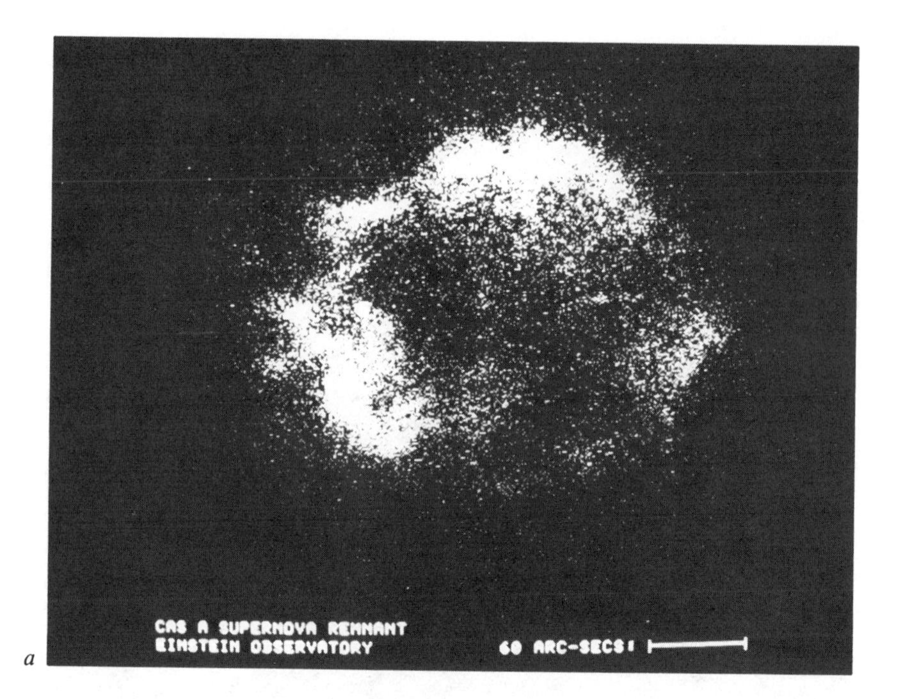

a

N

CAS A

W

60"

AUGUST 1973

b

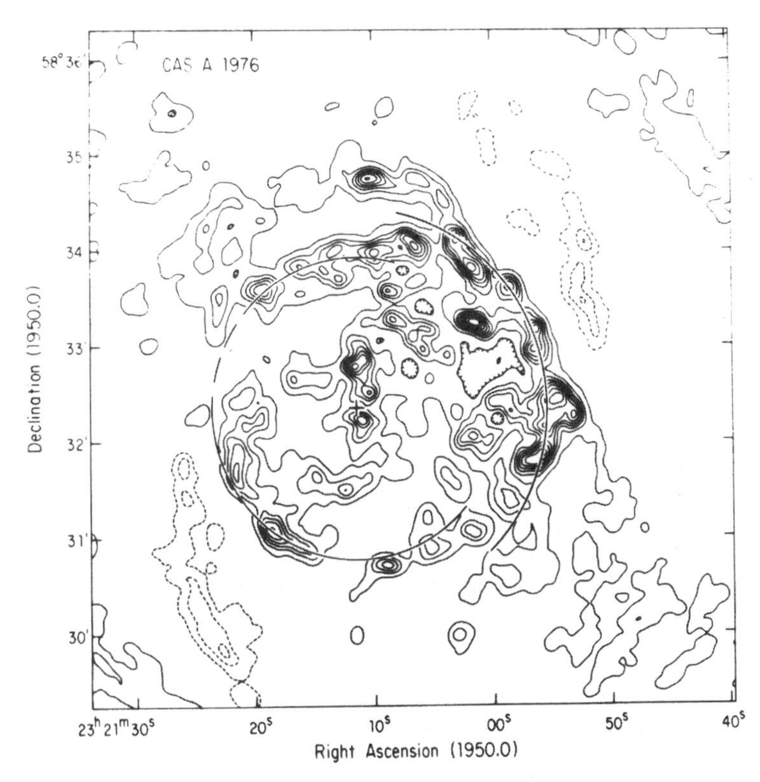

c

Figure 18 (*a*) X-ray, (*b*) optical, and (*c*) radio images of the Cassiopeia A supernova remnant.

anticipated (Helfand et al. 1980). Since the predicted rate of cooling of a neutron star depends on assumptions made about the importance of various nuclear processes, there is the very real and exciting possibility of doing nuclear physics with X-ray telescopes. The detected values or upper limits to the surface temperature of neutron stars as a function of the age for the supernova remnants observed by the Einstein Observatory can be compared to those predicted by various neutron star cooling models. The absence of detected emission from Cas A, Tycho, and SN1006, and the low result for the unpulsed emission from the Crab pulsar, suggest that at least "standard" neutron stars are not present in these supernova remnants (Tsuruta 1980).

The factor 100 increase in sensitivity for point sources should allow AXAF to either detect the remnants or to push the upper limits on the temperature down by more than a factor of three for the high latitude supernova remnants. This will severely constrain or eliminate most neutron star models.

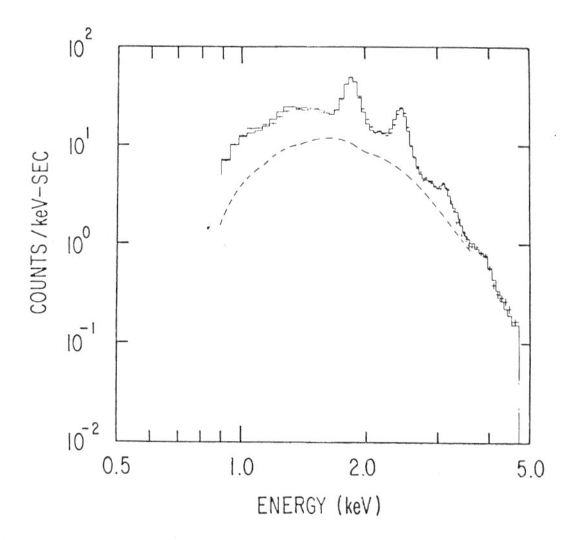

Figure 19 The spectrum of Cas-A as observed by the Solid State Spectrometer on the Einstein Observatory. Superimposed upon the data is the expected Solid State Spectrometer response to a two-component isothermal model. The lower dashed trace is the contribution to the spectrum from low z material (H, He, C, N, O, Ne).

6.5 *The Andromeda Galaxy*

Studies of normal galaxies are being carried out in order to compare the X-ray properties of these objects with their morphology and composition. In M31, it is possible to localize individual sources and obtain optical identifications of some of these. IPC observations have been used to map the entire galaxy at moderate resolution. HRI observations have been used to resolve the nucleus and to obtain precise positions for sources in the spiral arms. A major scientific goal has been to obtain the luminosity distribution of these galactic sources, for comparison with those in other nearby galaxies such as the Magellanic Clouds and the Milky Way.

Almost 100 individual sources have been detected in M31, each emitting more than about 10^{37} erg s^{-1} in the 0.5 to 4.5 keV band. These are comparable in luminosity to the brighter galactic sources in the Milky Way. The mean source luminosity of 10^{38} erg s^{-1} is somewhat higher than for galactic sources in the Milky Way.

Figures 8 and 20 show X-ray images of M31 (Van Speybroeck et al. 1979, Van Speybroeck 1980). The IPC images in Figure 8 have been superimposed on an optical photograph to show the correlations with specific features. The HRI data in Figure 20 show the nuclear region resolved into individual sources. One such source is coincident with the optical center and it is among the most luminous in the galaxy. The inset to Figure 20 is a second HRI image showing the existence of a variable source.

In general, the sources appear to correspond to optical features in the galaxy. There is a significant correlation between bright X-ray sources and other indicators of Population I activity, such as OB associations, H I, H II, CO, and dusty regions and a strong concentration of sources in the central bulge of the galaxy. The distribution of bulge sources in our Galaxy spans a region as much as ten times larger. No interpretation for this difference has been advanced, although the effect is large: in M31 a central region that contains only about 1.5% of the total mass appears to be responsible for about one third of the total number of bright X-ray sources.

Current observing plans for the Einstein Observatory include additional exploratory investigations of a large number of nearby elliptical, SO, spiral, and irregular galaxies which exhibit a wide variety of morphological types, colors, and luminosity.

In the future, we can expect the AXAF to greatly extend these studies beyond the exploratory phase of Einstein. The sample of galaxies for which the integrated emission can be detected will increase by a factor of one thousand. AXAF sensitivity will allow detection and study of individ-

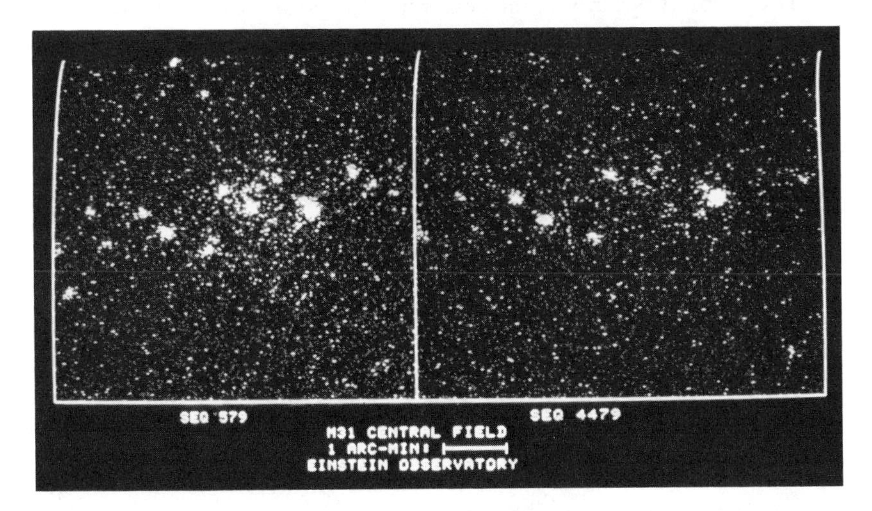

Figure 20 Two HRI images of the central region of M31. Sequence 579 was taken in January 1979, 4479 was taken in August 1979. Notice the absence of at least one strong source in August (4479).

ual high luminosity galactic sources in galaxies closer than 20 Mpc, thereby extending the type of results obtained with Einstein for M31 to well over a thousand galaxies.

6.6 *Clusters of Galaxies*

With the imaging capabilities of the Einstein Observatory, a new level of detailed studies of clusters of galaxies is possible. The observational program consists of (*a*) a detailed study of bright clusters with both the IPC and the HRI detectors; (*b*) a general survey to measure luminosity, surface brightness, and size for a large number of relatively nearby ($z \leq 0.2$) clusters; and (*c*) a survey of distant clusters to study evolutionary processes.

Among the results of this program thus far are (*a*) a measurement of the mass of M87 based on the distribution of X-ray–emitting as in the gravitational well of the galaxy; (*b*) detection of emission regions around M86 and M84 in the Virgo cluster; (*c*) observations of several distinct morphological X-ray structures of clusters; and (*d*) observations of very distant clusters ($z \geq 0.2$) which suggest evolution in the temperature and X-ray luminosity of these objects.

M87 lies at the center of the Virgo cluster. It is an X-ray source distinct from the general cluster emission and is by far the brightest individual source in the cluster, being about 100 times more luminous than other galaxies. Observations with the IPC imply the presence of a halo of visibly faint material containing a mass on the order of 3×10^{13} solar masses (Fabricant et al. 1980). This follows from a study of the radial surface brightness distribution, shown in Figure 21, and the assumption of hydrostatic equilibrium for the X-ray emitting gas in the gravitional potential well of M87. From the surface brightness distribution, the variation of the gas density with radius can be determined. Taking a temperature of the cluster equal to 2.5 keV and following the profiles out to a radius of 230 kpc where the X-ray emission from M87 merges into the overall cluster emission gives a value of about 3×10^{13} solar masses for the gravitational mass binding the hot gas to M87.

A similar analysis can be performed for clusters as a whole, particularly those that have smooth and azimuthally symmetric surface brightness profiles. The masses obtained are about 10^{14} solar masses, consistent with the virial mass estimates for bound systems. This lends strong support to the idea that massive halos similar to the one about M87 are common phenomena.

After M87, the brightest X-ray sources in the Virgo cluster are associated with the elliptical galaxies M86 and M84. In the IPC image shown in Figure 22, the extended source coincides with M86 and M84 is to the

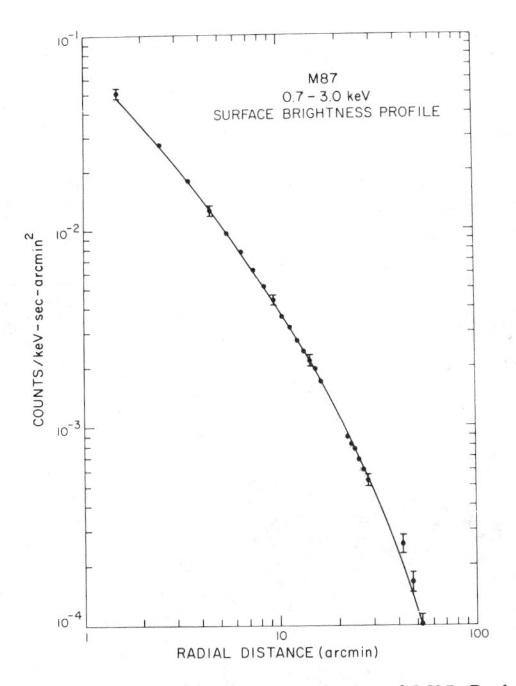

Figure 21 The 0.7 to 3.0 keV surface brightness profile of M87. Background from the Virgo Cluster and internal detector background have been subtracted.

east. An extended feature north of M86 has also been detected in a high resolution image of this region.

The emission directly associated with M86 is extended with an isothermal core radius of about 2.5 arcmin ($\sim$ 15 kpc) and a temperature of a few keV. The relatively high velocity of M86 (1500 km s^{-1}) with respect to the center of the cluster suggests that M86 is in an orbit which takes it far outside the cluster core. During intervals when M86 is far from the core, mass lost from stars within the galaxy can accumulate to form a halo. When the galaxy enters the core, the ram pressure of the intracluster gas with a 10-keV temperature and a density of 5×10^{-4} cm^{-3} strips the outer parts of the halo from the galaxy (Forman et al. 1979). The observation of a "plume" of hot gas near M86 may be evidence of this process in action. M84, on the other hand, is a much less extended source, with a core radius of only 20 arcsec (2 kpc). It also has a much lower velocity with respect to the cluster, and it probably never gets outside the core. It is constantly subjected to the ram pressure stripping of the hot core gas, and can retain only a tightly bound core of gas. A schematic summary of the X-ray structure of the Virgo cluster is shown in Figure 23.

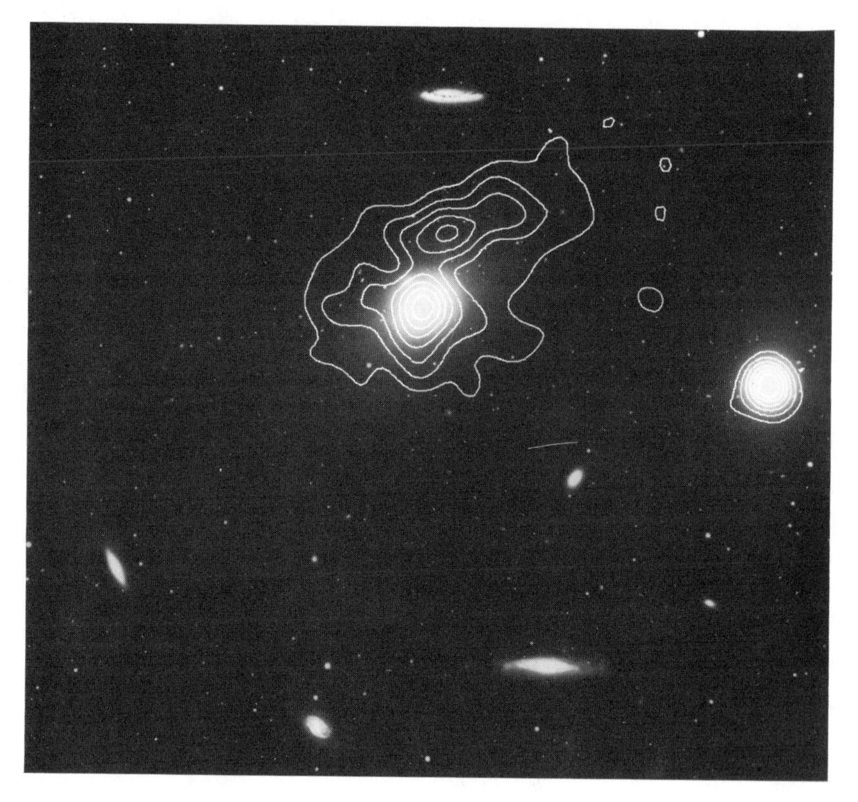

Figure 22 IPC image of the galaxies M86 and M84 in the Virgo Cluster. This is a plot of iso-intensity X-ray contours superimposed on a visible light photograph. M86 is near the center of the picture and an extension is seen in the X-ray contours to the north, which is interpreted as a plume of gas stripped from the galaxy.

The X-ray observations show a large diversity in the structure of clusters of galaxies, ranging from broad, highly clumped distributions to a smooth distribution building up to central peaks (Figures 24 and 25). Apparently these differences in morphology represent an evolutionary sequence in the dynamical evolution of the clusters (Jones et al. 1979). Thus, for example, in A1367, some X-ray peaks are associated with bright galaxies in the cluster, much as was the case for the Virgo cluster. In both these clusters, X-ray emission is dominated by relatively low temperature (few keV) gas which is still bound to the galaxies. Some gas has escaped, or has been stripped from this galaxies, and has been captured in the potential well of the cluster which is still relatively shallow.

As the cluster evolves, its gravitational potential well will deepen, the galaxies will move about more rapidly, and ram pressure stripping will pull more and more gas from the galaxies. The X-ray intensity of the intracluster medium will increase, as will its pressure, and the stripping of gas from the galaxies will accelerate until interactions of the galaxies with each other and with the gas produce a smooth, centrally peaked distribution of high temperature (10 keV) gas. The clusters A85, A2319, and Coma have presumably reached this highly evolved stage.

The clusters shown in Figure 26 may represent an intermediate stage in the evolutionary process. Their X-ray iso-intensity contours are characterized by a double structure (Forman et al. 1980). We could be seeing the final step in the merging together of galaxies into clumps of two, four, eight, sixteen, etc., galaxies until there remain only two subclusters, which will eventually merge to produce a relaxed, Coma-type cluster. The role of cD galaxies in the evolution of clusters is unclear. The data suggest that clusters can be divided into two main groups: those with centrally located X-ray emitting cD galaxies (Virgo, A85) and those without such galaxies (A1367, Coma). Within both of these categories the evolutionary tracks

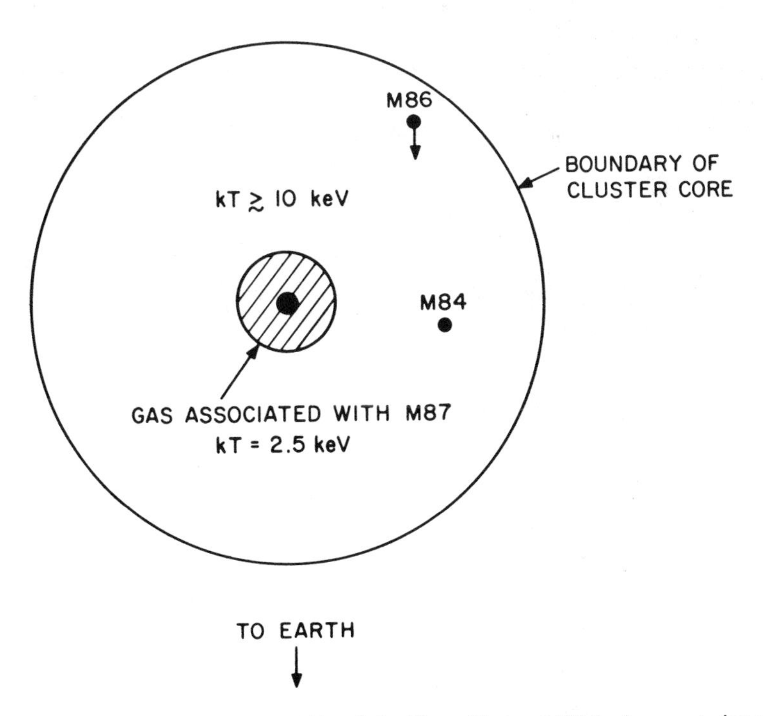

Figure 23 A schematic representation of the Virgo Cluster. M86 is shown entering the cluster core region. M84 is shown embedded in the high density gas of the core.

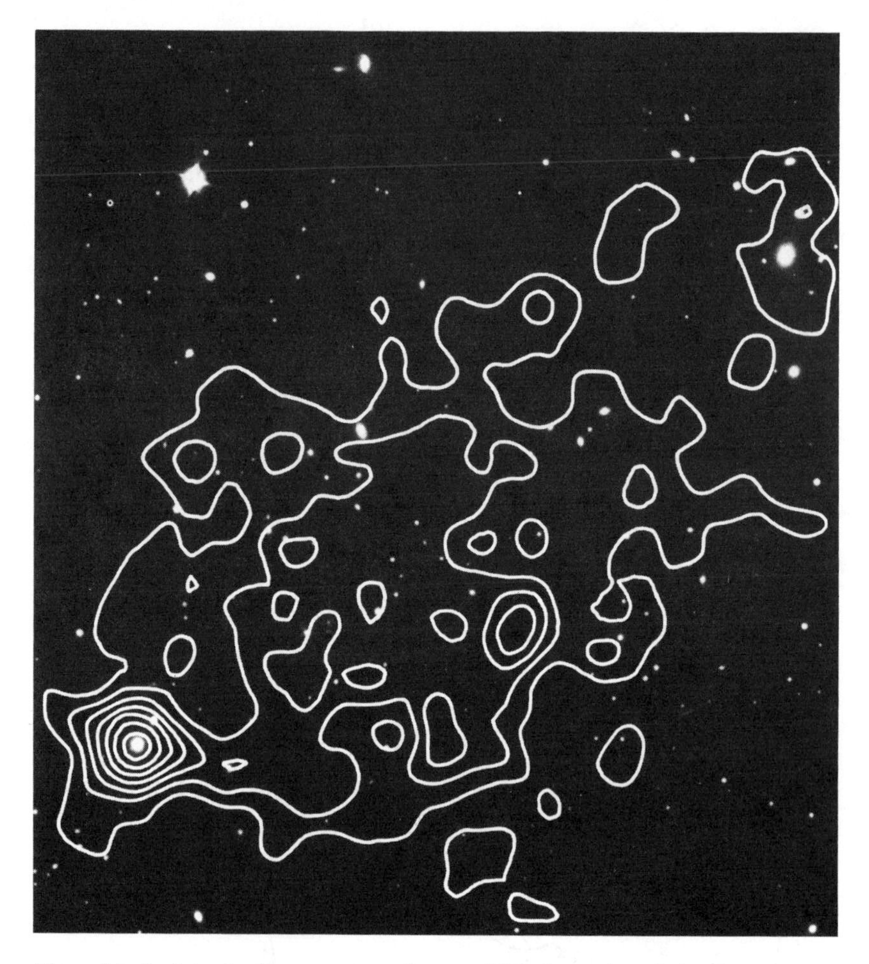

Figure 24 Iso-intensity X-ray contours from an IPC observation of the cluster A1367 superimposed on a visible light photograph. The X-ray data have been deconvolved and smoothed. The point source in the south is associated with 3C264 = NGC3864.

from irregular to centrally condensed distribution of mass are apparent. The existence of a dominant elliptical galaxy may be simply an initial condition that determines which branch the cluster will follow. Table 3 gives examples of clusters that fall into the two categories.

The evolutionary models for clusters of galaxies described above suggest a trend of increasing X-ray luminosity and temperature as a function of time, since the amount of gas available and the strength of the central gravitational potential are both increasing with time. This picture can be tested in a statistical manner by observing the X-ray emission from a

Figure 25 Iso-intensity X-ray contours from an IPC observation of the cluster A85 superimposed on a visible light photograph. The X-ray data have been deconvolved and smoothed.

number of distant clusters and making a comparison with the data for relatively nearby clusters. On the average, we should be seeing clusters at large distances at an earlier stage in their development and, hence, at lower temperatures and luminosities. The data obtained so far from the Einstein Observatory for clusters out to $z = 0.75$ are consistent with this picture (Perrenod & Henry 1980).

The issues of the masses of individual galaxies, the interaction of galaxies with the intracluster gas, the origin of the intracluster gas, the evolution of clusters, and the role of cD galaxies in this evolution cannot be

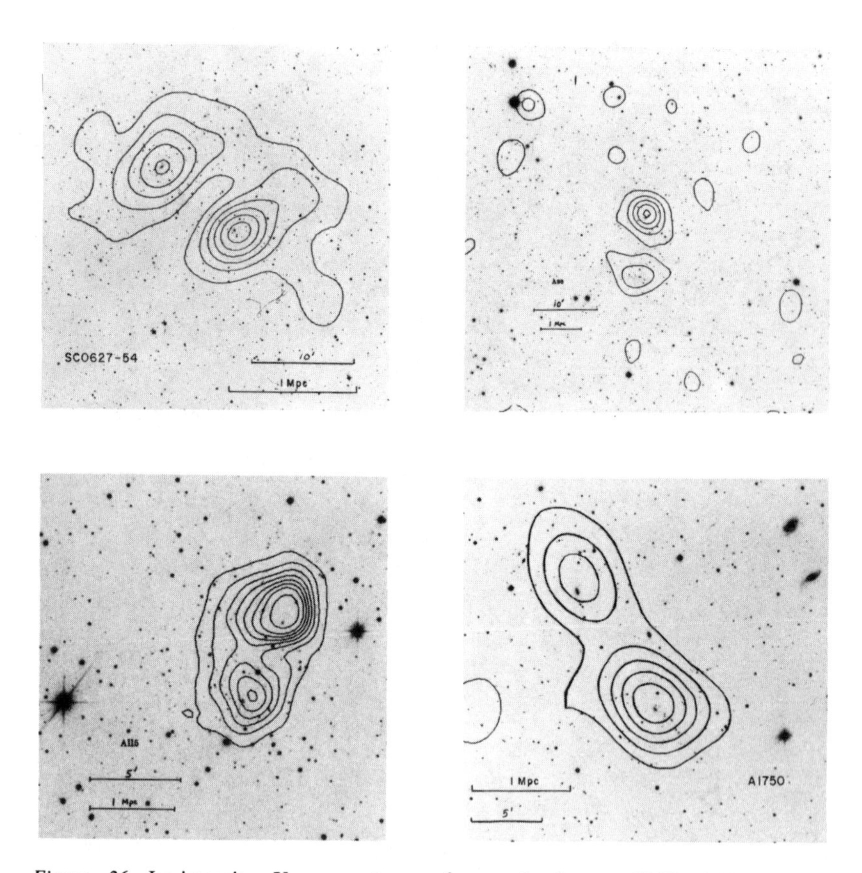

Figure 26 Iso-intensity X-ray contours of several clusters (A98, A115, A1750, SC0627-54) which show double structure. The X-ray data have been deconvolved and smoothed. Galaxy subclustering corresponding to the X-ray structure has been detected in these clusters.

fully resolved with Einstein. Extended observations with improved spectral capabilities and increased sensitivity are needed. Only then can the temperature, density, and abundance distribution across the clusters be determined. These results may give some indication of the origin of the gas and how it is heated. The extended AXAF bandwidth (0.1 to 8 keV) will be sensitive to a wider range of gas temperatures than Einstein, and it will cover the important iron K-shell region around 8 keV; this will allow processed material from galaxies to be detected directly and redshifts measured. It should be possible to detect and spatially resolve clusters out to $z = 3$ or greater and thus to study the origin and evolution of clusters.

Table 3 Examples of the two types of galaxy clusters

Clusters containing centrally located X-ray emitting galaxy	Clusters with no centrally located X-ray emitting galaxy	Evolutionary indicators
Virgo (M87), A262, A2199	A1367, A2634,	Low velocity dispersions. High fraction of spiral galaxies. Cool X-ray temperatures ($\sim$2 keV). Irregular galaxy distribution.
A85, A1795	Coma, A2256	High velocity dispersion. Small fraction of spiral galaxies. Hot X-ray temperatures ($\sim$8 keV). Centrally condensed, regular galaxy distribution.

6.7 *Active Galaxies and QSOs*

The Einstein Observatory has detected X-ray emission from virtually every type of object with an active nucleus: Types I and II Seyfert galaxies, radio galaxies, N-type galaxies, BL Lac objects, and QSOs. Perhaps the most significant results so far have come from QSO surveys and the resolution of jet structures in Centaurus A, M87, and 3C273.

For example, Einstein observations of the nearby radio galaxy Centaurus A have shown that a jet of sub-relativistic plasma connects the compact nuclear source and the extended inner radio lobes. Figure 27 shows the HRI iso-intensity contour map around the nucleus of Cen A. Also shown is the inner optical jet first reported by Dufour & Van den Bergh (1978). The X-ray jet is co-aligned with the optical jet and extends to the inner radio lobe several arcmin from the nucleus.

The interpretation of the X-ray jet as a stream of hot plasma may explain how the radio lobes are energized. The X-ray and optical emission from the jet are consistent with radiation from a hot plasma. The kinetic energy of the plasma jet provides sufficient power for the radio emission in the lobes, if a reasonable fraction (1–10 percent) of the streaming energy is converted into relativistic electrons through shock waves and magnetohydrodynamic turbulence (Schreier et al. 1979).

However, X-ray jets are also seen in M87 and 3C273. The relative X-ray, optical, and radio intensities of these jets are consistent with synchrotron emission. This mechanism cannot be ruled out for Centaurus A.

The spectral capability of AXAF will provide the most direct evidence bearing on the question of whether the radiation from the nucleus, jet, and lobes of a given source is thermal or nonthermal and what the connec-

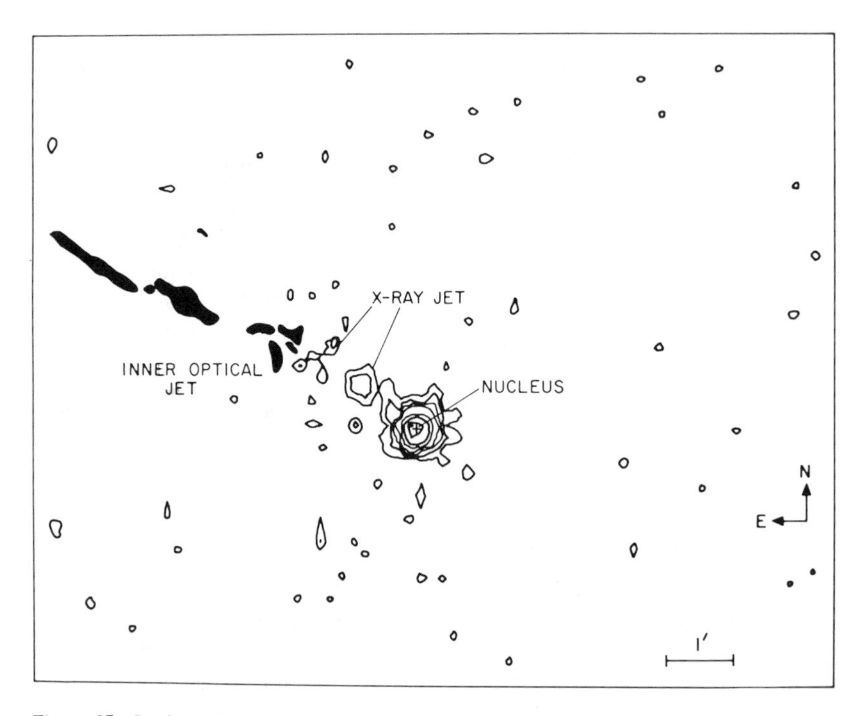

Figure 27 Iso-intensity X-ray contour map of the HRI image around the nucleus of Cen A. The jet to the northeast is clearly seen. The dark shapes further to the northeast show the positions of diffuse optical features discovered by Dufour & van den Bergh (1978).

tion between these various components may be. The 100-fold increase in sensitivity will allow complete samples of active galaxies of differing optical and radio classes to be detected. It should be possible to construct luminosity functions, to measure the evolution of more distant objects, and to clarify the apparently fundamental nature of the nuclear X-ray emission and its relation to the source of power in active galactic nuclei.

Observations of QSOs with the Einstein Observatory have shown that virtually all of these objects are strong X-ray emitters. In a continuation of the work of Tananbaum et al. (1979), Zamorani et al. (1980) have reported on observations of 107 QSOs of which 79 are detected. These results show a strong correlation between X-ray and radio emission: for QSOs of fixed optical luminosity, the X-ray luminosity of radio-loud QSOs is about 2 to 3 times that of radio-quiet QSOs. In addition, there is a suggestion that the ratio of X-ray to optical luminosity decreases with either increasing redshift or optical luminosity.

The QSOs detected so far span a luminosity range from 10^{43} to 10^{47} erg s^{-1} in the 0.5 to 4.5 keV band. As yet, no luminosity function for these objects has been determined since no statistically complete sample has been observed. As more data are accumulated, a sample of X-ray–selected QSOs can be established which will be free of the types of bias and selection effects found in optical and radio surveys. With the data currently available, the correlation of X-ray with optical and radio fluxes can be investigated. The relation between X-ray and optical luminosities has been determined separately for radio-loud and radio-quiet QSOs. The correlation, shown in Figure 28, of radio with X-ray luminosity is significant. Study of the integral distribution function for α_{ox}, the power-law index connecting the optical and X-ray flux densities at 2500 Å and 2 keV, for radio-loud and radio-quiet QSOs shows that the radio-loud ones are, on the average, three times brighter in X ray.

An average effective ratio of X-ray to optical luminosity can be determined from these two classes of QSOs by appropriately weighting the

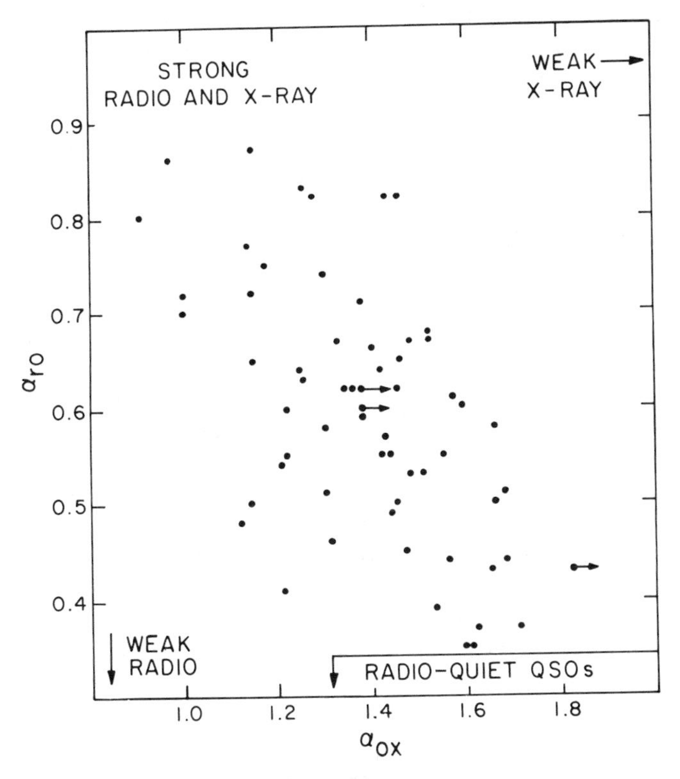

Figure 28 α_{ro} versus α_{ox} for 62 radio-loud quasars. Directions are indicated for decreasing radio emission and decreasing X-ray emission. More than 90% of the radio-quiet quasars have α_{ox} within the region bracketed at the bottom of the plot.

percentage of objects in each class: $\bar{\alpha}_{ox} = 1.45$. Zamorani et al. used this value to compute the contribution of QSOs to the X-ray background, and the number of QSOs expected at the limiting sensitivity of the Einstein deep surveys. Using the optical luminosity function of Braccesi et al. (1980) and taking a limiting magnitude of $m_B = 20.0$, corresponding to the X-ray deep survey flux limit of 2.6×10^{-14} erg cm^{-2} s, they find that QSOs account for 25 percent of the total X-ray background. One expects 13 objects per square degree, in good agreement with the Einstein results.

If the limiting magnitude is taken to be $m_B = 21.2$, with the same value for $\bar{\alpha}_{ox}$ then QSOs will account for 100 percent of the background. Either the optical luminosity function must flatten before $\mu_B = 21$, or the ratio of X-ray to optical luminosity must decrease beyond 20th magnitude. Independent evidence for a flattening in the optical luminosity function about 20th magnitude has been found from optical counts (Kron 1980, Bahcall & Soneira 1980, and Bonoli et al. 1980).

The X-ray observations of QSOs suggest a framework for QSO emission models. Both radio-loud and radio-quiet QSOs must have a central energy source that provides some optical and radio emission consistent with $\bar{\alpha}_{ox} \sim 1.5$. This source must have at least one variable parameter to allow for the large dispersion of the observed ratio of X-ray and optical fluxes. In radio-loud QSOs a second mechanism must operate to produce X-rays preferentially to optical emission. This could be radiation from a hot plasma, a relatively flat synchrotron spectrum, or synchro-Compton emission.

Detailed studies by AXAF of the intensity, variability, spectrum, and polarization should give direct information on the processes occurring at the source. For instance, if the X-ray emission process is Compton scattering of soft photons from a cloud of hot electrons, the measurement of an energy-dependent lifetime during flaring activity would yield a complete description of the input spectrum as well as of the scattering region. If QSOs produce X-ray emission or absorption lines, then modest resolution spectra offer the possibility of observing these lines at redshifts greater than $z = 3$. AXAF will also be used to obtain much larger complete X-ray samples of QSOs to be compared with optically and radio selected samples. Magnitude, redshift, and intrinsic luminosity distributions can be developed and refined, and the luminosity evolution can be examined with X-ray data.

6.8 *Surveys: Log* N–*Log* S

The search for X-ray sources and the extension of the sensitivity limit for detecting sources is the goal of the Einstein Observatory survey program. As was discussed in previous sections, the telescope and detectors combine to increase point source sensitivity by a factor of $\gtrsim 500$ compared with previous instruments. A systematic series of observations has been carried out using both the IPC and HRI detectors to detect sources to a limiting flux of $\sim 2 \times 10^{-14}$ erg cm^{-2} s^{-1} in the 1–3 keV energy band (Giacconi et al. 1979a and Murray et al. 1980). This corresponds roughly to $\sim 2 \times 10^{-3}$ Uhuru counts s^{-1} for sources with a spectrum similar to the X-ray background. In addition to these "deep" surveys which cover only a few square degrees of sky, a more modest sensitivity survey has also been carried out over a larger region of ~ 25 square degrees (Maccacaro et al. 1980, private communication). This "medium" survey utilizes observations with the IPC which were selected on the basis of exposure time and suitable targets that contained no complex or diffused primary sources of emission. Serendipitous sources within these IPC fields constitute the medium survey.

The key scientific objectives of these surveys are 1. the number-intensity distribution for X-ray sources and the composition of this distribution by type of object; 2. the contribution of discrete X-ray sources to the isotropic background; and 3. the discovery of new classes of X-ray sources.

Deep surveys have been carried out in several regions of the sky. Thus far a preliminary analysis for four survey regions has been carried out. We detect 15 sources at the 5σ level of significance or better to a limiting flux of 1.3×10^{-14} erg cm^{-2} s^{-1} in the 1–3 keV energy band. Assuming a power-law representation for the source distribution

$$\frac{dn}{dS} = A\, s^{-(\alpha+1)}, \qquad N(>S) = N, S^{-\alpha},$$

we have fit the observed source distribution to this function using a maximum-likelihood technique. This takes into account the effects of varying sky coverage and sensitivity for each of the survey regions. After making systematic corrections to the slope estimate (Murdock et al. 1973), for the effects of flux uncertainty and the small number of sources in the sample, the best corrected slope estimate is $\alpha_c = 1.63$, with a normalization of

$$N(S > 1.3 \times 10^{-14}) = 1.4 \times 10^5 \text{ Sr}^{-1}.$$

With these values for the number-intensity distribution of X-ray sources, the contribution of discrete sources to the isotropic background can be estimated. We assume that these sources on the average have a spectrum like the background, and find that in the 1–3 keV energy band sources with flux greater than 1.3×10^{-14} erg cm^{-2} s^{-1} contribute $\sim 24\%$ of the total background. If this number-intensity distribution continues to lower fluxes with the same parameters, then the total background would be

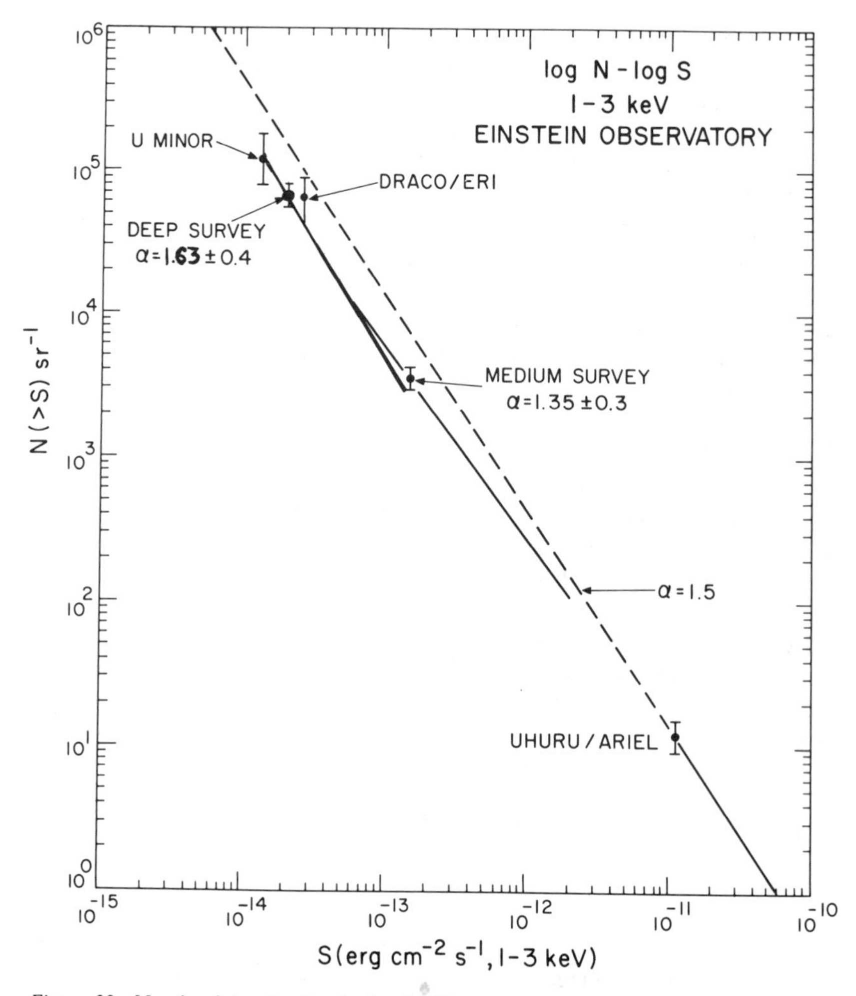

Figure 29 Number-intensity distribution for X-ray sources. The Medium Survey line represents the best maximum likelihood method fit over the flux range shown, similarly for the Deep Survey line. The error bars show uncertainty in normalization. The previously published Einstein Deep Survey result in Dracro and Eridanus is shown as well as the integral point obtained in the very deep Ursa Minor survey.

made up by sources with flux greater than 2.5×10^{-15} erg cm^{-2} s^{-1}, a factor of 10 more sensitive than the Einstein limit. The present limiting sensitivity was achieved with a 300,000-second exposure in a single HRI field of view ($\Omega \sim 3.4 \times 10^{-5}$ Sr). At this level of exposure the sensitivity increases only as the square root of time; further, systematic effects within the HRI detector such as nonuniform gain and quantum efficiency become important. Thus, a factor of 10 sensitivity increase is not possible with the Einstein Observatory and will require future telescopes such as the AXAF, which should easily achieve this level of sensitivity.

In addition to the X-ray observation and detection of weak sources in the survey program, a substantial effort is made to identify sources with optical and/or radio counterparts. In the set of 15 sources that have been used to study source counts, seven have likely identifications as follows:

4 previously unknown quasars (redshifts 0.5, 1.9, 1.4, and 2.6),
2 faint galaxies, which are also coincident with radio sources,
1 faint galaxy.

For the unidentified sources, six do not have an optical candidate that is brighter than 19^{m} and the remaining two have large error box sizes (~ 1 arcmin radius), which makes identifications difficult. We expect several of these sources will be quasars, but the required spectrographic observations have not yet been carried out.

The medium survey has thus far resulted in the detection of 16 sources at the 5σ or greater level of significance which are believed to be extragalactic. The resulting number-intensity distribution for this set of sources gives $\alpha \sim 1.35 \pm .3$ with 6.1×10^{3} sources Sr above a flux of 10^{-13} erg cm^{-2} s^{-1}. The sources detected in the deep survey range in flux from 1.3×10^{-14} to 2×10^{-13} erg cm^{-2} s^{-1}. The medium and deep survey source counts are consistent with one another over the region of overlap in flux (around 10^{-13} erg cm^{-2} s^{-1}). If one takes a composite number-intensity distribution made up of the results for the Einstein surveys then the total fraction of the background from discrete sources with flux above 1.3×10^{-14} erg cm^{-2} s^{-1} is $\sim 30\%$.

These results are shown in Figure 29, which is a plot of log $N(>S)$ vs log S. While the Einstein results thus far are preliminary, this figure is consistent with the following: For the strongest and nearest extragalactic sources, the expected distribution follows the classical Euclidean form with $\alpha = 1.5$. As weaker more distant sources are counted, cosmological effects become significant. For sources that do not evolve the effect is to lower the observed slope as occurs in the medium survey. At the faintest flux levels, the contribution for evolving sources, particularly quasars, starts to dominate the source counts and a steeper slope is observed. Ultimately, the source counts must turn over (i.e. the slope flattens) to avoid the background constraint. As discussed above, this must begin within a factor of 10 from the current-flux limit.

7. X-RAY OBSERVATORIES FOR THE FUTURE

The main requirement for the future of X-ray astronomy is the establishment of permanent observational capabilities. In a sense, X-ray astronomy has become too important to be left to X-ray astronomers. When observations dealt with restricted categories of objects, such as the galactic X-ray sources, it was quite appropriate to look at each X-ray astronomy observation as an enterprise in itself, conceived and executed as a single experiment. The lack of continuity intrinsic to the use of satellite instruments that could not be serviced was not of great concern to the astronomical community as a whole. Provided funding was available, it simply meant that different experimental groups alternated in the lead for brief periods of time.

The use of the Einstein Observatory by hundreds of guest investigators has, however, made quite clear the need for large-scale sophisticated instruments continuously available to astronomers in all branches of astronomy. The response of the X-ray astronomers to this obvious need has been the advocacy of permanent national X-ray observatories, such as AXAF, as the keystone of an X-ray program. These observatories are to be launched and serviced by the Space Shuttle and will be designed to respond to the observational requirements of all astronomers. They would be operated, in the same way as large ground-based national observatories or the Space Telescope, by an independent institute representing the interests of the entire community.

In addition to these major initiatives, X-ray astronomers plan to increase their participation in more specialized satellite missions which complement the national observatories. These missions, typically carried out by individual groups or by consortia of universities under the leadership of a single principal investigator, have been responsible for much of the vitality and rapid growth that has characterized X-ray astronomy over the past two decades. They must continue to be supported both because of their intrinsic scientific interest and because they will provide the scientific and technological basis for the observatories of the future. In this category of missions, the extension of X-ray observations to higher energies, the monitoring of individual sources with high time resolution over long periods of time, and detailed spectroscopic studies of individual sources appear particularly urgent.

In what follows, we give a brief description of two of the most prominent missions in the current NASA program: AXAF (Advanced X-ray Astrophysics Facility) and LAMAR (Large Area Modular Array of Reflectors).

7.1 *AXAF*

The successor to the Einstein Observatory will be the Advanced X-ray Astrophysics Facility (AXAF). This mission has been studied since 1977 by a NASA Science Working Group (appointed by the Office of Space Science), which includes representatives of all X-ray astronomy groups in the US, and X-ray astronomers from England, Germany, and Japan. Engineering studies were conducted by NASA's Marshall Space Flight Center and its subcontractors during the same period. The conceptual definition (Phase A) was completed in 1978 and a more detailed definition (Phase B) is under way. Current plans call for the final design and development activities to begin in Fiscal Year 1984 or 1985 with the launch of AXAF to take place in 1988 or 1989.

At first glance, AXAF seems just to be a larger Einstein; however, several important changes are planned, making a substantial improve-

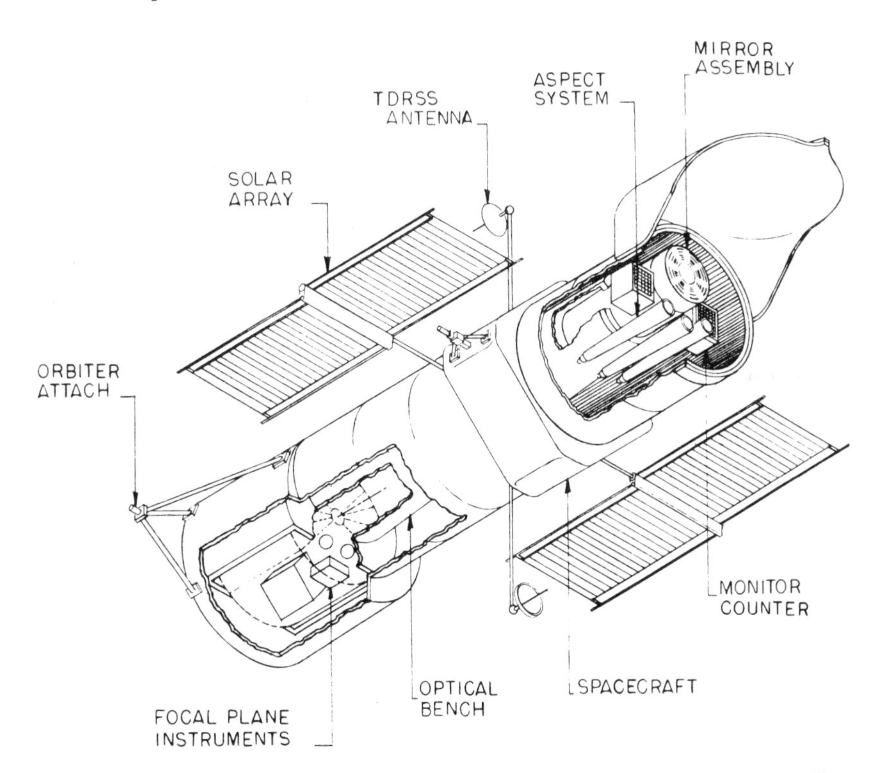

Figure 30 Diagrammatic representation of the Advanced X-ray Astronomy Facility (AXAF). The major subsystems are identified.

Table 4 Detection of point sources by AXAF (10^5 seconds observation time)

Z	Luminosity distance	Object	Minimum detectable X-ray luminosity (ergs s^{-1})	Corresponding linear dimension of 1″ detection cells
	150 pc	star	1×10^{27}	50 AU
	0.7 Mpc	point source in M31	2×10^{34}	3.5 pc
.003	19 Mpc	point source in Virgo Cluster	1×10^{37}	100 pc
.05	300 Mpc	normal spiral galaxy	3×10^{39}	1.5 Kpc
.16	1000 Mpc	active galaxy in Hydra cluster	3×10^{40}	5 Kpc
7	2×10^5 Mpc	quasar	1×10^{45}	1 Mpc
Examples of known X-ray sources				
	normal star:	sun	5×10^{27}	
	SNR:	Cyg Loop	2×10^{36}	
	point source:	Sco X-1	2×10^{37}	
	binary:	Cen X-3	4×10^{37}	
	normal spiral:	Milky Way	5×10^{39}	
	active galaxy:	M87	3×10^{43}	
	quasar:	3C273	5×10^{45}	

ment in capability. The present plans are centered on a 1.2-m diameter, 10-m focal length X-ray lens. Resolution in the center of the field is expected to be 0.5 arcsec; the mirror collecting area will be four times that of Einstein, and the mirror will reflect X rays up to energies of 10 keV. With some improvements in the focal plane instrumentation, a factor of 100 increase in sensitivity for detection of faint point sources over Einstein is expected. AXAF will be shuttle launched, free-flying, and expendables will be resupplied in orbit (Figure 30). It will be operated as a national facility and should achieve a lifetime of 10–15 years. Focal plane instruments will probably, as on the Einstein Observatory, include a high resolution imager, a high sensitivity, moderate resolution imaging instrument, a nondispersive solid-state spectrometer, and high resolution spectrometers. A polarimeter will give a new capability, and an objective grating 10 times as efficient as Einstein's can be made with present technology. Tables 4 and 5 give the calculated sensitivity for detection of sources and the off-axis performance of the proposed AXAF mirror.

The design of the AXAF is based upon experience with the fabrication, assembly, and testing of the Einstein mirrors. The design emphasizes the collecting area in the few keV region and also gives a useful response be-

Table 5 Off-axis performance of proposed AXAF mirror and Einstein mirror

	Off-axis angle in arcmin			
	0	5	10	20
Geometrically perfect AXAF mirror				
rms blur circle diameter in arcsec	0	2	—	—
half power diameter in arcsec	0	2	5	20
Einstein mirror				
half power diameter in arcsec (0.28 keV)	8	10	25	—
half power diameter in arcsec (3 keV)	20	25	40	—
FWHM in arcsec	3.5	3.5	5	—

yond $\sim$7 keV—the energy of K α X-ray lines from ionized Fe, an element of great astrophysical importance.

To achieve a sensitivity 100 times greater than that of Einstein for faint point sources, an improvement in the focal plane high resolution instrument is necessary. Current developments in detector technology give assurance that the AXAF high resolution detectors will permit the achievement of high angular resolution, high quantum efficiency, and moderate spectral resolution over the entire imaged field. A combination of detectors emphasizing one or the other of the requirements will probably be selected.

7.2 *LAMAR*

LAMAR is a Large Area Modular Array of Reflectors. It is an imaging instrument with high sensitivity (area) and moderate resolution which is to be built in modular form. Each module consists of a small telescope and detector, and a LAMAR with many identical modules can be packaged in a much smaller volume than a very large telescope with equivalent sensitivity. No new technology is needed; large area is of prime importance, and, therefore, one would build an array containing as many modules as physical and financial constraints would permit.

The approach currently followed by NASA is to start with the development of principal investigator experiments to be flown on the shuttle, to upgrade the telescopes and detectors on subsequent shuttle flights, and to eventually assemble in the late 1980s a free flyer having a very large area and a long lifetime which would be operated as a national facility. The observing time would probably be devoted to 1. an all-sky survey considerably more sensitive than any previous surveys; 2. mapping of soft X-ray

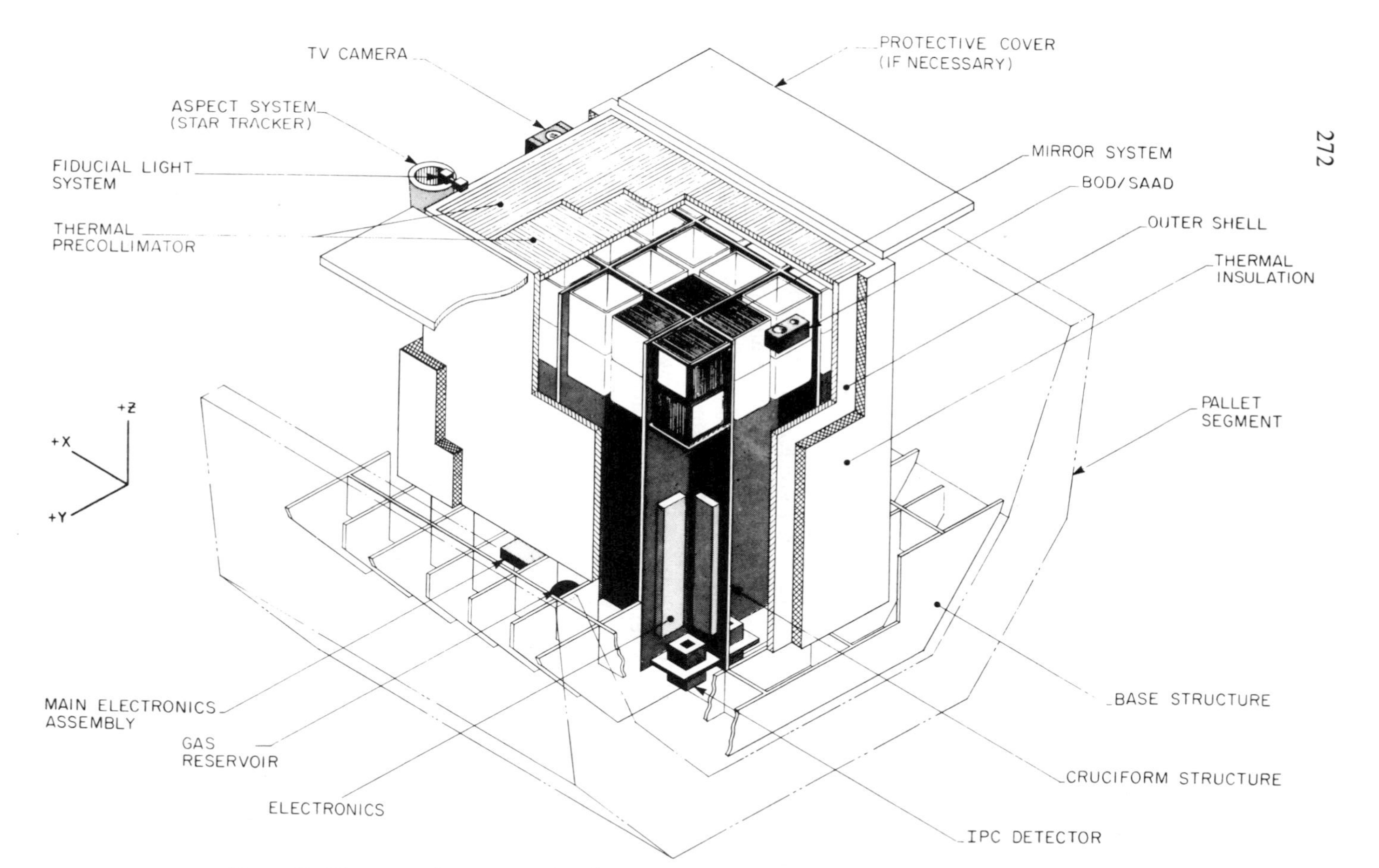

Figure 31 Diagrammatic representation of the shuttle-borne version of the (LAMAR) experiment.

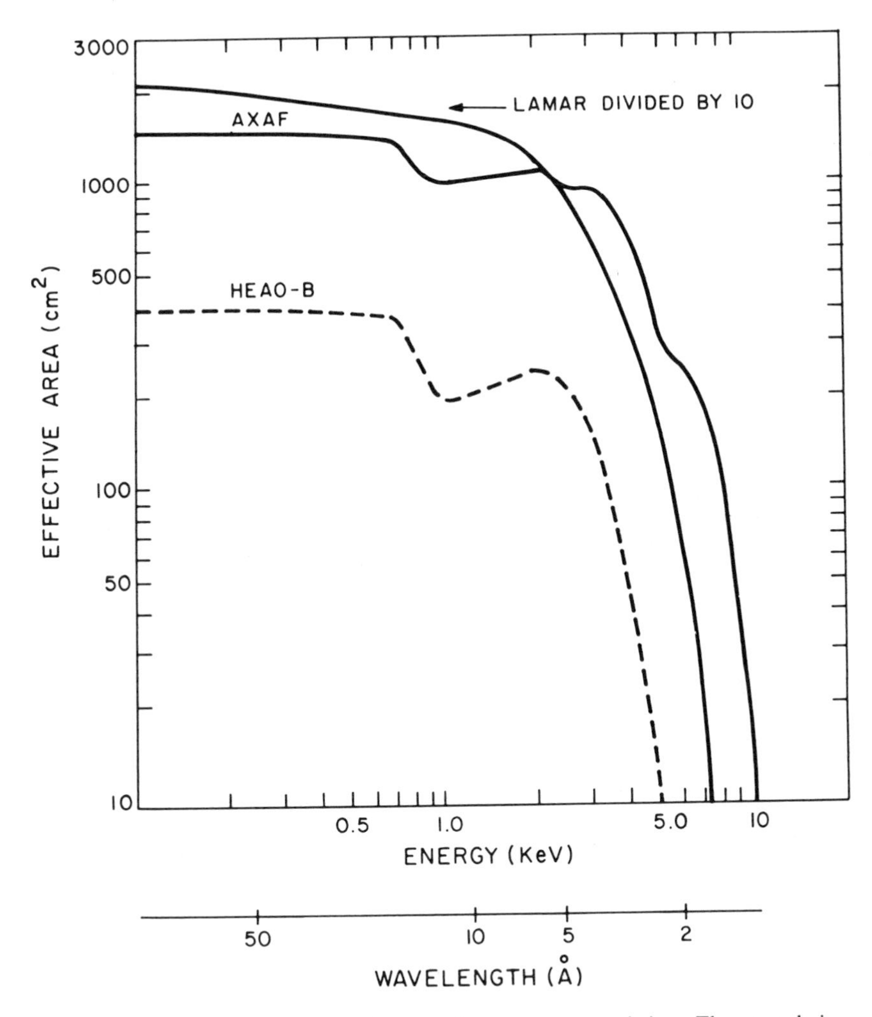

Figure 32 Effective area vs energy for three X-ray astronomy missions. The curve designated HEAO-B is the area actually achieved in the Einstein Observatory mission. The curve designated AXAF corresponds to the current baseline design for AXAF. Finally, the curve designated LAMAR corresponds to the area ultimately proposed for a LAMAR Free-Flyer Facility. The numerical values (for LAMAR only) have been divided by 10 to allow comparison with the other curves.

features of low surface brightness; and 3. the study of time-variable X-ray emission and spectra from distant sources.

Figure 31 shows a LAMAR package being designed for Spacelab. It is a 24-element array. Each element is a small telescope consisting of a Kirkpatrick-Baez nested plate-type lens and an imaging-proportional counter similar to that used on the Einstein Observatory. The modular design is flexible; it would not be difficult to expand the array to ultimately include 64 modules on a single pallet. The expected effective area of a 200-module free flyer with Wolter type 1 optics is shown in Figure 32.

An all-sky survey with this instrument, dwelling 100 seconds on each square degree of the sky and assuming a 50% duty cycle for useful observing, would take three months. The instrument would be capable of observing sources down to 0.016 Uhuru counts s^{-1} and, projecting the log N/ log S source distribution, we can predict that 20,000 extragalactic sources could be detected. The sources could be located to 0.5 arcmin, enough for unambiguous identification of most of these sources.

The most valuable use of LAMAR will probably be in the measurement of spectra and time variations of the faintest extragalactic sources for which AXAF is not well suited. An imaging instrument is necessary to observe the faint sources and avoid confusion, and a large area is required to measure spectra and to detect significant time variations in short times. Prime targets would be the nuclei of active galaxies, X-ray binaries in neighboring galaxies, and perhaps stellar coronae in our own Galaxy.

8. CODA

With the flight of the Einstein Observatory, X-ray astronomy has achieved parity with radio and optical astronomy. Figure 33 illustrates the sensitivity of major astronomical facilities in the radio, optical, and X-ray regions of the electromagnetic spectrum at different epochs.

Owing to the relative transparency of the interstellar medium to X rays and the relative ease of focusing and detection of high energy quanta in the X-ray band, X-ray observations can be extended to regions of high redshift ($1 < z < 10$) and they can be used to single out interesting objects, such as very distant QSOs, clusters of galaxies, and perhaps protogalaxies, for further studies at optical and radio as well as at X-ray frequencies. We believe that astronomical research for the remainder of this century must of necessity utilize observations in all wavelength bands to explore the still unanswered questions as to the mass density of the universe, the formation and clustering of galaxies, the evolution of stars and galaxies, the end points of stellar evolution, and the nature of the machine lurking in the core of galactic nuclei. To all these questions X-ray astronomy is making, and will continue to make, important and at times crucial contributions.

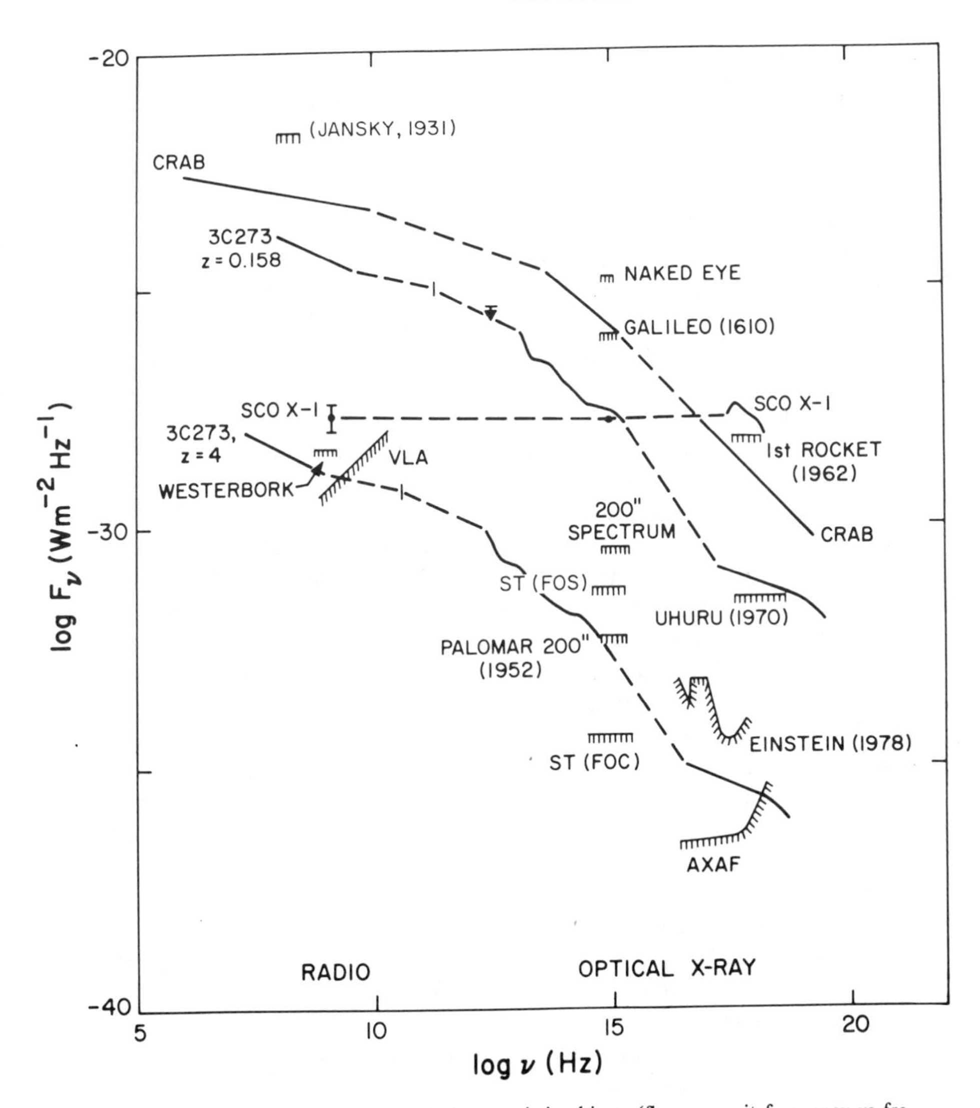

Figure 33 The spectrum of several characteristic objects (flux per unit frequency vs frequency) is shown. Plotted on the figure are the sensitivities of a number of different instruments to emphasize the progress that has occurred in different branches of astronomy over the years. The illustration is intended to emphasize the commonality of scientific objectives among radio, optical, and X-ray observations and the complementary nature of the AXAF mission to the Space Telescope and the Very Large Array.

ACKNOWLEDGMENTS

The research reported in this article is the result of the efforts of many scientists at the four Consortium institutions: Massachusetts Institute of Technology, Columbia Astrophysical Laboratory, Goddard Space Flight Center, and the Harvard/Smithsonian Center for Astrophysics. We wish to thank all our colleagues for their permission to utilize their results.

While it is not possible to acknowledge all the individuals associated with the development of the Einstein Observatory, we would like to recognize the help and support of Richard Halpern and his colleagues at NASA Headquarters, the guidance of Fred Speer and the management team at Marshall Space Flight Center, and the support by the Kennedy Space Flight Center and the Goddard Space Flight Center. Their efforts during launch and in the mission operations that followed were and continue to be essential to the success of the mission.

Also essential were the efforts of the industrial contractors: TRW for the spacecraft design and construction; AS&E for the experiment construction, integration, and testing; Perkin-Elmer for the fabrication of the X-ray mirror; Honeywell Electro-Optics Center for the construction and testing of the star trackers; Convair Division of General Dynamics for the construction of the optical bench; LND for the preparation of proportional counters; Parker-Hannifin Corporation, Lord Corporation, Aeroflex Laboratories, Inc., R&K Precision Machine Co., Astronautic Industries, Inc., for many of the mechanical assemblies; Spacetac, Inc., Matrix Research and Development Corporation, for low and high voltage power supplies; and, finally BASD (formerly Ball Brothers) for the construction of the cryostat.

We wish to acknowledge the constant support given to this project by the management and staff of the four institutions forming the scientific Consortium: The Harvard/Smithsonian Center for Astrophysics, Columbia Astrophysical Laboratory, Goddard Space Flight Center, and the Massachusetts Institute of Technology.

This research was sponsored under NASA contracts NAS8-30751, NAS8-30752, and NAS-30753.

References

Arnett, W. D. 1975. *Astrophys. J.* 195:727

ASE-106. January 1961. An X-ray Detector for High Resolution Mapping of the Sun from S-16A. Prepared for NASA

ASE-311. September 1962. Experiment to Measure Solar X-radiation from the Pointing Section of the S-57 Satellite. Prepared for NASA

ASE-623. 4 June 1964. A Proposal for the Initial Phase of a Program to Develop a High Resolution X-ray Telescope for AOSO. Prepared for NASA

ASE-2410-III. September 1970. Technical Proposal Summary for a Large Orbiting X-ray Telescope for the HEAO. Prepared for NASA

Bahcall, J. N., Soneira, R. M. 1980. Preprint

Becker, R. H., Holt, S. S., Smith, B. W., White, N. E., Boldt, E. A., Mushotzky,

R. F., Serlemitsos, P. J. 1979. *Astrophys. J.* 234:L73
Blake, R. L., Chubb, T. A., Friedman, H., Unricker, A. E. 1963. *Astrophys. J.* 137:3
Bonoli, F., Braccesi, A., Marano, B., Merighi, R., Zitelli, V. 1980. *Astron. Astrophys.* 90:L10–L12
Borkowski, C. J., Kopp, M. K. 1972. *IEEE Trans. Nucl. Sci.* NS-19:161
Bowles, J. A., Patrick, T. J., Sheather, P. H., Eiband, A. M. 1974. *Journal of Physics E; Scientific Instruments.* 7:191
Braccesi, A., Zitelli, V., Bonoli, F., Formiggini, L. 1980. *Astron. Astrophys.* 85:80
Canizares, C., Clark, G., Bardas, D., Markert, T. 1977. *Proc. Soc. Photo-Optical Instrum. Engrs.* 106:154
Chase, R. C., Van Speybroeck, L. P. 1973. *Appl. Opt.* 12:1042
Chevalier, R. A., Kirschner, R. P. 1978. *Astrophys. J.* 219:931
Compton, A. H., Allison, S. K. 1935. *X-rays in Theory and Experiment.* New York: Van Nostrant
Dietz, P. W., Bennett, J. M. 1966. *Appl. Opt.* 5:881
Donaghy, J., Canizares, C. 1978. *IEEE Trans. NS-25,* p. 459
Dufour, R. J., van den Bergh, S. 1978. *Astrophys. J.* 226:L73
Fabian, A. C., Willingale, R., Pye, J. P., Murray, S. S., Fabbiano, G. 1980. *Mon. Not. R. Astron. Soc.* In press
Fabricant, D., Lecar, M., Gorenstein, P. 1980. *Astrophys. J.* 241:522
Forman, W., Bechtold, J., Blair, W., Giacconi, R., Van Speybroeck, L., Jones, C. 1981. *Astrophys. J. Lett.* 243:1–133
Forman, W., Schwarz, J., Jones, C., Liller, W., Fabian, A. C. 1979. *Astrophys. J.* 234:L27
Giacconi, R., Bechtold, J., Branduardi, G., Forman, W., Henry, J. P., Jones, C., Kellogg, E., van der Laan, H., Liller, W., Marshall, H., Murray, S. S., Pye, J., Schreier, E., Sargent, W. L. W., Seward, F., Tananbaum, H. 1979a. *Astrophys. J.* 234:L1
Giacconi, R., Branduardi, G., Briel, U., Epstein, A., Fabricant, D., Feigelson, E., Forman, W., Gorenstein, P., Grindlay, J., Gursky, H., Harnden, F. R. Jr., Henry, J. P., Jones, C., Kellogg, E., Koch, D., Murray, S., Schreier, E., Seward, F., Tananbaum, H., Topka, K., Van Speybroeck, L., Holt, S. S., Becker, R. H., Boldt, E. A., Serlemitsos, P. J., Clark, G., Canizares, C., Markert, T., Novick, R., Helfand, D., Long, K. 1979b. *Astrophys. J.* 230:540
Giacconi, R., Harmon, N. F., Lacey, F. R., Szilagyi, Z. 1965a. *J. Opt. Soc.* 55(4): 345–47
Giacconi, R., Reidy, W., Vaiana, G. S., Van Speybroeck, L., Zehnpfennig, T. 1969. *Space Science Rev.* 9:3–57
Giacconi, R., Reidy, W., Zehnpfennig, T., Lindsay, J., Muney, W. 1965b. *Astrophys. J.* 142(3):1265
Giacconi, R., Rossi, B. 1960. *J. Geophys. Res.* 65:773
Gorenstein, P., Gursky, H., Harnden, F. R. Jr., De Caprio, A., Bjorkholm, P. 1975. *IEEE Trans. NS-22,* p. 616
Grindlay, J., Hertz, P., Steiner, J., Murray, S., Lightman, A. 1981. In preparation
Harnden, F. R. Jr., Buehler, B., Giacconi, R., Grindlay, J., Hertz, P., Schreier, E., Seward, F., Tananbaum, H., Van Speybroeck, L. 1980. *Bull. Am. Astron. Soc.* 11:789
Harvey, P., Sanders, J., Cabral, R., Morris, J., Bjorkholm, P., Ballas, J., Jagoda, N., Harnden, F. R. 1976. *IEEE Trans. NS-23,* p. 487
Helfand, D. J., Chanan, G. A., Novick, R. 1980. *Nature* 283 (5745):337–43
Henry, J. P., Kellogg, E., Briel, U., Murray, S., Van Speybroeck, L. P., Bjorkholm, P. 1977. *SPIE Proc.* 106:196
Humphrey, A., Cabral, R., Brissette, R., Carroll, R., Morris, J., Harvey, P. 1978. *IEEE Trans. NS-25,* p. 445
Jones, C., Mandel, E., Schwarz, J., Forman, W., Murray, S. S., Harnden, F. R. Jr. 1979. *Astrophys. J.* 234:L9
Joyce, R., Becker, R., Birsa, F., Holt, S., Noordzy, M. 1978. *IEEE Trans. NS-25,* p. 453
Kirkpatrick, P., Baez, A. V. 1948. *J. Opt. Soc. Am.* 38:766
Kron, R. G. 1980. In *Two Dimensional Photometry* (ESO Workshop), ed. P. O. Lindblad, H. van der Laan. Geneva
Lamb, S. A. 1978. *Astrophys. J.* 220:186
Ledger, A. M. 1979. *Proc. Soc. Photo-Optical Instrum. Engrs.* 184:176
Abbott, D. C., Bieging, J. H., Churchwell, E., Cassinelli, J. P. 1980. *Astrophys. J.* 238:196–202
Lindsay, J. C., Giacconi, R. 1963. High resolution (5 arcsec) X-ray telescope for advanced orbiting solar observatory. *Goddard Space Flight Cent. Doc. X-614-63-112*
Mangus, J. D., Underwood, J. H. 1969. *Appl. Opt.* 8:95
Mathur, D. P., Adamo, D. R., Bastien, R. C., Strouse, E. A. 1979. *Proc. Soc. Photo-Optical Instrum. Engrs.* 184:139
Murdock, H. S., Crawford, D. F., Jauncey, D. L. 1973. *Astrophys. J.* 183:1
Murray, S. S. 1980. In *X-ray Astronomy—Proc. HEAD/AAS, Jan. 1980, Cambridge, Mass. Meet.*, ed. R. Giacconi. Dordrecht, Netherlands: Riedel. In press

Murray, S. S., Fabbiano, G., Fabian, A. C., Epstein, A., Giacconi, R. 1979. *Astrophys. J.* 234:L69

NASA SP-213, July 1969. *A Long-Range Program in Space Astronomy.* Position Paper of the Astronomy Missions Board

NAS Publication 1403. 1966. *Space Research: Directions for the Future.* Report of a Study by the Space Science Board, National Academy of Sciences

Perrenod, S. C., Henry, J. P. 1980. Submitted to *Astrophys. J. Lett.*

Pravdo, S. H., Smith, B. W. 1979. *Astrophys. J.* 234:L195

Schreier, E. J., Feigelson, E., Delvaille, J., Giacconi, R., Grindlay, J., Schwartz, D. A., Fabian, A. C. 1979. *Astrophys. J.* 234:L39

Tananbaum, H., Avni, Y., Branduardi, G., Elvis, M., Fabbiano, G., Feigelson, E., Giacconi, R., Henry, J. P., Pye, J. P., Soltan, A., Zamorani, G. 1979. *Astrophys. J.* 234:L9

Tsuruta, S. 1980. International School of Astrophysics, 5th Course, X-ray Astronomy, Erice, Sicily, 1979

Vaiana, G. S., Cassinelli, J. P., Fabbiano, G., Giacconi, R., Golub, L., Gorenstein, P., Haisch, B. M., Harnden, F. R. Jr., Johnson, H. M., Linsky, J. L., Maxson, C. W., Mewe, R., Rosner, R., Seward, F., Topka, K., Zwaan, C. 1980. *Astrophys. J.* In press

Vaiana, G. S., Krieger, A. S., Timothy, A. F. 1973. *Sol. Phys.* 32:81–116

Vaiana, G. S., Reidy, W. P., Zehnpfennig, T., Van Speybroeck, L., Giacconi, R. 1968. *Science.* 161:564

Vaiana, G. S., Van Speybroeck, L., Zombeck, M. V., Krieger, A. S., Silk, J. K., Timothy, A. 1977. *Space Sci. Instrum.* 3:19

Van Speybroeck, L. 1973. In *Proc. X-ray Optics Symp., Mullard Space Sci. Lab. Univ. College London, April 1973,* ed. P. W. Sanford

Van Speybroeck, L. 1980. In *X-ray Astronomy, Proc. HEAD/AAS, Jan. 1980, Cambridge, Mass. Meet.,* ed. R. Giacconi. Dordrecht, Netherlands: Riedel

Van Speybroeck, L., Chase, R. 1972. *Appl. Opt.* 11:440

Van Speybroeck, L., Epstein, A., Forman, W., Giacconi, R., Jones, C., Liller, W., Smarr, L. 1979. *Astrophys. J.* 234:L45

Winkler, P. F., Canizares, C. R., Clark, G. W., Markert, T. H., Kalata, K., Schnopper, H. 1981. *Astrophys. J.* 246:In press

Wolter, H. 1952. *Ann. Physik* 10:94

Young, P. S. 1979. *Proc. Soc. Photo-Optical Instrum. Engrs.* 184:131

Zamorani, G., Henry, J. P., Maccacaro, T., Tananbaum, H., Soltan, A., Avni, Y., Liebert, J., Stocke, J., Strittmatter, P. A., Weymann, R. J., Smith, M. G., Condon, J. J. 1980. Submitted to *Astrophys. J.*